From the
AMERICAN SYSTEM *to*
MASS PRODUCTION
1800–1932

STUDIES IN INDUSTRY AND SOCIETY
Glenn Porter, General Editor

Published with the assistance of
The Eleutherian Mills-Hagley Foundation

From the
AMERICAN SYSTEM *to*
MASS PRODUCTION
1800–1932

The Development of Manufacturing
Technology in the United States

DAVID A. HOUNSHELL

THE JOHNS HOPKINS UNIVERSITY PRESS/Baltimore and London

This book has been brought to publication with the generous assistance of the Eleutherian Mills-Hagley Foundation.

The Johns Hopkins University Press, Baltimore, Maryland 21218
The Johns Hopkins Press Ltd, London

Library of Congress Cataloging in Publication Data

Hounshell, David A.
  From the American system to mass production, 1800–1932.

  (Studies in industry and society; 4)
  Bibliography: p. 385
  Includes index.
  1. Mass production—United States—History. I. Title.
II. Series.
TS149.H68 1983   338.6′5′0973   83–16269
ISBN 0–8018–2975–5

Title page illustration: Detail of South Wall, *Detroit Industry,* Diego Rivera, Detroit Institute of Arts, 1932–33. (Founders Society Purchase, Edsel B. Ford Fund & Gift of Edsel B. Ford. Courtesy of Detroit Institute of Arts.)

To Eugene S. Ferguson
Teacher, scholar, and friend

Handicrafts and methods of production that follow
the precedent of handicrafts, serve best an aristocracy
of consumers, while factories serve best the
consumption of a democracy.

—Victor S. Clark, *History of Manufactures in the United States* (1916)

If I could give every Russian just one American
book, I would choose the Sears, Roebuck catalogue.

—Attributed to Franklin Delano Roosevelt

Contents

List of Figures

List of Tables

Foreword

This volume marks an important departure in the Industry and Society series of the Johns Hopkins University Press and the Eleutherian Mills–Hagley Foundation. It has been our intent from the first that most of the studies in this series focus on the economic and social history of the Mid-Atlantic states yet be pertinent to wider issues and topics. Burton Folsom's *Urban Capitalists,* John Bodnar's *Worker's World,* and Paul Paskoff's *Industrial Evolution* led the way in fulfilling that goal; all three deal closely with aspects of industrial history in a single state, but all approach their subjects in a way that makes them of interest to the broad community of historians. The publication of David Hounshell's *From the American System to Mass Production* represents another kind of work that we also hoped to include in the series—studies that make significant contributions to important historical questions about the relationship between industry and society but do not necessarily deal exclusively with the Mid-Atlantic region.

Hounshell's work is pathbreaking in many respects. He gives us, for the first time, a comprehensive analysis of the evolution of the most significant and best-known set of mass production technologies in American history. Beginning with their origins in the federal armories of the early nineteenth century, Hounshell traces those interrelated technologies through a number of industries—the sewing machine, the woodworking industries, the reaper, the bicycle, and the automobile. The techniques were spread and improved by a close-knit network of key mechanics, who moved out in concentric circles from the armories to the machine tool companies, the sewing machine manufacturers, and the rest, all the way to the automobile makers. Mechanics who had learned parts of the growing body of ideas about how to create highly productive factories passed those ideas on to others, who, in turn, expanded and spread them. The extent of the linkages between the critical firms and individuals was astonishing. Over time, these men steadily widened the range of methods for manufacturing items of wood and metal. Factory organization, specialized machines, precision manufacture, interchangeable parts, carefully coordinated work sequences and materials flows, and new methods for stamping and welding metal eventually become standard items in the repertoire of America's production engineers.

By the early twentieth century some industries had become so adept at turning out masses of parts that they encountered a bottleneck at the point of assembly. The mechanics and engineers responsible for the revolution in production proved equal to this new challenge as well. On the eve of World War I the Ford Motor Company came up with the answer—the assembly line. The last obstacle had been removed. Seemingly limitless numbers of virtually identical goods could now be produced. A perpetual cornucopia machine had emerged, fully realizing the promise of true mass production that began in the armories at Springfield and Harpers Ferry.

Almost as quickly as the new order had dawned, however, it ran into another bot-tleneck. For the first time in the story of the industries Hounshell examines, the major problem ceased to be the challenge of how to produce enough goods to meet an ever-growing demand and became, instead, how to dispose of these goods. Although creative marketing and heavy advertising had been central elements in the success of firms such as Singer and McCormick, their factory superintendents had usually been called on again and again to expand output. In the auto industry, however, after Ford's assembly innova-tions at Highland Park things began to change. For the first time in the firms studied here, serious difficulties arose as the volume of manufactured goods exceeded the demand. The solution to these problems came not from Henry Ford, perhaps the greatest figure in the history of mass production, but from Alfred Sloan, Charles Kettering, and others at General Motors. They pioneered a revolutionary approach to marketing in which they continued to introduce real mechanical improvements in their products but, in addition, they now emphasized style and superficial annual model changes. Furthermore, GM began to create individual products consciously aimed at different income groups.

The General Motors strategy succeeded so well that the Ford Motor Company plunged into decline and losses. Ford and his production engineers eventually were forced to follow the lead of GM. Hounshell traces in detail the story of the painful transition at Ford to the new technology of "flexible mass production." Ford had, with the Model T, taken American mass production to its most extreme form. He had also led his company into an economic and technical dead end. The dream born in the federal armories a century earlier became the nightmare of crushing inventories of unsold cars and a rigid production system with enormously costly and inflexible plants.

The tragedy of Henry Ford pointed to a cruel dilemma that had long troubled other manufacturers. In a sense it is a fundamental problem of technically advanced capitalism itself—manufacturers could produce ever greater quantities of any given item, but eventually they reached the point at which they could no longer sell them at a price that would yield what they considered an acceptable profit. The system is built on endless growth, however, and for any given firm, continuing growth can be achieved only through change. No solution is ever final, no product ever so successful that its growth phase continues endlessly. Even the Model T—the perfect car for the masses—fell victim to this hard fact. Ever since American consumers have been free to choose new goods in the marketplace, they have done so sooner or later, no matter how useful, durable, or inexpensive the existing product might have been. Firms committed to growth exist in a treadmill universe; the machine of growth must never stop.

In many industries, though generally not in the ones detailed in this study, the problem of overcapacity had existed long before Ford pioneered the assembly line. Manufacturers tried various ways of dealing with the problem, from trade associations and pools to the mergers that created so many big businesses before World War I. Oligopolistic competi-tion and heavy reliance on modern marketing techniques provided a way out for some industries. When the problem of overcapacity struck the auto industry, however, it hit the Ford Company especially hard because of the company's wider-ranging commitment to the mass production of a single product.

David Hounshell skillfully traces the evolution of the most important set of production technologies in American industrial history. In the process, he inters a number of myths that have grown up over the years, such as the ideas that the use of interchangeable parts was widespread in the nineteenth century, that interchangeability always meant lower costs, and that technical barriers prevented mass production in the woodworking industries. Perhaps most important, Hounshell shows that the most successful American firms relied on the most careful attention to both marketing and production. The story he tells is one in which a complex and difficult set of new technologies came to fruition with the benefit of initial government subsidy and only when business leaders made long-term commitments to innovation and excellence. At a time when Americans are worrying about a decline in the performance of their industrial corporations, it is both timely and instructive to have this history of the era in which U.S. firms rose to a dominant position in world markets.

This is not, however, entirely a paean to progress. Although most of the book is written from the perspective of the factory superintendents and engineers who perfected mass production, the gradual spread of mechanized production technology meant more routinized work for the people who tended the machines and whose workday came to be controlled and defined by the production engineers. In its final chapter, the study assesses the ambiguous meaning for American society of the coming of mass production.

Few topics are as central to the interests of the Regional Economic History Research Center at Hagley as David Hounshell's subject. Since he has served for some years as a member of the Center's Academic Advisory Board and as curator of technology at the Hagley Museum, I am particularly pleased to have his work appear in this series.

GLENN PORTER, Director
Regional Economic History Research Center
Eleutherian Mills–Hagley Foundation

Acknowledgments

The debts I incurred in writing this study are staggering. First and foremost, I am indebted to the Eleutherian Mills–Hagley Foundation, which supported me as a Hagley Fellow while I was a graduate student at the University of Delaware and later as a curator in the Hagley Museum. The foundation's generous travel budget allowed me to carry out an important part of my manuscript research, and its Eleutherian Mills Historical Library held resources that I could not obtain elsewhere. Among numerous members of the library staff who helped me, Carol Hallman deserves special thanks for handling countless interlibrary loans and solving other problems for me. The library's Pictorial Collections Department also made a substantial contribution to the illustrations of this book. I deeply appreciate the patience shown me by my museum colleagues and superiors through the process of refining the manuscript. Robert Howard, Hagley's curator of engineering, contributed to this book through his excellent drawings.

I remain flattered that my colleague, Glenn Porter, wanted to have this book as part of the Regional Economic History Research Center's series with the Johns Hopkins University Press. He played a dual role, serving as my greatest critic and as my constant supporter. Those who have read Glenn Porter's own work know the precision of his scholarship; I consider myself fortunate to have experienced at firsthand the benefit of his outstanding editorial skill. He contributed significantly to this book in more ways than even he will ever know. For each of them and for his constant support, I am most grateful. Two other members of the foundation staff helped to make this book possible. Dora Mae Blake sustained me during the manuscript's most trying moments, that of typing. She cheerfully turned a manuscript that looked worse than a rotten sow's ear into a real silk purse. Mary Meyers rendered great help with the index.

By awarding me a predoctoral fellowship in the history of science and technology, the Smithsonian Institution made possible a major portion of the work on my doctoral dissertation, which provides the basis for this book. I wish to thank Edward Davidson, Gretchen Gayle, and Elsie Bliss for their help on administrative matters. The National Museum of History and Technology, now the National Museum of American History, and its staff were critical in my research and writing. I am especially indebted to Robert C. Post, who served as my Smithsonian supervisor and who continually went beyond the call of duty. His own work, his criticism of mine, his careful editorial markings, and his abiding friendship will always be gratefully remembered. Carlene E. Stephens answered hundreds of my questions and solved more than a few perplexing historical riddles. I also thank Lu Rosignol and Charles Burger of the Smithsonian Library, who often rendered heroic service. Other Smithsonian staff who helped me in important ways are Brooke Hindle, Silvio Bedini, Joyce Ramey, Nancy Long, Robert Friedel, Otto Mayr, Robert Vogel, Rita

Adrosko, Barbara Suit, William Henson, Joanna Kofron, and Hazel Jones. I sincerely appreciate their help and support.

I am also indebted to Harvey Mudd College in Claremont, California, for funding my research on woodworking technology. The college not only awarded me a summer faculty research fellowship but it provided other nonpecuniary support. Much of that support came from B. Samuel Tanenbaum, dean of the college, and Richard Olson, my colleague and oftentimes pedagogue.

Thanks also go to the College of Arts and Science of the University of Delaware for helping to defray the costs of obtaining and publishing many of the illustrations in this book.

I wish to thank the State Historical Society of Wisconsin and its staff in the manuscripts reading room for their help with my work in the McCormick Collection and the papers of the Singer Manufacturing Company. I am indebted to Joy Levien, assistant secretary of the Singer Company, for permission to use the Singer papers.

Henry D. Sharpe, Jr., chairman of the Brown & Sharpe Manufacturing Company, North Kingstown, Rhode Island, gave me permission to use the historical material on Brown & Sharpe which still survives. I sincerely appreciate his help and encouragement.

To the staff of the manuscripts section of the Connecticut State Library I extend my thanks for their help with the surviving papers related to the Pope Manufacturing Company. These papers gave me an unexpected and important perspective on manufacturing problems in late nineteenth-century America.

The Ford Archives of the Edison Institute will always hold a special place in my debt list. Douglas A. Bakken, director of the Ford Archives, has made this archive into a first-rate research facility, and his interest in my project greatly aided my work there. Reading manuscripts in the Ford Archives was pure pleasure. I profited enormously from the help of David Crippen, reference archivist. His familiarity with the vast Ford Motor Company collection and with Ford history is delightfully rare and is deeply appreciated.

My study was aided by Jane McCavitt of the MIT Institute Archives. Her quick responses to my numerous inquiries faciliated my work on Foster Gunnison. Also, I must acknowledge my debt to Foster Gunnison, Jr., for providing me with important information on his father's life and work and for allowing me to use the photographs of Gunnison houses and the Gunnison factory. In addition, my thanks go to the Department of Special Collections, Case Western Reserve University Libraries, for its help with the Fred Colvin Papers, which aided my analysis of the Ford assembly line.

My wife, Nancy Eddy, contributed in countless ways to the making of this book. Her patience with my long hours and her help when little time and much work remained will never be forgotten. Our daughter Jennie was born at the same time this book was being conceived. She has grown up with her daddy always either talking about or working on his book, and she has made her special contributions, as has her younger brother, Blake. His passion for Tin Lizzies has been exceeded only by his father's own passion for how they were made. In short, we have sustained each other.

The editorial team of Johns Hopkins University Press deserves special thanks. Henry Tom gave his support early on, and he demonstrated great patience in working with me. Trudie Calvert performed the difficult task of copyediting the manuscript of this book with impressive care and thoroughness and real grace. To them and others not mentioned, I offer my sincere thanks.

I could never have realized, much less addressed, conceptual problems without the work and help of John J. Beer, Merritt Roe Smith, and Eugene S. Ferguson. When I began this study, I set as my goal to equal the quality of John Beer's dissertation and hoped my resulting published work would have the same importance as his. I am not qualified to judge whether I have reached these ends, but I know that Beer's work and his excitement for learning have contributed to my intellectual makeup.

One need only peruse the first chapter of this book to see the magnitude of the debt I owe Merritt Roe Smith. His writings on arms production technology form the basis of my work. Smith's careful criticism of the manuscript and his probing questions have sharpened my work at every turn.

My greatest debt, both intellectual and otherwise, is to Eugene S. Ferguson. He was one of the first historians of technology to stress the importance of the American system of manufactures and the development of mass production technology. In his class on American technology, Ferguson demonstrated the necessity of understanding this historical phenomenon if we are to comprehend fully the nature of technology in nineteenth-century America. This need is the reason I undertook my study. Ferguson continually directed me to important sources and raised questions that never occurred to me. No student ever had a better teacher, and no person could ever have a more devoted friend.

These many individuals and institutions have contributed to the makeup of this book, but the weaknesses that remain are mine and mine alone.

From the
AMERICAN SYSTEM *to*
MASS PRODUCTION
1800–1932

Introduction

Mass production became the Great American Art.
—Paul Mazur, *American Prosperity*

Since the 1920s the term *mass production* has become so deeply ingrained in our vocabulary that we seldom question its origin. The purpose of this study is to determine how the term arose and to provide historical background on the development of mass production in America. Manufacturing in the United States developed along such distinct lines in the first half of the nineteenth century that English observers in the 1850s referred to an "American system" of manufactures. This American system grew and changed in character so much that by the 1920s the United States possessed the most prolific production technology the world has ever known. This was "mass production."

In 1925 the American editor of the *Encyclopaedia Britannica* wrote to Henry Ford asking him to submit an article on mass production for the three-volume supplement to the *Britannica.* Apparently Ford's office, if not Ford himself, responded favorably and promptly set Ford's spokesman, William J. Cameron, to work on the article. Cameron consulted the company's chief production planner about how to state the principles of mass production for the "general reader." When Cameron completed the article, he placed Henry Ford's name beneath it and sent it to the *Britannica's* New York office.[1]

Although Cameron would later say that he "should be very much surprised to learn that [Henry Ford had] read it,"[2] this article played a fundamental role in giving the phrase *mass production* a place in the English vocabulary. Even before the article appeared in the *Britannica,* the *New York Times* had published it as a full-page feature in a Sunday edition, titled "Henry Ford Expounds Mass Production: Calls It the Focussing of the Principles of Power, Economy, Continuity and Speed," and distributed it through the wire service. An editorial on the subject appeared in the same edition.[3] The *Britannica* editor's reason for choosing the term *mass production* is unknown, but there is little doubt that the ghost-written Ford article led to its widespread use and identification with the assembly line manufacturing techniques that were the hallmark of automobile production.[4] After 1925, the term appeared in reference works such as the *Reader's Guide to Periodic Literature* and the *New York Times Index.* It soon superseded the previously popular expression *Fordism.* Thus the article signed by Ford endowed the expression *mass production* with a certain universality despite its ambiguity and its status as poor grammar.*

*Mass was—and most grammarians would say still is—a noun rather than an adjective. The term *mass production* raises the question of whether this is production aimed at the "masses" or merely quantity production.

1

FIGURE 0.1.   A Day's Output of Ford Model T's, Highland Park Factory, 1915. (Henry Ford Museum, The Edison Institute. Neg. No. 0-716.)

Much more important than the story of how *mass production* entered the English vocabulary are the developments that lay behind the production system described in the article. Commenting in 1940 on Henry Ford and the *Britannica* article in his *Engines of Democracy,* Roger Burlingame raised the essential questions:

> With [Ford's] great one-man show moving toward a dictatorship of which any totalitarian leader might well be proud he was ready for what he calls [Mass] Production. [Mass] Production, Ford believes, had never existed in the world before. With the magnificent contempt of men immune to history, he disregards all predecessors: Whitney, Evans, Colt, Singer, McCormick, the whole chain of patient, laborious workers who wrought his assembly lines and all the ramifications of his processes out of the void of handicrafts. In a colossal blurb printed in the *Encyclopaedia Britannica* under the guise of an article on mass production, he writes: "In origin, mass production is American and recent; its notable appearance falls within the first decade of the 20th century," and devotes the remainder of the article and two full pages of half-tone plates to the Ford factory.

Burlingame asked rhetorically, "What are those production methods in use today in every large automobile plant with scarcely any variation? They are simply the methods of Eli Whitney and Samuel Colt, improved, coordinated and applied with intelligent economy— economy in time, space, men, motion, money and material."[5]

FIGURE 0.2. Ford Motor Company, Highland Park Factory Employees, 1915. (Henry Ford Museum, The Edison Institute. Neg. No. 833-700.)

Since the establishment of the history of technology as an academic discipline in the United States, the assertions contained in both Ford's *Britannica* article and Burlingame's popular work have received close study. Indeed, the American system of manufactures, which describes the methods of Eli Whitney, Samuel Colt, Oliver Evans, Isaac Singer, Cyrus McCormick, and others, has become one of the most productive areas of American scholarship in the history of technology.[6] Portions of that new scholarship, combined with the research in this study, indicate that the Ford article came much nearer the truth than did Burlingame and his followers. "In origin," the Ford piece suggested, "mass production is American and recent"—what Whitney et al. did in the nineteenth century was not true mass production. The title of this study suggests that mass production differed in kind as well as in scale from the techniques referred to in the antebellum period as the American system of manufactures, as can be seen most clearly by first considering the American system itself.

Two decades of research on this topic have yielded a number of conclusions, particularly concerning a basic aspect of modern manufacturing, the interchangeability of parts. The symbolic kingpin of interchangeable parts production fell in 1960 when Robert S. Woodbury published his essay "The Legend of Eli Whitney and Interchangeable Parts" in the first volume of *Technology and Culture*. Woodbury convincingly argued that the parts of Whitney's guns were not in fact constructed with interchangeable parts. In 1966 the artifactual research of Edwin A. Battison solidly confirmed Woodbury's more traditional document-based research findings. Eugene S. Ferguson later wrote of Woodbury's pioneering article, "Except for Whitney's ability to sell an undeveloped idea, little remains of his title as father of mass production."[7]

With Eli Whitney reinterpreted as a promoter rather than as a pioneer of machine-made interchangeable parts manufacture,[8] it remained for Merritt Roe Smith to identify conclusively the personnel and the circumstances of this fundamental step in the development of mass production.[9] Smith demonstrated that the United States Ordnance Department was the prime mover in bringing about machine-made interchangeable parts production of small arms. The national armory at Springfield, Massachusetts, played a major role in this process, especially in its efforts to coordinate its operations with those of the Harpers Ferry Armory and John Hall's experimental rifle factory, also at Harpers Ferry. Although these federally owned arms plants were central to its efforts, the Ordnance Department also contracted with private armsmakers. By specifying interchangeability in its contracts and by giving contractors access to techniques used in the national armories, the Ordnance Department contributed significantly to the growing sophistication of metalworking and woodworking (in the case of gunstock production) in the United States by the 1850s.[10] British observers found these techniques sufficiently different from their own as to allude to them as the "American system," the "American plan," and the "American principle."[11]

Although British visitors to the United States in the 1850s, especially Joseph Whitworth and John Anderson, were impressed with every aspect of American manufacturing, small arms production received their most careful and detailed analysis. Certainly this was Anderson's job, for he had been sent to the United States to find out everything he could about small arms production and to purchase armsmaking machinery for the Enfield Armoury. In his report, Anderson indicated that the federal armory at Springfield had indeed achieved what the Ordnance Department had sought since its inception: true interchangeability of parts. Anderson and his committee had designed a rigorous test to verify this achievement, and when they had completed it, they were no longer doubters.[12]

What Anderson was not likely to have known was the extraordinary sum of money that the Ordnance Department had expended over a forty- or fifty-year period, "in order," as an Ordnance Officer wrote in 1819, "to attain this grand object of uniformity of parts." Nor was Anderson necessarily aware that the unit cost of Springfield small arms with interchangeable parts almost certainly was significantly higher than that of arms produced by more traditional methods. He should, however, have known that the Ordnance Department could annually turn out only a relatively small number of Springfield arms manufactured with interchangeable parts. Despite the high costs and limited output, Anderson pointed out that the special techniques used in the Springfield Armory as well as in some private armories could be applied almost universally in metalworking and woodworking establishments.[13] In fact, by the time Anderson reached this conclusion, the application of those techniques in other industries was already under way.

The new manufacturing technology spread first to the production of a new consumer durable, the sewing machine, and eventually it diffused into such areas as typewriters, bicycles, and eventually automobiles. Nathan Rosenberg has provided economic and technological historians with an excellent analysis of a major way in which this diffusion occurred.[14] Rosenberg identified the American machine tool industry, which grew out of the small arms industry (notably Colt's Patent Firearms Manufacturing Co. in Hartford, Connecticut, and Robbins & Lawrence Co. in Windsor, Vermont, and Hartford, Connecticut) as the key agent for introducing armsmaking technology into the sewing machine, bicycle, and automobile industries. The makers of machine tools worked with manufacturers in various industries as they encountered and overcame production problems relating to the cutting, planing, boring, and shaping of metal parts. As each problem was solved, new knowledge went back into the machine tool firms, which then could be used for solving production problems in other industries. Rosenberg called this phenomenon "technological convergence." In many industries that worked with metal, the final products were sold in vastly different markets—the Springfield Armory, for example, "sold" its products to a single customer, the government, whereas sewing machine producers marketed their products among widely scattered individual consumers. Nevertheless, these products were technologically related because their manufacture depended upon similar metalworking techniques. These common needs "converged" at the point where the machine tool industry interacted with the firms that bought its machine tools.

Although he did not emphasize the point, Rosenberg recognized that individual mechanics played an equally important role in diffusing know-how as they moved from the firearms industry to sewing machine manufacture to bicycle production and even to automobile manufacture. Examples of such mechanics abound. Henry M. Leland is an obvious example: he worked at the Springfield Armory, carried this knowledge to the Brown & Sharpe Manufacturing Company when it was manufacturing both machine tools and Willcox & Gibbs sewing machines, and created the Cadillac Motor Car Company and finally the Lincoln Motor Company.[15]

But the process of diffusion was neither as smooth nor as simple as Rosenberg and others would have it. New research suggests that the factories of two of the giants of nineteenth-century manufacturing, the Singer Manufacturing Company and the McCormick Harvesting Machine Company, were continually beset with production problems. Previously, many historians attributed the success of these two companies to their advanced production technology. But it now appears that a superior marketing strategy, including advertising and sales techniques and policies, proved to be the decisive factor.[16]

Although the Singer sewing machine was the product of the colorfully scandalous Isaac Singer, the successful enterprise known as I. M. Singer & Co. (incorporated in 1863 as Singer Manufacturing Company) was primarily the handiwork of lawyer Edward Clark. Clark's success rested on marketing, not on production techniques. The Singer company initially held no technical advantages and no decisive patent monopoly over major competitors because in order to construct a workable sewing machine, four organizations (including Singer) had been forced to pool their patents. In fact, one member of the pool, the Wheeler and Wilson Manufacturing Company, took an early and wide lead until Singer surpassed its production in 1867 (forty-three thousand Singer machines versus thirty-eight thousand Wheeler and Wilson). After 1867, Singer dominated the industry and eventually absorbed Wheeler and Wilson. Wheeler and Wilson had based its production on what contemporaries called ''armory practice,'' that is, the production techniques used at leading armories, such as Springfield. Its manufacturing system was established by three former armsmaking machinists, one trained at Colt's Hartford armory, one who worked at Nathan Ames's armory and for eight years at the Springfield Armory, and the other who had been a contractor at the Robbins & Lawrence–Sharps rifle factory at Hartford.

Unlike Wheeler and Wilson, Singer initially built its machines in a Boston scientific instrument maker's shop and later it rented ''rooms'' in a New York manufacturing district. Not until 1862 did the Singer company hire any mechanic familiar with arms production technology, and then it chose a man whose experiences had been gained in the small, New Jersey–based Manhattan Firearms Company, rather than in one of the great advanced armories of New England. As the company's leader, Edward Clark had emphasized marketing rather than production. In 1855 he wrote a high-level company employee that ''a large part of our own success we attribute to our numerous advertisements and publications. To insure success only two things are required: 1st to have the best machines and 2nd to let the public know it.''[17] ''To have the best machines'' implied not only excellence in design but also quality in manufacture. There was no question in Clark's mind that the Singer approach to manufacture, called the European method because it depended largely on skilled machinists, provided this quality essential for commercial success.

A notable aspect of Singer's marketing strategy—as well as Cyrus McCormick's—was that the Singer machine was deliberately sold at the top of the price list for the industry throughout the nineteenth century. Moreover, Singer maintained its high price for most of this period despite significant growth in production and sales. Its marketing strategy, which in addition to advertising eventually included retail dealerships and service centers and an installment purchasing plan, allowed the company to continue to sell more and more machines at the same price level.[18]

Singer's business continued to expand both in the United States and abroad. By 1880 the firm's world output had reached five hundred thousand machines annually. Singer's factory superintendent, who had been hired away from the Manhattan Arms Company, had gradually introduced special-purpose machinery and had striven toward production of more uniform parts. Yet for a long time, as B. F. Spalding pointed out in the *American Machinist* in 1890, Singer "compromised with the European method [of manufacture] by employing many cheap workmen in finishing pieces by dubious hand work which could have been more economically made by the absolutely certain processes of machinery."[19] The records of the company show conclusively that Spalding was right. In fact, despite the increasing use of a rational jig, fixture, and gauging system* (a hallmark of arms production technology), parts of Singer sewing machines were hand-fitted together by skilled fitters as late as 1883. The inability of Singer's major U.S. factory to meet the continually growing demand for sewing machines finally led the president of the company, who had "worked his way up from the bench," to establish an ad hoc production committee. This committee, which included the president, the factory superintendent, and the superintendent's chief assistants, resolved in March 1883 that "each piece commenced in a department shall be finished there to gauge [sic] ready for assembling and no part shall be made in the department where it is assembled into the machine."[20]

This resolution clearly indicates that extensive hand-fitting and custom machining were done during the process that Singer publicly called "assembly." Try as they would to attain interchangeable parts on Singer machines, however, a Singer official noted almost two years later that the factory was "no further ahead than we were two years ago" in perfecting interchangeable parts manufacture.[21] Whereas Springfield Armory had turned out arms numbering into the thousands constructed with perfectly interchangeable parts, the Singer Manufacturing Company could not achieve the goal at a time when it made a half million sewing machines annually. Singer simply could not afford to lavish the same amount of care in machining and inspection on its sewing machine parts as Springfield did on its muskets. In this connection, one cannot help but notice a central requirement for mass production stated by Ford in the *Encyclopaedia Britannica:* "In mass production there are no fitters."[22] Despite its grand successes in both sales and production, the Singer Manufacturing Company left the development of mass production unfinished

---

*It is important to understand what is meant by "a rational jig, fixture, and gauging system," because it was with this system that firearms makers in the antebellum period were able to produce weapons with interchangeable parts. The system was "rational" because it was based on a model, which in one sense can be interpreted as a kind of Platonic model in that armsmakers viewed the model weapon as an ideal form. All production arms were but imperfect imitations of this ideal (but real) model. Jigs and fixtures are devices to fix or mount workpieces in machine tools. How a workpiece is fixed in a machine tool determines (in part) its accuracy, especially when more than one machining operation is involved. If several operations are performed on a workpiece which requires several different fixtures to hold it in a machine tool or a series of machine tools, accuracy becomes problematic unless the fixtures are designed on some common, rational basis. In the nineteenth century, the model provided this basis. All fixtures were designed with reference to the model, thereby ensuring uniformity. In addition, gauges to verify this uniformity were also constructed. Where dimensions and fits were critical, gauges were made based on the model, or ideal form. With such designed gauges and fixtures, parts produced in machine tools approximated comparable parts of the model.

because it continued to rely upon fitters. The same was even more true at the McCormick Harvesting Company.

Perhaps no major American manufacturing establishment has been more misunderstood than the McCormick reaper works in the nineteenth century. Throughout popular literature of the nineteenth century[23] and in secondary historical literature of this century, the McCormick works is described as a model manufacturing establishment and credited with advanced production techniques. This certainly may have been true when compared to other agricultural implement makers, but when viewed alongside the Springfield Armory or even the Singer Manufacturing Company, the production technology at the McCormick works appears crude. It has long been asserted that Cyrus McCormick adopted the manufacturing techniques developed in New England armories when he established the reaper works in Chicago in 1848. But the firm's founder never took a serious interest in the manufacture of his reaper. He left that to his youngest brother, Leander J. McCormick, who had learned only the craft of blacksmithing before he left the family's Virginia homestead to superintend Cyrus's Chicago factory.

Between that date, 1848, and 1880 there is little evidence that Leander expanded his technical horizons to encompass the developments that have become known as the American system of manufactures. The McCormick factory employed almost no special- or single-purpose machinery, and there is little evidence that Leander knew of the techniques of special gauges, jigs, and fixtures which distinguished the arms industry. Handwork and skilled machine work appear to have prevailed during this period. Moreover, the output of reapers and mowers remained surprisingly small. In 1873, slightly more than 10,000 machines were produced while in 1880, 21,600 machines were made, including some 5,000 of the smaller but more mechanically sophisticated binder attachments. Compared to the half million sewing machines Singer made that year and to the half million Model T automobiles Ford Motor Company produced in 1916, McCormick was manufacturing on a small scale.

Like the Singer machine, the price of McCormick's products was top of the line. From the outset, Cyrus McCormick had marketed his machine aggressively, spending what his brother William considered "enormous" sums of money on advertising. As the years proceeded, McCormick changed his initial agent or distributor system of sales to franchised dealerships supervised by regional office managers.[24] Although these changes resulted in greater sales and the potential for even more, Leander steadfastly refused to allow significant increases in the factory's output. For this reason and for related personal ones, Cyrus McCormick finally fired his brother as superintendent of the factory in 1880 and replaced him with a mechanic who was familiar with the latest production technology. This person, Lewis Wilkinson, had been employed at the Colt armory, the Connecticut Firearms Company, and the Wilson Sewing Machine Company.

The arrival of Wilkinson and his tutelage of Princeton-educated Cyrus McCormick, Jr., played a major role in bringing about radical change in McCormick production methods. Drawing on his experience in small arms production, he introduced the principles of armory practice into the McCormick factory. Although Wilkinson stayed at McCormick for only one year, Cyrus McCormick, Jr., who served as his assistant during that year, learned the principles well. Cyrus, Jr., carried the new approach forward in his "new regime" as superintendent and soon as the chief executive officer of the company. Output under the new regime expanded rapidly.

Despite the introduction of production methods commonly used in American small arms plants, the McCormick company continued to be plagued by the farm implement industry's propensity for what could be termed annual model changes. Indeed, these changes may have been the principal reason Leander McCormick wanted to maintain the more flexible but less productive traditional approach to manufacture during his tenure as superintendent from 1848 to 1880. The perceived necessity to make annual changes in order to keep the McCormick machines attractive in the market imposed severe production limitations on the McCormick factory. In fact, they made it impossible for the McCormick works to become the birthplace of mass production.

At about the same time that the McCormicks were adopting important elements of the American system, a new product was being born that would serve as a bridge between that system and mass production. That new product was the bicycle. The American bicycle industry played a transitional role in the development of mass production for a number of reasons.[25] The physical nature of the product itself clearly provided a stepping stone to the automobile. With important exceptions, early automobile chassis consisted of bicycle tubing and tires, and many early automobile makers were also manufacturers of bicycles. In addition, the safety bicycle introduced the American public to the wonders of personalized transportation, which was probably used more for recreation than for transportation to the workplace. During the 1890s, with more and more Americans riding bicycles (sales in 1896 exceeded 1.2 million bicycles), speed in personalized transportation came to be looked upon as a virtue and as a necessity for a mobile nation, and this attitude hastened the day of the automobile. Furthermore, with the American bicycle industry, advertising grew in importance and sophistication. During this period several commercial artists became famous for their bicycle advertising poster work and advertising layouts in popular journals. But it was in production technology above all that the bicycle left its mark as a transitional industry to mass production.

Joseph Woodworth, author of *American Tool Making and Interchangeable Manufacturing* (1907), argued that the "manufacture of the bicycle . . . brought out the capabilities of the American mechanic as nothing else had ever done. It demonstrated to the world that he and his kind were capable of designing and making special machinery, tools, fixtures, and devices for economic manufacturing in a manner truly marvellous; and has led to the installation of the interchangeable system of manufacturing in a thousand and one shops where it was formerly thought to be impractical.''[26] Clearly the bicycle industry as a staging ground for the diffusion of armory practice cannot be overemphasized. Rosenberg's idea that the machine tool industry played a leading role in this diffusion applies even more clearly to the bicycle than to the sewing machine. The bicycle boom of the 1890s kept the machine tool industry in relatively good health during the serious depression that began in 1893, and it was accompanied by changes in production techniques.

Entirely new developments occurred in bicycle production—sheet metal stamping and electric resistance welding techniques. These new techniques rivaled in importance the diffusion of older metalworking technologies. During the 1890s, bicycle makers located principally but not exclusively in areas west of New England began to manufacture bicycles with many components (pedals, crank hangers, steering heads, joints, forks, hubs, and so forth) made from sheet steel. Punch pressing or stamping operations were combined with the recent invention of electric resistance welding to produce parts at significantly lower costs. This technology would become fundamental to the automobile industry.

Albert A. Pope is regarded as the father of the American bicycle industry because he first imported English ordinary or high-wheel bicycles to the United States and then began to make them here. Pope initially built an effective patent monopoly for his high-wheel Columbia bicycle (the bicycle with the big front wheel), but his patent position faded during the first years of the safety bicycle era (the chain-driven bicycle we know today). For this reason and because no single manufacturer gained a strong patent position, the industry became highly competitive during the bicycle boom which began about 1892–93 and ended abruptly in 1896–97. Nevertheless, Pope had created a large enterprise during the high-wheel era and had (because of his virtual patent monopoly) sold his Columbia at the high price of $125–135. Through aggressive marketing and advertising, he managed to maintain for his safety bicycle both the prestige and the price of the high-wheel Columbia, whose name was also used for the Pope safety bicycle. The Columbia, which was made by methods growing directly out of New England armory practice and refined by sewing machine manufacture, was decidedly the most expensive bicycle manufactured in America. Despite the price, the Pope Columbia, like the Singer machine and the McCormick implement, dominated its industry. At the peak of the boom, Pope Manufacturing Company produced sixty thousand Columbias in a year, each carefully hand assembled and adjusted.

Bicycle makers such as Pope who used traditional armory production techniques looked with disdain at those who manufactured bicycles with parts made by the new techniques in pressing and stamping steel. An executive at the Columbia works called them cheap and nasty.[27] Despite such views, the one manufacturer that outstripped Pope's production at the peak of the bicycle boom was the Western Wheel Works of Chicago, which made a "first class" bicycle out of pressed steel hubs, steering head, sprocket, frame joints, crank hanger, fork, seat, handlebar, and various brackets. Although slightly less expensive than the Columbia, the Western Wheel bicycle ranked high in the top price category among some two to three hundred manufacturers. Production of this bicycle reached seventy thousand in 1896, an output that was significantly less than that of the Ford Model T in 1912, the last full year of its manufacture before introduction of the assembly line.

Singer, McCormick, Pope, and the Western Wheel Works all held one characteristic in common. Although they sold the most expensive products in their respective industries, they were the dominant firms. This fact raises serious questions about the widely held notion that American-made products succeeded in the market because they were cheaply made and low priced. Only Singer annually produced numbers of products ranging into the hundreds of thousands—figures that conjure up in our own minds an image of "mass production." But the techniques used by Singer near the end of the nineteenth century proved problematic. As late as 1883 Singer was still using many fitters, and the manuscript records end before resolution of these problems is apparent. In terms of production, it is only with the rise of the Ford Motor Company and its Model T that there clearly appears an approach to manufacture capable of handling an output of multicomponent consumer durables ranging into the millions each year.

Moreover, the rise of Ford marks an entirely new epoch in the manufacture of consumer durables in America. The Ford enterprise may well have been more responsible for the rise of "mass production," particularly for the attachment of the noun *mass* to the expression, than we have realized. Unlike Singer, McCormick, and Pope, Ford sought to manufacture the *lowest* priced automobile and to use continuing price reductions to produce ever greater demand. Ford designed the Model T to be a "car for the masses." Before the era of the Model T, the word *masses* had carried a largely negative connotation, but with such a clearly stated goal and his company's ability to achieve it, Ford recognized "the masses"[28] as a legitimate and seemingly unlimited market for the most sophisticated consumer durable product of the early twentieth century. Whether Henry Ford envisioned "the masses" as "the populace or 'lower orders' "[29] of late nineteenth-century parlance or merely as a large number of potential customers hardly matters, for the results were the same. Peter Drucker long ago maintained that Ford's work demonstrated for the first time that maximum profit could be achieved by maximizing production while minimizing cost. He added that "the essence of the mass-production process is the reversal of the conditions from which the theory of monopoly was deduced. The new assumptions constitute a veritable economic revolution." Drucker saw mass production as an economic doctrine as well as an approach to manufacture. For this reason if for no other, the work of the national armories, Singer, McCormick, Pope, et al., differed substantially from Ford's. But Ford was able to initiate this new "economic revolution" because of advances in production technology, especially the assembly line.[30]

Before their adoption of the revolutionary assembly line in 1913, Ford's production engineers had synthesized the two different approaches to production that had prevailed in the bicycle era. First, Ford adopted the techniques of armory practice. All of the company's earliest employees recalled how ardently Henry Ford had supported efforts to improve precision in machining. Although he knew little about jig, fixture, and gauge techniques, Ford became a champion of interchangeability within the Ford Motor Company, and he hired mechanics who knew what was required to achieve that goal. Certainly by 1913, most of the problems of interchangeable parts manufacture had been solved at Ford. Second, Ford adopted sheet steel punch and press work. Initially he contracted for stamping work with the John R. Keim Company in Buffalo, New York, which had been a major supplier of bicycle components. Soon after opening his new Highland Park factory in Detroit, however, Ford purchased the Keim plant and promptly moved its presses and other machines to the new factory. More and more Model T components were stamped out of sheet steel rather than being fabricated with traditional machining methods. Together, armory practice and sheet steel work equipped Ford with the capability to turn out virtually unlimited numbers of components. It remained for the assembly line to eliminate the remaining bottleneck—how to put these parts together.

The advent of line assembly at Ford Motor Company in 1913 is one of the most confused episodes in American history. Although a detailed version of those events is recounted in Chapter 6, some general observations are needed here. The assembly line, once it was first tried on April 1, 1913, came swiftly and with great force. Within eighteen months of the first experiments with moving line assembly, assembly lines were used in almost all subassemblies and in the most symbolic mass production operation of all, the final chassis assembly. Ford engineers witnessed productivity gains ranging from 50 percent to as much as ten times the output of static assembly methods. Allan Nevins accurately called the moving assembly line "a lever to move the world."[31]

There can be little doubt that Ford engineers received their inspiration for the moving assembly line from outside the metalworking industries. Henry Ford himself claimed that the idea derived from the "disassembly lines" of meatpackers in Chicago and Cincinnati. William Klann, a Ford deputy who was deeply involved in the innovation, agreed but noted that an equally important source of inspiration was flour milling technology as practiced in Minnesota. Klann summarized this technology in the expression "flow production."[32] Of course, early twentieth-century flour milling technology had clear antecedents in the automatic flour mill developed by the Delawarean Oliver Evans. For this reason, one might agree with Roger Burlingame that Ford's mass production owed much to Oliver Evans, a debt never recognized in Ford's *Britannica* article. Although there is merit to this view, it should be recalled that Evans's flour mill, especially its flour handling machinery, represented a brilliant synthesis of existing components, not an entirely new technology.[33] Similarly, although there may have been a clear connection in the minds of Ford engineers between "flow production" and the moving assembly line, there is little justification for saying that the assembly line came directly from flour milling. Both materials and processes were too different to support such a view.

The origins of the Ford assembly line are less important than its effect. While providing a clear solution to the problems of assembly, the line brought with it serious labor problems. Ford's highly mechanized and subdivided manufacturing operations already imposed severe demands on labor. Even more than previous manufacturing technologies, the assembly line implied that men, too, could be mechanized. Consequently, during 1913 the Ford company saw its annual labor turnover soar to 380 percent and even higher.[34] Henry Ford moved swiftly to stem this inherently inefficient turnover rate. On January 5, 1914, he instituted what became known as the five-dollar day. Although some historians have argued that this wage system more than doubled the wages of "acceptable" workers, most recently the five-dollar day has been interpreted as a plan whereby Ford shared excess profits with employees who were judged to be fit to handle such profits.[35] In any case, the five-dollar day effectively doubled the earnings of Ford workers and provided a tremendous incentive for workers to stay "on the line." With highly mechanized production, moving line assembly, high wages, and low prices on products, "Fordism" was born.

During the years between the birth of "Fordism" and the widespread appearance of the term *mass production,* the Ford Motor Company expanded its annual output of Model Ts from three hundred thousand in 1914 to more than two million in 1923. In an era when most prices were rising, those of the Model T dropped significantly—about 60 percent in current dollars. Throughout the Model T's life, Henry Ford opened his factories to technical journalists to write articles, series of articles, and books on the secrets of production at Ford Motor Company. Soon after the appearance of the first articles on the Ford assembly lines, other automobile companies began putting their cars together "on the line." Manufacturers of other consumer durables followed suit. Ford's five-dollar day forced automakers in the Detroit vicinity to increase their wage scales. Because Ford had secured more than 50 percent of the American automobile market by 1921, his actions had a notable impact on American industry.

Ford's work and its emulation by other manufacturers led to the establishment of what could be called the ethos of mass production in America. The creation of this ethos marks a significant moment in the development of mass production and consumption in America. Certain segments of American society looked at Ford's and the entire automobile industry's ability to produce large quantities of goods at surprisingly low costs. When they did so, they wondered why, for example, housing, furniture, and even agriculture could not be approached in precisely the same manner in which Ford approached the automobile.

Consequently, during the years that the Model T was in production, movements arose within each of these industries to introduce mass production methods. In housing, an industry always looked upon as one of the most staid and preindustrial, prefabrication efforts reached new heights. Foster Gunnison, for example, strove to become the "Henry Ford of housing" by establishing a factory to turn out houses on a moving assembly line,[36] and Gunnison was only one among many such entrepreneurs. Furniture production also saw the influence of Ford and the automobile industry. In the 1920s a large number of mechanical engineers in America banded together within the American Society of Mechanical Engineers (ASME) in an effort to bring the woodworking industry into the twentieth century—the century of mass production.[37] Consequently, the ASME established in 1925 a Wood Industries Division, which served to focus the supposed great powers of mechanical engineering on all aspects of woodworking technology. In agriculture, Henry Ford argued that all problems could be solved simply by adopting mass production techniques.[38] Ford conducted experiments in this direction, but he was no more successful than the mechanical engineers and housing fabricators were in bringing about mass production in their respective industries. One could argue, however, that today such an agricultural product as the hybrid tomato, bred to be picked, sorted, packaged, and transported by machinery, demonstrates that mass production methods have penetrated American agriculture. But furniture and housing seem to have no equivalent to the hybrid tomato.

A conclusive exploration of why mass production in housing, furniture, and some other industries failed to take hold must be the subject of another study but is worthy of speculation. One hypothesis has been explored in recent seminars at the University of Delaware on material culture, economic history, and the history of technology. In housing, furnishings, and clothing, Americans for some reason refused to allow their tastes to succumb to mass production techniques and its concomitant standardization. Certainly technology itself was not the limiting factor. Gunnison actually assembled houses on a line in a factory. Yet he sold few houses in comparison with the number of on-site, traditionally built houses in the United States. Singer Manufacturing Company built two large woodworking factories that produced cabinets and tables for its entire U.S. and European output of sewing machines. But the production of a phenomenally large number of sewing machine cabinets failed to lead to a true mass production furniture industry.[39] American furniture manufacturers continued to operate relatively small factories employing around 150 workers, annually turning out between five thousand and fifty thousand units. Beliefs that automotive production technology holds the key to abundance in all areas of consumption persist today. As recently as 1973, Richard Bender observed in his book on industrial building that ''much of the problem of industrializing the building industry has grown out of the mistaken image of the automobile industry as a model.''[40] In many areas, the panacea of Fordism will continue to appeal to those who see in it solutions to difficult economic and social problems. The ethos of mass production, established largely by Ford, will die a hard death, if it ever disappears completely.

Yet the very timing of the rise of this ethos along with the appearance of the *Encyclopaedia Britannica* article on mass production shows how full of paradox and irony history is. Although automotive America was rapidly growing in its consumption of everything under the sun and although Ford's achievements were known by all, mass production as Ford had made it and defined it was, for all intents and purposes, dead by 1926.[41] Ford and his production experts had driven mass production into a deep cul-de-sac. American buyers had given up on the Ford Model T, and the Ford Motor Company watched its sales drop precipitously amid caustic criticism of its inability to accept and make changes. In mid-1927, Henry Ford himself finally gave up on the Model T after 15 million of them had been produced. What followed in the changeover to the Model A was one of the most wrenching nightmares in American industrial history. Designing the new model, tooling up for its production, and achieving satisfactory production levels posed an array of unanticipated problems, which led to a long delay in the Model A's introduction.

In some respects, the Ford Motor Company never recovered from the effects of its first big changeover. Changes in consumers' tastes and gains in their disposable incomes made the Model T and the Model T idea obsolete. Automobile consumption in the late 1920s called for a new kind of mass production, a system that could accommodate frequent change and was no longer wedded to the idea of maximum production at minimum cost. General Motors, not Ford, proved to be in tune with changes in American consumption with its explicit policy of "a car for every purpose and every purse," its unwritten policy of annual change, and its encouragement of "trading up" to a more expensive car. Ford learned painfully and at great cost that the times called for a new era, that of "flexible mass production."

The Great Depression dealt additional blows to Ford's version of mass production. As dramatic decreases in sales followed the Great Crash, Ford and the entire industry began laying off workers. As a result, Detroit became known as the "beleaguered capital of mass production."[42] Mass production had not prevented mass unemployment or, more properly, unemployment of the masses but seemed rather to have exacerbated it. Overproduction had always posed problems for industrial economies, but the high level of unemployment in the Great Depression made mass production an easy culprit for critics as they saw hundreds of thousands of men out of work in the Detroit area alone.

Americans may have had concerns about the ill effects of mass production, but they by no means were willing to scrap it. Already their desire for style and novelty, coupled with increased purchasing power in the 1920s, had forced even Henry Ford to change his system of mass production. When pushed by the Depression, the greater part of Americans looked for solutions in the sphere of mass consumption. The 1930s witnessed the publication of extensive literature on the economics of consumption.[43] As history would have it, the prophets of mass consumption were proven at least temporarily correct as the United States pulled itself out of the Depression by the mass consumption of war materiel and, after the war, by the golden age of American consumption in the 1950s and 1960s.

Today, however, when we live in a period labeled variously as the spage age, the information era, the nuclear age, the computer society, and postindustrial civilization, mass production and mass consumption have lost much of their centrality as concerns shared by Americans. There are few discussions about mass production today that mirror those of the late 1920s and early 1930s. Nevertheless, one still reads about our nation's "productivity dilemma"—the problem of choosing between frequent product changes and lower productivity or no change and higher productivity.[44] This so-called dilemma is by no means new. It was born with the establishment of the ethos of mass production and the new consumption patterns of the late 1920s. Henry Ford, whose company brought mass production into being, well knew the productivity dilemma, even though he seems never to have been able to resolve it. Indeed, the dilemma itself may be insoluble. Yet the origins of the dilemma in the "American system" of manufactures in antebellum America provide an important starting point for more clearly understanding its dimensions.

CHAPTER 1

The American System of Manufactures in the Antebellum Period

[Americans] call in the aid of machinery in almost every department of industry. Wherever it can be introduced as a substitute for manual labour, it is universally and willingly resorted to.
—Joseph Whitworth, *Special Report* (1854)

The two national armories of Springfield and Harper's Ferry, the private establishments of Colonel Colt, Robbins and Lawrence, and the Sharpe's Rifle Company, are all conducted on the thorough manufacturing system, with machinery and special tools applied to the several parts. . . . Besides the machinery and tools . . . there are hundreds of valuable instruments and gauges that are employed in testing the work through all its stages, from the raw material to the finished gun, others for holding the pieces whilst undergoing different operations, such as marking, drilling, screwing, etc., the object of all being to secure thorough identity in all parts.
—Report of the Committee on Machinery of the United States of America (1855)

Most American historians know the American system of the antebellum period as the political program put forward by Henry Clay in his 1824 tariff speech before the United States House of Representatives. Clay sought political measures to maintain and promote American industry and to eliminate foreign competition. As he outlined it, Clay's plan "consist[ed] in modifying our foreign policy, and in adopting a genuine American system." One contemporary succinctly summarized Clay's aims: "Internal improvement, and protection of American interests, labor, industry and arts, are commonly understood to be the *leading* measures, which constitute the American system."[1] Clay's program became a major part of Whig politics during the party's heyday.

To historians interested in nineteenth-century American technology, however, the American system denotes an entirely different phenomenon. To them, the American system (often called the American system of manufactures or the American system of manufacturing), as defined by Eugene S. Ferguson, means manufacturing involving "the sequential series of operations carried out on successive special-purpose machines that produce interchangeable parts."[2] Unfortunately, the American system of manufactures had no Henry Clay; the expression cannot be traced to a single proponent who made or wrote something analogous to Clay's tariff speech of 1824. In fact, during the last three decades the term has taken on an almost legendary or mythical character. Without citing any specific sources, historians writing about the American system of manufactures usually say that the expression's origins trace to England in the 1850s. As Nathan Rosenberg noted in his introduction to *The American System of Manufactures,* an edition of three British reports on American manufacturing written in 1854 and 1855, "It was at the Crystal Palace Exhibition [1851] that many Englishmen were first familiarized, through an examination of American products, with productive methods which seemed so novel and original that they were promptly dubbed 'The American System of Manufacturing.' " (See Figure 1.1.) Rosenberg did not indicate any appearance in the Crystal Palace Exhibition literature of the expression. It did not appear. Moreover, in the *Special Reports* of George Wallis and Joseph Whitworth (1854) and the *Report of the Committee on the Machinery of the United States of America* (1855), the complete term never appears. John Anderson, the author of the latter report, did once use the words "American system" in his discussion of musket production in the United States. But given the number of times he used "system" and his choice of such words as "principle," "plan," "arrange-ment," and "mode," it is doubtful that "American system" meant anything more to him than did "American principle" or "American method." A close reading of the report indicates that Anderson employed "system" frequently and in a variety of usages.[3] Anderson and his contemporaries used "system" in ways that generally conformed to several definitions given in the *Oxford English Dictionary.* Unlike historians speaking of the "American system of manufacturing," they did not consciously endow "system" with great significance or with transcendent qualities.

FIGURE 1.1.   United States Exhibit at the Crystal Palace Exhibition in London, 1851. Samuel Colt's display of revolvers was given a central place in the *Illustrated London News*'s pictorial account of the American exhibit. (*Illustrated London News,* December 6, 1851. Library of Congress Photograph.)

Among historians, the expression probably gained currency because of Joseph Wickham Roe's early and still influential *English and American Tool Builders,* first published in 1916. Roe wrote that the ''system of interchangeable manufacture is generally considered to be of American origin. In fact, for many years it was known in Europe as the 'American System' of manufacture.'' In another passage he argued that the interchangeable system ''was known everywhere as 'the American system.'" To document this assertion, the Yale mechanical engineer–historian relied upon the autobiography (1883) of James Nasmyth, the English inventor and machine tool builder who was a keen observer of British and American manufacturing practice. Writing about the British government's decision in 1853–54 to change its small military arms and arms procurement procedures, Nasmyth noted, ''It was finally determined to improve the musketry and rifle systems of the English army. The Government resolved to introduce the American system, by which arms might be produced much more perfectly, and at a great diminution of cost.'' Roe italicized Nasmyth's words, ''introduce the American system,'' to emphasize and to document his earlier claim that interchangeability was ''known everywhere as 'the American system.'"[4] Relying on Roe's analysis, historians from John E. Sawyer, who in 1954 published the seminal article ''The Social Basis of American System of Manufacturing,'' through H. J. Habakkuk, Nathan Rosenberg, Paul Uselding, and others, have all spoken of a clear, unambiguous, distinct, and meaning-laden ''American system of manufacturing.''[5] With notable exceptions, however, British contemporaries, who are supposed to have invented and commonly used the expression, saw a great deal of ambiguity in the originality, procedures, capability, and potential of American manufacturing methods. These ambiguities were genuine. Understanding them provides an excellent way to begin a critical discussion of the American system of manufactures in the antebellum period. (For a brief chronological narrative of the words and expressions used to describe manufacturing methods in nineteenth-century America, see Appendix 1.)

British Study of American Manufacturing Methods in the 1850s

In 1854 Parliament established a select committee ''to consider the Cheapest, most Expeditious, and most Efficient Mode of providing *Small Arms* for *Her Majesty's Service.*'' Britain was in the midst of the Crimean War, and during the three previous years the Board of Ordnance had had difficulty in procuring enough muskets for its troops. To deal with this problem, the Board of Ordnance proposed that the British government establish a small arms factory modeled after the United States federal armories at Springfield, Massachusetts, and Harpers Ferry, Virginia. One of the select committee's duties, therefore, was to judge the political, economic, and technological feasibility of the Board of Ordnance's proposal. Naturally, the committee was also to study in detail the ''existing system'' by which the Board of Ordnance secured small arms.[6]

The British obtained their military small arms through a system of contracting with private manufacturers located principally in the Birmingham and London areas. The fabrication of individual musket parts and the fitting together and finishing of the parts that had passed the government's inspection were contracted for separately. Although significant variations occurred, almost all of the contractors manufactured parts or fitted them through a highly decentralized, putting-out process using small workshops and highly skilled labor. In small arms making as in lock production, the ''workshop system'' rather than the "factory system" was the rule.[7] One alternative to the existing system of procurement was to mechanize armsmaking in a centralized, government-owned establishment.

The Select Committee on Small Arms summarized the Board of Ordnance's position on such an establishment. Advantages of machine-made muskets included "cheapness in the manufacture, an exact similarity in the several parts, so that they may be readily interchanged and replaced, and above all, the facility of rapidly producing muskets." The committee noted that two factories in the United States—the Springfield Armory and the Harpers Ferry Armory—had clearly demonstrated these advantages. Each, it reported, was "capable of producing 30,000 muskets a year.[8] The Board of Ordnance also called the committee's attention to the London armory of Samuel Colt, established shortly after the Crystal Palace Exhibition and employing American armsmaking techniques similar to those employed at his Hartford, Connecticut, armory. (See Figure 1.2.)

The Board of Ordnance and in turn the select committee had obtained most of their information on American armsmaking technology from Colt and from Joseph Whitworth. A noted Manchester machine tool manufacturer, Whitworth had toured many U.S. arms plants, including Springfield and Harpers Ferry, while serving as a commissioner in 1853 to the New York Crystal Palace Exhibition. His *Special Report* on American manufacturing proved to be of major interest to the British. Another commissioner, George Wallis, had also seen the Springfield Armory and had written a report on his U.S. tour. He, too,

FIGURE 1.2.   Samuel Colt's London Armory, 1854. Located beside the Thames, in the Millbank area near Vauxhall Bridge, Colt's factory began production on January 1, 1853. Note that both the Union Jack and Old Glory were flown. (*Household Words*, May 27, 1854. From Charles T. Haven and Frank A. Belden, *A History of the Colt Revolver*.)

served as an informant to the Board of Ordnance. Colt, Whitworth, and Wallis were among the experts called to testify before the committee. Other witnesses included the machine tool maker, James Nasmyth; a former superintendent at Colt's armory, Gage Stickney; a noted English machine builder, Richard Prosser; and the Board of Ordnance's technical expert, John Anderson.[9] Anderson was the author of the board's proposal to establish a small arms plant. After a four-month study trip to the United States, he was the most ardent advocate of American small arms production technology and probably the most knowledgeable British engineer on the subject.[10] The testimony of all of these men, however, suggests the ambiguities of the system adopted in the United States armories to produce muskets.

Beyond its study of the existing system of procuring small arms for the British government, the select committee sought primarily to obtain information about American methods of arms production. For the purposes of analysis, the committee's investigation can be classified into five categories of questioning: (1) whether small arms could be produced by machines; (2) whether mechanics and machine tool builders could produce a weapon; (3) what effect mechanized production would have; (4) whether arms made by machines would contain interchangeable parts (a corollary question was what ''interchangeable'' meant); and (5) whether the Americans indeed had pioneered in this approach to production. These lines of questioning will be discussed in turn.

Although it may seem strange to the modern observer, the parliamentary select committee first explored the question of whether any small arm, particularly that of the British service, could be made by machinery. Some members were interested in the question of technical feasibility while others wondered about economic feasibility. For example, John Anderson was asked whether he was ''aware that the same principle has been tried in the Netherlands of manufacturing guns by machinery, and given up, because it was more expensive.'' Anderson answered that he had not heard of this occurrence, but he argued throughout his testimony that small arms could and should be made by machinery. Samuel Colt's London armory, Anderson said, provided abundant evidence that such an approach was feasible. He quoted a report he had written on Colt's factory, which he said had been ''reduced to an almost perfect system.'' Colt himself assured the committee that ''there is nothing that cannot be produced by machinery,'' including the British musket. Nasmyth, Whitworth, and Prosser concurred, and each argued—Nasmyth most strenuously—that the present system of arms manufacture was by no means the most rational allocation of resources but merely a patchwork of ancient customs and practices. In fact, Nasmyth implied that hitherto the government had rejected certain articles simply because they were made by machinery.[11]

Their questions showed that some members of the committee doubted whether Nasmyth and his colleagues, as machine builders, could manufacture muskets, rifles, or even machines for their manufacture. None had any inkling of the phenomenon Nathan Rosenberg has called ''technological convergence,'' whereby the machine tool industry could serve a critical role in developing the convergent technology of light manufacturing.[12] As will be seen later in this chapter, Samuel Colt's dictum that ''there is nothing that cannot be produced by machinery'' and the very organization of his American armory suggests that he understood the crucial role of toolmaking in manufacturing even though members of the select committee may not have.

Given the overwhelming evidence that the British musket could indeed be made by machinery, there remained the question of what effect mechanized production would have. The select committee had been established in part to consider the impact of a mechanized government armory on private manufacturers. While important to the committee, this essentially political question need not concern us. Much more pertinent was the inquiry about the effect of mechanization on the cost and the quality of the arms and the extent to which labor would be employed. These questions brought to the fore the ambiguities of the American system in the antebellum period. Expert opinion on that system differed significantly.

Those experts who advocated the establishment of a mechanized government armory argued consistently that the end result of this system would be a lower-cost arm. Each spoke of cost reduction resulting from mechanization as an article of faith because none presented any conclusive evidence that this would indeed be the case. John Anderson's testimony typified the proponents' position on cost. When asked, "Do you think you will really make [the musket] cheaper?" Anderson replied, "There is no certainty in anything that is not done; I am sure the Government could not lay out money to better advantage; I cannot prove it very well."[13] Earlier a member of the select committee had suggested to Anderson that the British arms industry, by its very existence and enormous annual output, had found the most efficient way to allocate resources. Anderson replied that he had spoken to some English gunmakers who "acknowledge that it would have been better if they had had more machinery to assist them." Yet many of the armsmakers who testified thought that extensive mechanization would not pay in England. London gunmaker Charles Clark took a strong position, arguing that Samuel Colt had sought an extension of his United States patent "in order to cover the losses that he has sustained by the machinery that he has put up here." Clark quoted at length the report of the Committee on Patents to the House of Representatives, which baldly stated that Colt should receive a seven-year extension because of his activities in England.[14]

Colt had convinced the representatives on the Committee on Patents that he had established his London revolver factory solely to protect himself from the introduction of spurious imitations of his pistol in England and Europe. Only by employing American-made machinery operated by American mechanics, Colt argued, could he maintain the necessary quality of his firearm. Yet, as the House committee report noted, the London armory was "a constant drain on the resources and energies of the inventor." Thus Colt believed that his missionary activities in London justified his request for a patent extension.[15]

Clark argued before the select committee that Colt had "recklessly advanced capital in England, and set up expensive machinery that was not paying but was a great loss."[16] The discussion of costs of machine-made arms revolved around the question of how much labor would be saved. Would the employment of machinery eliminate labor in the manufacture of small arms?

Samuel Colt's affirmative answer was implicit in his suggestion that anything could be made by machinery. James Nasmyth stated explicitly that "with properly contrived machinery you might reduce the employment of manual labour, I may say down to zero." Since Nasmyth of course recognized the necessity for machine operators, he was referring to skilled hand labor used in the fabrication of musket parts. For reasons that will be discussed below, Nasmyth's more famous colleague, Joseph Whitworth, disagreed with such a radical position.[17] Not everything could be mechanized, Whitworth argued. He was joined in this view by Richard Prosser. Hand labor—*skilled* hand labor—would always be required. Prosser argued that "labour-saving machinery" would probably save only 50 percent of the cost of labor required under the existing system. His calculation included the wages or rates of machine operators as well as skilled hand laborers. A report of the Board of Ordnance, which had been quoted extensively in the select committee's hearings, had said that "from what Colonel Colt stated, . . . the amount of skilled and unskilled labour did not exceed 20 per cent [of the cost of the revolver]."[18]

Gage Stickney's testimony, however, confirmed Prosser's estimation of 50 percent savings with data from Colt's showplace London armory and from the Springfield Armory. Stickney certainly possessed impressive credentials to judge the question. For six years he had worked for Samuel Colt, and for fifteen months he had served as superintendent at Colt's London armory. "As near as I can judge," Stickney testified, "I should say about 50 per cent" labor was required in mechanized arms manufacture. But when questioned further, Stickney made it clear that machinery had not eliminated the need for skilled labor. Even the Colt pistol required "first-class labour and the highest price is paid for it." Replying to further questions, Stickney stressed that each part of Colt's arm was finished by skilled labor "if it is done properly." No arm of quality, he generalized, could be produced without skilled hand labor, and the more skilled labor used, the higher would be the quality.[19] (See Figure 1.3.)

Such testimony raised the important question of how the supposedly machine-made arms of the Americans compared in quality to those of the British. William Richards, a Birmingham barrelmaker who had developed a new process for barrelmaking, believed Colt's arm to be of lower quality than the British musket.[20]

Joseph Whitworth agreed. When asked if "the work in Colonel Colt's pistols is as good as that in the Minié rifle," he replied, "No, I do not consider that the parts are as well finished." The committee asked Whitworth if it would be possible to produce a machine-made Minié rifle that was as good as one made by hand. He replied, "Yes; but it would have to be finished by hand; the quality would depend entirely upon the finish after it came from the machine." As in other instances, Nasmyth's opinion was opposite to Whitworth's. He told the committee that he considered the Colt pistol superior in workmanship to the British rifle.[21] Colt's testimony suggests that he vehemently disagreed with the gunmakers and with Whitworth. His reasons provide an important insight into why American armsmaking technology had followed the path that it had.

When addressing the committee's question of whether "muskets manufactured by machinery in America are as well fabricated as the Minié rifle," Colt stated, "There is none so badly made at *our* national armouries as the Minié rifle shown to me; that arm would not pass one of our inspectors." Asked why he held this view, Colt said that American-made arms were "more uniform" than their English counterparts. Lack of uniformity was the only fault he found with English-made weapons: "There is more difference between one and another where they are made by hand, than there can possibly be when they are made by machinery." In America, Colt argued, "it is the uniformity of the work that is wanted."[22]

Colt's reasoning, his prior claims about the uniformity of his pistol parts, and his very presence in England raised fundamental questions about the interchangeability of parts made, in the words of one committee member, "upon the Springfield principle."[23] The committee's interrogation of John Anderson on this issue suggests their doubts and, more important, Anderson's own uncertainties about the interchangeability of parts. Seeking clarification of an earlier statement by Anderson, a committee member asked him if he meant "to say that they will be so exactly alike, that of the 150,000 muskets any one part from one musket, if it failed, could be replaced by another part." Anderson hedged his reply: "That is the point that I would aim at; I think that we would come to that ultimately; but that is a very difficult point to get to; that would be the great difficulty, for that would

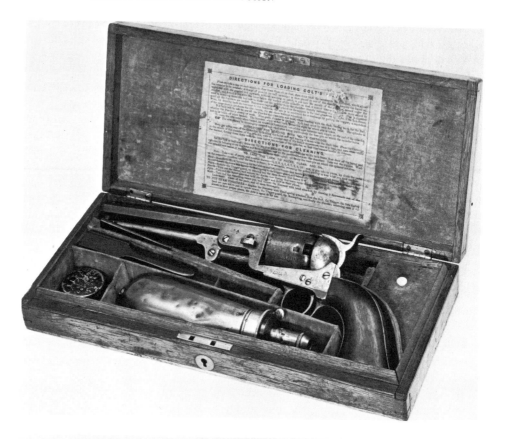

FIGURE 1.3.   London-Made Colt Revolver. This revolver was one of ninety-five hundred 1851 Navy Model revolvers that Colt's London armory produced for the British navy during its short history of operations. The arrow on the barrel is the British proof mark. The lower photograph identifies the place of manufacture. (Private collection. Eleutherian Mills Historical Library.)

be perfection." Nevertheless, he said optimistically, "I think we shall come very nearly to it, but to say that we should come altogether to it, would be saying that it would be perfect." The committee asked if the Springfield Armory muskets were truly interchangeable, did that mean they were perfect? Anderson could only reply, "They say that is the case, and I have heard that it is the case; I can imagine things so perfect, and it would be perfection if they were so." James Nasmyth presented a complementary and more certain view. He argued that by implementing a mechanized "system of manufacture" for a musket, one could achieve "the unerring precision of mechanical tools." Complete mechanization, Nasmyth stressed, "would result in impressing absolute identity on the parts," whereby they would be "so perfect as to fit promiscuously." He noted that this would be the "inevitable result of the introduction of machinery."[24]

While Anderson and Nasmyth asserted that in principle perfect interchangeability could and would be achieved by mechanization, Joseph Whitworth argued—convincingly as far as the committee was concerned—that in principle perfect interchangeability would be impossible. Under any circumstances, he testified, machine-made gun parts would never be so accurate that they did not have to be fitted together by hand labor.[25]

Whitworth explained his reasons for this assessment. The process of hardening parts made interchange impossible because iron always changed its shape during and after this process (parts were worked or machined in a soft state and then hardened for their final use). Even if parts fitted together nicely before hardening, they would not do so after it. They had to be "restored." The eminent machine tool builder and master of precision argued that this could be done "only by hand labour." "Whenever we want great perfection of parts we must do it by hand labour," Whitworth concluded.[26]

Having received significant differences in opinion about the possibility in principle of perfect interchangeability, the committee sought information about how the idea fared in practice. Whitworth expressed definite opinions. He did not think that parts of Colt revolvers could be taken indiscriminately and assembled into a complete weapon, "nor do I think it is possible in an American musket, although there is such care taken." Although not perfectly interchangeable, American-made parts possessed an important uniformity. "I think this," Whitworth argued, "that on a field of battle, if a great number of muskets were disabled, made by the American mode of manufacture, that the armourer would be able to put a greater number of parts together; whereas by the mode of manufacture in this country he could not do that."[27]

Far different than Whitworth, James Nasmyth steadfastly maintained that parts of Colt revolvers made in London were interchangeable. Basing his case on Colt's gauging methods, Nasmyth said that "the means which he employs would attain that object [of interchangeability]."[28]

Colt himself hedged on the issue. When asked if "interchange of parts is possible," Colt replied, "Yes, it is so near that it would not cost you very much to interchange them." A member then queried Colt if "this interchange" had been realized in America. "Yes, many thousands," Colt snapped back and then told the committee about service of his arms in Oregon and California. When asked earlier whether his revolver parts could be assembled indiscriminately, however, Colt had testified that "they would do that a great deal better than any arms made by hand." Later he said, "In my own arms one part corresponds with another very nearly," and blamed English workers for producing less uniform parts than those made in his American factory. William Richards of Birmingham had purchased a half dozen Colt revolvers while the committee's hearings were proceeding. He testified categorically that "there are no two parts that will change round."[29]

Colt's former superintendent had already said that he had never seen a case when the parts of one Colt pistol would interchange with another. Gage Stickney added much additional confusion to the committee's deliberations on interchangeability when a member asked him what a U.S. congressional committee had meant by the word "interchangeability" in a report on Colt's application for a seven-year extension of his revolver patent. "I do not know what they mean," Stickney replied. The committee then asked if the congressional committee had been mistaken in its report. Stickney could only respond that by "interchangeable" the report meant that parts were "very near alike." But, he argued, it required a skilled armorer to make parts interchange, if they would at all. From such divided opinion on what "interchange" was, and whether perfect uniformity could be achieved in principle and in practice, it was difficult for the committee to derive any solid conclusion about the interchangeability aspect of the "Springfield principle." Ambiguities existed.[30] The committee also heard conflicting testimony about the originality of this "almost perfect system," as John Anderson had called it in his testimony.

Although James Nasmyth had appeared as the most ardent champion of the mechanization of production, particularly as exemplified at Colt's London armory, he nevertheless hastened to point out to the committee that Colt and the Americans were simply "carrying out those systems that were originated by Sir Samuel Bentham, Mr. [Henry] Maudslay, and [Marc Isambard] Brunel, in the block machinery." Nasmyth was alluding to special-purpose machinery constructed in the first decade of the 1800s to produce ships' blocks for the British navy. (Details of this machinery will be discussed later in this chapter.) On this issue, Whitworth agreed with Nasmyth. He, too, believed that the blockmaking machinery had demonstrated conclusively the merits of using special-purpose machinery in a manufacturing process.[31] Of all the witnesses, however, Richard Prosser was by far the most opinionated on the issue of American originality.

Prosser suggested that the Americans were not the first to introduce "labour-saving machinery" in gunmaking. That honor fell to a Birmingham firm, which in the 1810s had equipped an armsmaking plant for the Russian government at Tula. Prosser showed the committee a Russian book on the plant which contained forty-two illustrations of the machinery. He told the committee about the Tula operation, stating that every part of the Russian musket, except the stock, was made "by means of labour-saving machines." Prosser noted that in 1822 a Mr. Fairy [John Farey?] toured the Russian plant and witnessed twelve soldiers disassemble twelve muskets, put the parts in a basket, then reassemble the randomly selected parts, and fire the resulting muskets, all within two minutes. These guns, Prosser said, had been manufactured by English machinery made by James & Jones. Later Prosser emphasized the point by alluding to mechanized arms production not as the "American system of manufacturing" but as the "Russian plan."[32] Prosser showed his contempt for American technology:

> At the Great Exhibition the place was filled with inventions; for instance American reaping machines. I said, "This is an old invention." I thought the Yankees were going to eat me, and they got a council medal for it [i.e., the reaper]. . . . It proves now to be an English invention brought back from America, and put up here as an American invention, and they are constantly doing that whereas all the improvements in the working of wood that I have ever seen are due entirely to General Bentham's patent of 1791 and 1793.

Prosser was obviously alluding to the much-hailed gunstocking machinery, which was at the heart of the "Springfield principle."[33]

Despite the questions of priority and despite the ambiguities concerning quality and interchangeability, the select committee recommended that "manufacturing of Small Arms under the Board of Ordnance should be tried to a limited extent."[34] Tacitly it approved of the Board of Ordnance's desire not only to model its armory after the Springfield and Harpers Ferry armories but to equip it with American-made machinery. The committee's hearings had raised most of the important questions concerning the American system of arms production and consequently had brought out most of the ambiguities of the system. But the committee failed in one important respect. It had not carefully proved nor even begun to appreciate what Hyde Clark would soon after call the "artificial circumstances" that had given rise to the American system of arms production.[35] This system had come about only after a forty- to fifty-year period of relentless effort on the part of the United States government to realize in practice a technical-military ideal that was born in Enlightenment France.

The Origins of the American System of Arms Production

Those who seek to explain the development of the American system of arms production purely as a matter of economic causation—a unique blend of land, labor, and capital—or of social bases such as literacy and attitudes toward innovation are apt to overlook the fundamental role of the Enlightened military mind in this development.[36] The quest for interchangeability of parts—or, in the American parlance of the early nineteenth century, the "uniformity system"—grew out of eighteenth-century French military rationalism. Recently, General Jean-Baptiste de Gribeauval has been convincingly identified as the principal originator of this plan, which in fact had been known for a long time in France as "le système Gribeauval."[37] Beginning in 1765 Gribeauval sought to rationalize French armaments by introducing standardized weapons with standardized parts. The uniformity system, Gribeauval reasoned, would allow complete interchangeability in the French military; parts of small arms and artillery pieces could be easily interchanged, and arms themselves could be interchanged as readily as soldiers could be switched.

Although Lewis Mumford long ago recognized the importance of the military as an agent of mechanization and particularly of standardization and large-scale production, his notions have not been fully appreciated by historians until recently.[38] Perhaps one reason is that the process by which the Gribeauval system was worked out took a long time. The ideal sought by Gribeauval and his followers was not easy to achieve in practice. Its realization took almost a century of work and commanded high monetary and human costs. What is important is that Gribeauval envisioned a rationalized world of standardized, interchangeable parts. He and his followers—particularly the Americans—were willing to bear the costs of its achievement.

Gribeauval's ideas exerted influence in the United States in two different ways. First, Americans learned about the efforts of the Frenchman Honoré Blanc to achieve uniformity of musket parts—one of the ultimate goals of the Gribeauval system. Second, because of the tremendous influence of French military thought and practice, the United States War Department became an aggressive champion of the uniformity principle.

It is fitting that the American most often associated with the Enlightenment in France, Thomas Jefferson, should have introduced the idea of interchangeability of parts into the

United States. He did so in a latter to John Jay, written in 1785 while he was minister to France:

> An improvement is made here in the construction of the musket which it may be interesting to Congress to know, should they at any time propose to procure any. It consists in the making every part of them so exactly alike that what belongs to any one may be used for every other musket in the magazine. The government here has examined and approved the method, and is establishing a large manufactory for this purpose. As yet the inventor [Honoré Blanc] has only completed the lock of the musket on this plan. He will proceed immediately to have the barrel, stock and their parts executed in the same way. Supposing it might be useful to the U.S. I went to the workman. He presented me with the parts of 50 locks taken to pieces and arranged in compartments. I put several together myself taking pieces at hazard as they came to hand, and they fitted in the most perfect manner. The advantages of this, when arms need repair, are evident.[39]

The system's "inventor," Blanc, served as general inspector of three French arsenals. He had been experimenting with producing uniform musket locks. For this work and his design of the French model '77 musket he received approbation and patronage from General Gribeauval, who was inspector general of artillery. The latter provided Blanc with funds to establish in 1786 an armory at Vincennes to produce muskets with uniform locks. Gribeauval's death in 1789 and the Revolution brought an end to Blanc's patronage. Between then and his death in 1801, Blanc sought to apply his system as a civilian contractor, but his factory went severely into debt. His contemporaries argued that his methods were not practicable in the civilian sphere.[40]

Three years after his initial letter to Jay, Jefferson discussed with Blanc the possibility of removing his operations to the United States. Jefferson wrote to Secretary of War Henry Knox that Blanc's "method of forming the firearm appears to me so advantageous, when repairs become necessary, that I have thought it my duty not only to mention to you the progress of this artist, but to purchase and send you half a dozen of his officers' fusils [light muskets]." Jefferson also suggested that Knox ought to "give the idea of such an improvement to our own workmen." The following year he sent Knox a copy of Blanc's *Mémoire important sur la fabrication des armes de guerre, à l'Assembleé Nationale* (1790), which was Blanc's unsuccessful petition to the National Assembly for continued government support of his efforts. Jefferson pointed out to Knox that the *Mémoire's* author was the gunsmith who had produced the six muskets sent to him earlier. Blanc's work and Jefferson's enthusiasm for it apparently failed to impress Knox because he never responded to Jefferson's communications.[41] Nevertheless, within a decade, Jefferson's enthusiasm about uniformity manifested itself in the rhetoric of Eli Whitney, who in 1798 received a contract with the War Department for ten thousand muskets.

French military thought and practice played a paramount role in the United States military during the early national period. One need hardly point out how instrumental French military personnel and equipment had been during the American Revolution. After the war, French practice was studied and emulated, an easy task because a number of French officers continued to serve in the United States.[42] Crucial among them was Major Louis de Tousard.

A military engineer and artillerist who had served under Lafayette in the American Revolution, Tousard became the champion of le système Gribeauval in the American War Department. Tousard had left France in 1793 because of the French Revolution. When in 1795 the U.S. Corps of Artillerists and Engineers was created, Tousard joined it as a major in the First Regiment. For the next few years, he performed a wide variety of duties, from fortification construction and conducting courts-martial, to teaching fellow officers principles of artillery and engineering he had learned in France. His penchant for teaching may have led him to write in 1798 a thorough proposition titled "Formation of a School of Artillerists and Engineers," which he sent to Secretary of War James McHenry. This proposal has been called the "blueprint" for West Point, and it was clearly based on French military-technical experience.[43]

Even earlier, Tousard began a work on military artillery suggested by President George Washington. After Tousard's book was published in 1809 as the *American Artillerist's Companion* (3 volumes), le système Gribeauval—although not identified as such—became deeply ingrained in the American military mind. Tousard's book became the standard textbook for military officers in the United States at West Point and elsewhere. In his treatise, Tousard steadfastly argued for the creation of "a system of uniformity and regularity." "Want of uniformity," he stressed, "impeded for a long time the progress of the French military, as it will that of America, unless a similar system is adopted." The military was an ideal institution to realize the uniformity system, Tousard argued, for its officers in the armories could be *"made accountable"* for failure to introduce, perfect, and maintain the system.[44]

The importance of Tousard's book, as well as his informal teaching of officers in the Corps of Artillerists and Engineers, cannot be overemphasized. The first U.S. chief of ordnance, Colonel Decius Wadsworth, who became a powerful champion of interchangeability of small arms, had been a member of the Corps of Artillerists and Engineers and had served with Tousard and other Frenchmen. In the spirit of Gribeauval, Wadsworth adopted the motto *"Uniformity Simplicity and Solidarity."*[45]

Thomas Jefferson's enthusiasm for Honoré Blanc's experiments with the manufacture of interchangeable musket parts and the influence on the American military of the rationalism of General Gribeauval and his followers firmly established the intellectual and institutional basis for the rise of the American system of arms production. The pure rationalism of "system and uniformity" provided an adequate incentive for the pursuit of this goal. The United States War Department soon found the idea of interchangeability irresistible, and through its own armories and through private arms contracts it encouraged and supported attempts to achieve this end. Eventually the War Department *demanded* interchangeability. Ordnance officers elevated the idea of interchangeability to an ideal and helped to transform it into a reality.

Likewise, the government's role in the development of mechanized production, the other main thread in the history of American arms manufacturing, is noteworthy. It is important to keep in mind that interchangeability of parts does not necessarily imply nor require production by machinery. Through proper care, firearms parts can be made by hand (with hand tools) such that they will interchange. Blanc had attained his results in essentially this way. In his testimony before the Select Committee on Small Arms Richard Prosser argued that hand filing with jigs to gauge would result in interchangeable parts. Conversely, despite James Nasmyth's views on the matter, machine production does not in any way imply interchangeability of parts, for in countless instances in the nineteenth century products made in large quantities by highly specialized machine tools did not contain interchangeable parts.

The government armories and private arms factories that were given large government contracts were important sites for the development of machines used in metal and wood fabrication during the first half of the nineteenth century. Excited and encouraged by possibilities and existing developments in interchangeable parts manufacture and production by machines, the War Department, through the extensive resources of the public treasury, wove these two threads into a well-formulated manufacturing system. The history of the "private" efforts of Eli Whitney and Simeon North and of the federal arms establishments at Springfield, Massachusetts, and Harpers Ferry, Virginia, clearly demonstrates this development.

Private Contractors and Interchangeable Manufacture: Simeon North and Eli Whitney

Although it had established a federal armory at Springfield in 1794 (and would begin a second one at Harpers Ferry, Virginia, in 1798), the imminence of war growing out of the XYZ Affair led the United States War Department beginning in 1798 to issue cash-advance contracts with private armsmakers for small arms (muskets and pistols). Two of these contractors, Simeon North and Eli Whitney, not only secured the earliest War Department cash-advance contracts but also were attracted by the idea of producing arms with interchangeable parts. Both had also considered mechanizing some of their production operations. Of the two, Whitney was the more vocal, but the most recent historical scholarship suggests that North more fully developed his ideas and perhaps realized in practice the ideal of interchangeability.

Before Simeon North was awarded his government contract in October 1798 for five hundred horse pistols, he had made scythes and other small agricultural implements at his Berlin, Connecticut, workshop.[46] In February 1800, shortly before he completed the first contract, the War Department granted North another contract for an additional fifteen hundred pistols. Felicia Deyrup points out that the early arms contractors, including North, relied upon the federal government advancing sizable percentages of the total contract value as their sole source of capital. Exactly how North used this capital is not clear, but in 1808, obviously engaged in additional contracts with the War Department, he wrote, "To make my contract for pistols advantageous to the United States and to myself I must go to a great proportion of the expense before I deliver any pistols. I find that by confining a workman to one particular limb of the pistol until he has made two thousand, I save at least one quarter of his labor, to what I should provided [that] I finished them by small quantities; and the work will be as much better as it is quicker made."[47]

This letter suggests that North found savings from a division of labor more noteworthy than any machinery he may have built. North continued to refine his methods and in 1813 signed a contract that marks a minor milestone in the development of armory practice. This contract for twenty thousand pistols, delivered over a five-year period, specified that "the component parts of pistols, are to correspond so exactly that any limb or part of one pistol may be fitted to any other pistol of the twenty thousand."[48] No previous War Department contract for arms is known to have contained this stipulation. Its insertion is indicative of the growing enthusiasm within the War Department for firearms constructed with interchangeable parts, though North himself may have been the motivating force behind this clause.

In 1816, problems arose that led North to request an additional $50,000 advance on the 1813 contract above the $30,000 already advanced by the War Department. This request caused the department to question North's ability to execute the contract. The secretary of war ordered an investigation of North's factory, which was made by Roswell Lee and James Stubblefield, then superintendents of the Springfield Armory and Harpers Ferry

Armory, respectively. Their report not only judged North capable of fulfilling his contract but also pointed out how the New Englander accomplished uniformity of parts for the pistol locks. "By fitting every part to the same lock," North obtained "a more rigid uniformity than they [Lee and Stubblefield] have heretofore known."[49] It is clear that North proceeded from a well-reasoned principle: If each part of the lock were constructed so that it could take the place of the corresponding part of the model or pattern (or standard) lock, then corresponding parts would in principle interchange within the entire lot of locks so fashioned. This may seem to be an elementary geometric principle, if $B = A$ and $C = A$ then $B = C$, but its application is an intellectual leap, for it means deciding to make parts to a standard gauge (in this case, the model lock) rather than shaping parts similar to the pattern and then fitting the component parts together to form a complete—and unique—lock. It represents an entirely different approach to arms manufacture. This principle and its corresponding procedures are vital to interchangeability; according to Lee and Stubblefield, Simeon North made the innovation.[50]

North also pursued the other fundamental idea that lay behind the development of the American system of manufactures, producing components by special-purpose machinery. Merritt Roe Smith recently demonstrated "that the earliest known milling machine in America originated at North's factory [about 1816] and that the owner devised it." Until the milling machine was developed, the principal way to remove metal in shaping it was by filing, using either hand or rotary files. As developed, milling technology allowed flat or curved iron surfaces to be cut by simply passing a revolving hardened steel cutter over the iron, resulting in uniform and quickly executed parts. As Smith points out, in many ways the development of this powerful technology "presents an encapsulated overview of the rise of the American System."[51] In his execution of a huge contract for pistols, Simeon North helped set American manufacturing on the road toward mass production. North attempted to complete only the first few hundred pistols to the specification of uniformity, however, because, he wrote, the War Department made changes in the model. (See Figure 1.4.)

Eli Whitney, who in common wisdom is hailed as the pioneering father of the American system, has not fared well under recent historical scrutiny. The first private citizen to obtain from the United States government a contract with a monetary advance clause, Whitney had gained some notoriety through his claims to have invented the cotton gin. He had also acquired a substantial debt as a result of efforts to enforce his claim to the gin and his disastrous attempts to manufacture it in New Haven, Connecticut. Largely through the graces of the secretary of the treasury, a fellow Yale graduate, Whitney won an unprecedented contract in 1798 for ten thousand muskets with a guaranteed advance of $5,000 upon closing the contract and another $5,000 when Whitney presented evidence that the initial advance had been "expended in making preparatory arrangements for the manufacture of arms." Whitney informed a friend that this contract "saved me from ruin." As Merritt Roe Smith wrote, "Whitney's entry into the arms business undoubtedly was an act of desperation; the thought of 'bankruptcy and ruin' drove him into making a very rash proposal."[52]

When Eli Whitney originally suggested to the secretary of the treasury that he "should like to undertake to manufacture ten to fifteen thousand stands of arms," he could not possibly have demonstrated that he possessed the financial or technical resources or the labor force required to carry out such an unprecedented feat; his performance as a manufacturer of cotton gins had been disastrous both in his ability to make them and in the quality of the product. But Whitney's idea greatly intrigued not only the secretary of the

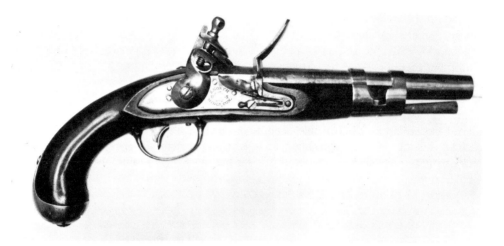

FIGURE 1.4.   North-Made Horse Pistol, Model 1813. In 1813, Simeon North received a contract from the War Department to manufacture twenty thousand of these pistols with parts that would be interchangeable. North encountered difficulties in executing this contract, which led to a formal investigation of his armory. Although the report of the investigation makes clear how North approached the technical problem of interchangeability, it is not conclusive about whether North had achieved perfect interchangeability. North abandoned his goal of interchangeability after the first couple of hundred pistols. (Edwin W. Bitter Collection. Photograph by Mr. Bitter.)

treasury but also the secretary of war. "I am persuaded," Whitney wrote, "that machinery moved by water, adopted to this [musket] business, would greatly diminish the labor and facilitate the manufacture of this article. Machines for forging, rolling, floating, boring, grinding, polishing, etc., may be made use of to advantage."[53]

Had Whitney's unfortunate experience with making cotton gins led him to this conclusion about the manufacture of muskets? Although this is unlikely, it seems logical that anyone proposing to make ten to fifteen thousand muskets in slightly over two years would have to suggest an alternative to the handicraft methods being practiced at the Springfield Armory, whose largest annual output in its first ten years was fewer than five thousand muskets. Any reasonable person knew that such an output would be a major achievement even employing a small army of gunsmiths. But Americans were not ignorant of the potential benefits from machine production, for by 1798 they had seen or heard about nailmaking machines and other metalworking devices being used in the United States.[54]

Whitney scholars have suggested that by the date of his contract Eli Whitney had heard or read about Honoré Blanc's ideas and methods for the manufacture of muskets in France. He may have learned about them from Jefferson himself or from someone who knew about Jefferson's enthusiasm for Blanc's work. Blanc's approach, however, as detailed in a report of the Académie des Sciences of March 19, 1791, did not involve machine production methods.[55] Outside of an "end miller" (a hand tool), the dies, molds, and jigs Blanc used can hardly be termed machines, and they did not resemble the "machines for forging, rolling, floating, boring, grinding, polishing, etc." that Whitney envisioned. If Whitney had heard of Blanc, it is strange that he did not tout the idea of interchangeable musket construction (which is the essence of Blanc's work in France) in his proposal for musket manufacture. As the historical record now stands, not until after October 1798—ten months after the contract began—when the secretary of the treasury sent Whitney a foreign pamphlet on arms manufacturing techniques (perhaps the Académie report or Blanc's *Mémoire* to the National Assembly) did Whitney begin to advocate the "uniformity principle."[56] The reason he began to espouse this idea is simple. By the middle of 1800, when his contract was to have been fulfilled, Whitney had not delivered a single weapon.[57] He needed an excuse, for government officials had become anxious about what had become of the advances on the contract and the lack of any deliveries. Whitney pleaded that he was attempting a unique and highly rational way of manufacturing muskets—with uniform parts made by machinery. This took time, he argued, and great amounts of advance capital.

In hindsight, as well as at the time, Whitney's plea for patience appears reasonable because it *does* take time and money to begin machine production of such an item as a firearm. But in recent years it has been shown that Whitney did not, in fact, ever try to carry out the system which he had implied he was hard at work on. Gene Cesari has suggested that in 1798 Whitney intended to subcontract fabrication of all the musket parts and merely fit and assemble them at his New Haven factory. Other authors have argued that Whitney used the contract, with its $10,000 advance, simply as a reprieve from his debts and as sustenance in recovering losses through damage suits for widespread infringement of his cotton gin patent. Indeed, between 1798 and 1801 Whitney spent considerable time in the South attending to cotton gin litigation.[58] Certainly forging a system of interchangeable parts made by machinery required more time then Whitney ever gave to it.

Nevertheless, Whitney kindled the flame of rational arms manufacture while biding for time and lenience on his contractual obligations. In January 1801, eight months before he had delivered a single arm to the War Department, Whitney staged a demonstration in Washington on the interchangeable character of his product. Whitney had made sure that the right people were invited. In addition to congressmen and government officials, President John Adams and President-elect Thomas Jefferson witnessed Whitney, using only a screwdriver, assemble ten different locks to the same musket. Whitney took care, however, to interchange only the assembled locks, not the lock parts. Nevertheless, as Smith noted, "the intimation was clear: Whitney was making firearms with uniform or interchangeable parts." Jefferson, perhaps because he had personally interchanged parts of locks made by the Frenchman Blanc, jumped to the conclusion Whitney sought. He later informed Governor James Madison of Virginia that Whitney had "invented moulds and machines for making all the pieces of his locks so exactly equal, that take 100 locks to pieces and mingle their parts and the hundred locks may be put together as well by taking the first pieces which come to hand." So enthusiastic was Jefferson about the uniformity he had long advocated that he reached this conclusion about Whitney's musket lock parts without any real evidence. Whitney's demonstration won for him government concessions both in time and in money.[59]

The story of Eli Whitney and armory production is the story of a man who espoused the two principal ideas that lay behind the system—interchangeability and mechanization— but who never understood, much less developed, its basic principles let alone its complex subtleties. That Whitney survived as an armsmaker until his death in 1825 is a tribute to his entrepreneurial abilities, but he was certainly not the heroic innovator pictured in, say, Constance Green's *Eli Whitney and the Birth of American Technology*. Whitney was a publicist of mechanized, interchangeable parts manufacture, not a creator.[60]

The inventor completed the 1798 contract in 1809, and, had it not been for the war emergency of 1812, the War Department would probably never have given him another contract because of the poor delivery record and especially the wretched quality of his arms.[61] By 1812, Whitney had begun to realize that wealth from the cotton gin had eluded him. In some respects, this date marks the beginning of his career as a legitimate arms-maker rather than an opportunistic entrepreneur. In executing the 1798 contract Whitney may have adopted the die-forging and jig-filing methods used by Blanc in France, but unlike the Frenchman's products, his firearms did not contain interchangeable parts. Nor did the Yankee develop the machines he had envisioned in 1798. As Cesari points out, in comparison with other private arms factories of the period, Whitney's armory consisted of "only the simplest, least expensive equipment."[62]

In neither interchangeability nor mechanization can Whitney be seen as a pioneer or even as comprehending the principles that lay behind the systems. Unlike Whitney, Simeon North elucidated his ideas on how he could achieve interchangeability of parts, and he developed a milling machine, one of the fundamental machines of the American system. North's clarity of statement, which comes about only through understanding, was absent in Whitney. Throughout his life, the inventor continued to talk about interchangeability and mechanization in musketmaking but always at a superficial level.

The Federal Establishments

When under the armory bill of 1794, the War Department established its first armory for the production of small arms (muskets, at this date) at Springfield, Massachusetts, even the most optimistic observer could not have hoped or dreamed that within fifty years the armory would be considered a world leader in the manufacture of firearms and a showcase of production techniques. Springfield Armory began making muskets using current methods, which were almost wholly the handicraft techniques developed through generation after generation of gunsmiths.[63]

David Ames, the first superintendent of the armory, and Robert Orr, the first master armorer, had been trained as gunsmiths through customary craft channels, and they approached gunmaking at Springfield with this familiar system by bringing together a group of gunsmiths and setting up an apprenticeship system. Problems of producing a satisfactory number of muskets with this system continually plagued Ames, who left the armory in 1802.[64] Although he had only recently been employed as master armorer, Henry Morgan took charge as superintendent and promptly instituted a series of what no doubt seemed like radical changes. Morgan, formerly a United States inspector of muskets, exercised the well-enunciated principles of division of labor by splitting the workers into four main divisions: barrelmakers and forge men; filers; stockers and assemblers; and grinders and polishers. Among Morgan's changes were the introduction of new managerial procedures and a move toward the elimination of the apprenticeship program. But production during Morgan's tenure decreased significantly without much change in the labor force, so in 1805 the paymaster dismissed him for reasons of inefficiency.[65]

Springfield Armory's next ten years present a checkered history marked by discord between the superintendents and both the War Department and the workers. Benjamin Prescott, Morgan's successor, introduced the piece-rate system of wages soon after taking over. This system, which lends itself to routinized production, did not please the craftsmen, who had already been stripped of much of their wide-ranging craftsmanship through the division of labor. Yet Prescott was relieved of his duties in 1813 as a result of a dispute with the War Department. After a poor performance by Henry Lechler, a Pennsylvania gunmaker, Prescott again took charge in 1815 as interim superintendent while the armory awaited the arrival of Colonel Roswell Lee, who would hold this position until his death in 1833 and to whom much of the credit has been given for the armory's rise to prominence.[66]

About the time of Lee's death, the British philosopher of machinery and economy, Charles Babbage, argued that the division of labor in manufacturing, especially when accompanied by a piecework system, naturally and almost inevitably would lead to improvements by the workers in the various techniques of the subdivided processes.[67] It might be tempting to interpret the developments at Springfield during Lee's tenure as proving Babbage's observation, arguing that the division of labor at the armory and the subsequent introduction of piecework led to development in mechanization of processes in the 1815–40 period, but the historical record does not support such an interpretation. The significance of events at the Springfield Armory from 1794 to 1815 is that here armsmaking was transformed from a craft pursuit into an industrial discipline and the weapon from a shop creation into a factory product. Without question, these changes were a precondition for the rise of the system of production that so greatly interested the British during the 1850s. Babbage's observation, however, is not borne out at Springfield, for many of the seminal ideas and machines developed at the armory originated outside this government establishment. Roswell Lee became a pivotal figure in the armory's history not because of his own particular know-how but because he cultivated productive relationships between the armory and private arms contractors and because he was captivated by the idea of interchangeability of parts.

Lee's appointment as superintendent of the Springfield Armory in 1815 followed an important change within the War Department. James Monroe, then secretary of war, pushed through Congress legislation that gave control of the Springfield and Harpers Ferry armories to the Ordnance Department, thus removing them from the immediate supervision of the secretary of war. An army bureau, the Ordnance Department had been created in 1812 to inspect and distribute military stores. Colonel Decius Wadsworth, who had worked with Louis Tousard in the Corps of Artillerists and Engineers, was the first chief of ordnance. To assist him, Wadsworth recruited a group of West Point junior officers. When the department gained jurisdiction over the armories in February 1815, these West Pointers moved immediately toward the institution of an American version of "le système Gribeauval," which Tousard had championed in his *American Artillerist's Companion*. From their experience in the War of 1812, when a vast number of arms had been damaged beyond repair in the field but could have been fixed had parts simply interchanged, the ordnance officers believed that uniform parts manufacture (proven technically possible by Blanc, North, and possibly others) would be worth almost any price. Lieutenant George Bomford, Wadsworth's chief assistant and his successor (1821–42), proved to be instrumental in the department's efforts to achieve uniformity, but as Merritt Roe Smith points out, "both men became zealous advocates of the 'uniformity system' and relentlessly pursued the idea of introducing it at the national armories."[68] Soon they would also demand it of private armories doing work for the government.

Bomford and Wadsworth provided more than zeal and administrative demands for uniformity; they also furnished much of the basic thought that went into its achievement. In June 1815 Wadsworth organized a meeting in New Haven, Connecticut, at which he,

Roswell Lee of Springfield, James Stubblefield of Harpers Ferry, Benjamin Prescott, the former Springfield superintendent, and Eli Whitney, who had long been a friend of Wadsworth's, considered the possibility of interarmory standardization in musket man-ufacture. This select group concluded that the initial requirement was the production and testing at the national armories of a new model musket, which would be standard to both armories and eventually to private contractors.[69] Bomford charged Lee and Stubblefield with the preparation of the new Model 1816 musket, but after a year of effort Lee learned that Bomford "regrets to observe a total disagreement between your pattern and that lately received from Harpers Ferry. It was hoped and expected that there would have been great coincidence with regard to their construction." One year later Lee complained, "It is difficult . . . to *please everybody. Faults* will *realy* [sic] exist & *many imaginary ones will be pointed out.* . . . It must consequently take some time to bring about a uniformity of the component parts of the Musket at both Establishments." Yet Bomford kept insist-ing that rigid uniformity in pattern pieces with parts "so critically alike as scarcely to be distinguished" was the only basis upon which to manufacture uniform weapons.[70] The Ordnance Department finally solved to some degree the problem of making uniform model muskets when it gave Roswell Lee the duty of preparing the models as well as inspection gauges for both national armories and later for armories contracting with the War Department.

When considering inspection devices, Lee and the Springfield Armory become vastly important in the development of interchangeable arms manufacture. Reaching beyond the preparation of uniform models, Lee moved toward perfecting a system by which to gauge parts while in the process of manufacture as well as in the final inspection. Lee's goal was to achieve greater uniformity in the armory's products. Hitherto, inspection at the federal armories had been done primarily with the eye (a subjective process) rather than by use of a gauge (a more objective procedure). Simeon North had used the model lock as a receiving gauge by which to fit each of the lock components, but by 1819 gauging at Springfield had become an elaborate system. Major James Dalliba of the Ordnance Department argued that the Springfield gauging system was the only means "to attain this grand object of uniformity of parts." The basis of the Springfield system was, according to Dalliba, the pattern or model weapon. Alongside the model, the master armorer maintained a set of master gauges that verified critical dimensions of the model. All shop foremen and inspectors also kept a set of gauges corresponding to the parts made or inspected in their departments. Finally, the workmen responsible for the production of particular parts were issued gauges to check dimensions of those parts while they were being made. In theory all gauges corresponded to the model and master gauges. Although made of hardened steel, the production and inspection gauges were subject to wear and therefore were checked regularly against the model or master gauges.[71]

At this early date in the development of gauging techniques, the complexity of the system and the demands it placed on those who adopted it were already evident. Yet in spite of much effort, perfection had not yet been achieved. Dalliba was optimistic, however, that with further refinement and more faithful adherence to it, the Springfield gauging system would eventually lead to the "grand object of uniformity of parts." Roswell Lee was equally sanguine. With this system, Lee pointed out to George Bom-ford, "our Muskets are now substantially uniform," although "I am sensible that consid-erable improvements are yet to be made to complete the system of uniformity throughout all the Establishments."[72]

In developing this system Roswell Lee readily understood the new demands dictated by precision. James Stubblefield, however, and no doubt many armory workers at Harpers Ferry and Springfield were "at a loss" in grasping the immediate and long-term results to be obtained from such gauging methods.[73] The goal of interchangeability, still very elusive, Lee believed, became an exacting exercise that imposed a bureaucratic system upon the armory in its attempt to prevent any deviation from the standard pattern. In terms of the number of gauges used and the tolerances allowed, gauging methods would become even more rigorous during the next score of years. Yet by 1822, Lee still had not fulfilled his 1818 orders that "the component parts of the musket may be made to fit *every musket*," and he began to argue that "the uniformity of the parts of the musket is believed to be as perfect as is practicable without incurring unreasonable expense and perhaps more attention to that particular point would not prove beneficial."[74] At this juncture, however, Lee had only begun to realize the powerful effects obtained from full mechanization of work processes, for Thomas Blanchard had just negotiated a contract to make gunstocks at the Springfield Armory for a specified price using his newly patented battery of stocking machinery. In many ways, this machinery is a microcosm of the American system of manufactures.[75]

The invention and development of stockmaking machinery by Thomas Blanchard set American manufacturing firmly on the road toward mechanized production. The so-called Blanchard "lathe," the fundamental machine in a battery that eventually included fourteen machines, was a truly elegant invention that reproduced in wood the irregular shape of a gunstock (or any other irregularly shaped object such as an ax handle or a shoe last).[76] (See Figure 1.5.) But this machine alone would not have so easily pointed the way had Blanchard not linked it sequentially with additional and more special-purpose machines that carried out the remainder of operations on the stocks such as recessing for the barrel and lock and mortising for the trigger mechanism. Hand labor was virtually eliminated. It is this sequential operation of special-purpose machines which characterized mechanization in American manufacturing.[77] The logic and the success of Blanchard's operation suggested to those both inside and outside the Springfield Armory that such mechanization was the path to pursue in other areas and with other materials.

Blanchard's ideas for stockmaking machinery grew immediately out of a machine he built in 1818 to turn musket barrels. Although several New England armsmakers had attacked the problem of turning musket barrels instead of grinding them, none had devised a way to machine the breech end of the barrel where it was no longer cylindrical but was flat and oval-shaped. Asa Waters, a prominent arms contractor who had tried in vain to turn this irregular shape, asked Thomas Blanchard, a local mechanician, for advice on how this object might be met. In a fashion characteristic among many American inventors of the nineteenth century, the solution to the problem of turning irregular forms, according to Asa Waters II, "flashed through his mind."[78] As worked out in wood and metal, Blanchard's inspiration centered around using the desired irregular form as a cam.

Louis Pasteur once declared that discovery favors the prepared mind; such was the case with Blanchard. The Diderot *Encyclopédie* contained descriptions of devices for turning objects through the use of cams. Blanchard had apparently seen the French work as well as the description of Marc I. Brunel's Portsmouth blockmaking machinery published in the *Edinburgh Encyclopaedia*. Smith notes that Blanchard "readily acknowledged" these sources of information in his patent specifications. According to Blanchard, this information had been available long enough to become everyman's province. Thus, as Smith points out, "Blanchard based his claim not on general principles, *per se*, but on their systematic arrangement and specific method of operation."[79] Perhaps Blanchard did not

FIGURE 1.5. Blanchard's "Lathe" to Manufacture Gunstocks, 1822. Built in 1822 by Thomas Blanchard for the Springfield Armory, this profile-tracing lathe became the fundamental machine in a battery of fourteen machines that produced all-but-finished gunstocks from sawn lumber. Blanchard's gunstock operations demonstrated the potential of special-purpose, sequentially used machinery. (Smithsonian Institution Neg. No. 24437.)

FIGURE 1.6.   Portsmouth Blockmaking Machine. This shaping engine, designed by Marc I. Brunel and built by Henry Maudslay, was one of some twenty-two special-purpose machines used sequentially to produce wooden blocks for the British navy. (British Crown Copyright. Science Museum, London.)

recognize his own originality. Diderot describes and illustrates methods of turning *regular* figures with the use of cams. It is difficult to understand how Blanchard derived inspiration to turn *irregular* forms from the techniques covered in the *Encyclopédie*. The use of a cam as Blanchard employed it is not obvious; his arrangement appears to have been entirely novel.

More important for the development of the American system of manufactures, however, Blanchard realized the promise of Brunel's special-purpose machines used sequentially for the production of rope blocks for the Royal Navy. Brunel designed and Henry Maudslay built twenty-two different kinds of machines to saw, drill, mortise, recess, turn, and shape the shells and sheaves of the wooden blocks. (See Figure 1.6.) When Brunel and Maudslay had finished their work in about 1807, forty-five machines produced three different ranges of blocks (4 to 7 inch, 7 to 10 inch, and 10 to 18 inch) with an annual output of 130,000 blocks. Because the *Britannica, Penny, Edinburgh,* and *Chambers'* encyclopedias described the Portsmouth machines and operation in detail (Ree's *Cyclopedia* of 1819 devoted eighteen pages and seven plates to it), it may safely be said that Brunel's work was a showpiece of British production technology.

Although Blanchard clearly did not draw inspiration from Brunnel's machinery for his fundamental gunstock-turning lathe (because the blocks were not irregularly shaped), it is entirely possible that he used Brunel's ideas for mortising and recessing. Yet even this theory of origination becomes muddled when it is realized that Brunel's interest in blockmaking machinery had first arisen while he was living in the United States at the end of the eighteenth century.[80] Perhaps Brunel drew his ideas from as yet unknown American inventions.

In any case, the Portsmouth blockmaking operation showed the vast productive benefits of using a number of machines, each designed to carry out a single operation and so arranged as to complete sequentially all the necessary operations upon a product. Blanchard firmly grasped the principle of sequentially arranged, single-purpose machinery, and since British producers failed to apply this principle to the manufacture of other goods, Blanchard can be credited with the rediscovery of Brunel's production methods. As noted above, the British machine builder Richard Prosser expressed his belief in 1854 that Blanchard's gunstocking operations (with the exception of the eccentric turning lathe) derived from the Portsmouth blockmaking machinery. Nevertheless, Blanchard's fundamental machine, in addition to being a brilliant invention, initiated an unprecedented movement in the construction of special-purpose machine tools that would be used sequentially.[81]

Between October 1818, when he had completed a barrel-turning machine for the Harpers Ferry Armory, and February 1819, Blanchard developed his lathe for turning gunstocks. By March he had demonstrated its operation at the Springfield Armory. Soon after this demonstration Blanchard built additional machinery to cut out space in the stocks for musket locks and for barrels. After 1822, the inventor worked out fully his ideas on mechanized gunstock production as an inside contractor at the Springfield Armory. The national armory provided him with shop space, free use of its tools and machinery, water power, and the necessary raw materials. Blanchard granted the armory free use of his patented machinery in return for a contracted price of thirty-seven cents per musket stock. By the end of 1826 he had perfected his battery of machines, now fourteen in number, and had eliminated the use of skilled labor in stockmaking.[82] Blanchard had shown Americans the meaning of mechanization. John H. Hall would combine this notion with the Ordnance Department's quest for interchangeable firearms, further developing and elucidating the basic principles of the American system of arms production.

Although recognized by his contemporaries as a major contributor to the American system, John H. Hall escaped the attention of modern historians until recently. Merritt Roe Smith's *Harpers Ferry Armory and the New Technology* has provided an outstanding study of Hall's achievements.[83] Sometime before 1820 Hall set for himself a mandate to produce his patented breechloading rifle with interchangeable parts. Drawing from best-practice techniques used at North's armory and the national armories, he radically improved upon them largely by thinking abstractly about interchangeability and how this goal might be achieved with machines. North and others had already considered this same object and had made progress, but their thought lacked Hall's precision and their commitment to interchangeability of parts was moderated far more by practical or economic considerations than was Hall's. It is important to keep in mind that throughout much of the nineteenth century the ideal of interchangeability, despite its powerful appeal and seeming rationality, was considered a somewhat irrational pursuit because it continually flew in the face of experience, which had borne out time and again that the system could not survive practical application. Roswell Lee, it will be recalled, was acutely aware of these problems. Yet in the United States, mechanics continued to pursue the dream, and the government allowed them generous financial support for their efforts. John Hall, who drew on his own wealth and the government's, clearly stands as a pivotal figure in the transformation of an ideal into a reality.

Originally a Portland, Maine, cooper, cabinetmaker, and boatbuilder, Hall invented a rifle that loaded at the breech rather than the customary and almost universal method of loading from the muzzle or barrel end. When Hall attempted to patent the breechloader in 1811, the commissioner of patents, William Thornton, claimed to have invented a similar weapon in 1792 and promised to block the patent unless it was issued jointly to Hall and himself. Secretary of State James Monroe refused to intercede on behalf of Hall, and the inventor was left with no recourse but to accept Thornton as co-inventor, which also meant granting the commissioner one-half of all income gained through the patent. The two inventors never agreed on how best to maximize profits from the patent; Thornton advocated a quick sale of rights and collection of royalty while Hall desired to hold onto the patent and sell his own rifles. Despite continued harassment from Thornton, Hall prevailed and worked to convince the government that it needed his invention. He sold eight breechloaders (five rifles and three muskets) to the War Department, which field-tested the weapons. With a favorable report, the War Department asked Hall in late December 1814 to deliver two hundred of his rifles by April 1, 1815. Hall, still in Portland, became acutely aware of production problems when he was forced to turn down the request because he lacked the necessary workmen and facilities to meet such a large order.

Hall had not given up, however, for during 1815 he laid some of the groundwork for a method and facility to produce "any number" of breechloaders for the War Department. With the same missionary zeal displayed by Eli Whitney and Simeon North, Hall told the secretary of war that he had "spared neither pains nor expence [sic]" in building tools and machinery. He noted, however, that "only one point now remains to bring the rifles to the utmost perfection, which I shall attempt if the Government contracts with me for the guns to any considerable amount viz., to make every similar part of every gun so much alike that . . . if a thousand guns were taken apart & the limbs thrown promiscuously together in one heap they may be taken promiscuously from the heap & will all come right." At the same time, Hall expressed his abiding faith that this plan was practicable, "although in the first instance it will probably prove expensive." The ultimate economic advantages of such a system, clearly professed by Hall, intrigued George Bomford and Decius Wadsworth, yet even these advocates of uniformity could not accept the high price of $40 each in Hall's offer to manufacture one thousand rifles nor his revised bid of five hundred weapons at $25 apiece. Ordnance chief Wadsworth persuaded the secretary of war that if Hall made one hundred rifles, this would constitute a "pretty extensive Experiment."[84] Perhaps Wadsworth was more concerned with testing the viability of the Hall breechloading rifle than with obtaining an idea of the bottom limits of the price per rifle by realizing economies of scale. With little promise of being able to reduce his cost through high volume, Hall reluctantly accepted the War Department's offer for one hundred rifles, which he completed in Maine by November 1817.

Despite initial skepticism about the viability of breechloading weapons in the federal service, Hall's one hundred rifles received much praise. One officer suggested that "they ought & must eventually supercede [sic] the common rifle."[85] Praise notwithstanding, the War Department seemed reluctant to negotiate a larger contract. Hall therefore resorted to asking political friends in Washington to influence the secretary of war, who in late 1818 asked the inventor to journey to Harpers Ferry to manage the production of a small number of breechloaders in order to perfect the model. Four such weapons were made there. Thorough testing by the superintendent of the armory, James Stubblefield, again proved Hall's claims for the weapon. But as Smith points out, Stubblefield and possibly others in the War Department maintained that, because these four weapons had cost $200 each to make, the government could not afford to adopt the breechloader. This decision indicates either a rationalized prejudice against the weapon or a profound ignorance of cost-reducing benefits obtained through high-volume production. Smith tacitly suggests that, incredible as it may seem to twentieth-century minds, Stubblefield had no idea about the possibilities of "economies of scale." (This is by no means an indictment against Stubblefield. But as will be seen later, John Hall had a very clear idea of the cost-reducing tendencies of high-volume production.) An additional three-month test of Hall's rifles pitted against the common muzzleloading rifle, checking for endurance, rapidity of loading and firing, and accuracy, yielded results that the Ordnance Department—which above all sought to arm the nation with a single, standard firearm—could no longer deny or ignore. Hall at last won a contract for one thousand weapons.

Hall's 1819 contract seems to have been a new departure for the War Department because the inventor agreed to produce his rifle for a monthly salary of $60 and a royalty of $1 per arm. The War Department would pay the cost of manufacture at Harpers Ferry. Whereas the government had in effect subsidized Whitney and North through cash advances, it truly subsidized Hall, who was, as Smith points out, "a private manufacturer at a public armory."[86]

John Hall began his work in earnest in April 1820 at the Rifle Works, a unit separate from the armory at Harpers Ferry. Originally believing he would fulfill the contract by September 1821, the inventor finally finished the one thousand breechloaders in December 1824. Although this delay may seem reminiscent of Whitney's first contract, the contrast is striking. Hall's rifle components *did* interchange. As he wrote to Secretary of War John Calhoun, "I have succeeded in an object which has hitherto completely baffled (notwithstanding the impressions to the contrary which have long prevailed) all the endeavors of those who have heretofore attempted it—*I have succeeded in establishing methods for fabricating arms exactly alike, & with economy, by the hands of common workmen, & in such a manner as to endure a perfect observance of any established model, & to furnish in the arms themselves a complete test of their conformity to it.''* Although one must beware of Yankee brag, Hall's claims were borne out in 1827 by a three-man committee charged by the United States House of Representatives with investigating Hall's Rifle Works at Harpers Ferry. This committee also confirmed that the inventor produced his interchangeable rifle components with a series of special-purpose machine tools of his own design and make. The committee's report constitutes one of the foremost documents in the early history of American manufacturing technology.[87]

Using this report as well as a wide array of published and manuscript material, Merritt Roe Smith has provided a comprehensive picture of Hall's machine tools and operations at Harpers Ferry. A number of devices and procedures distinguish Hall's work. Drawing upon the prior work of Simeon North, Hall developed three classes of milling machines, which he used to finish parts after they had been initially shaped in the dies of large and small drop forges. In France, Honoré Blanc had used hardened steel dies to fashion heated iron into various shapes for the lock parts of his musket, and Hall's drop forges may be seen as an extension of Blanc's work. Of course, coins have been made with the use of dies for centuries, but drop-forging was different. Moreover, Hall developed a method to eject the newly forged piece from the die in order to prevent heat buildup which quickly removed the die's temper. He also recognized that because of shrinkage in cooling a forged piece did not retain the exact shape of the die, and he developed procedures to deal with this problem. In addition to these metalworking machines, Hall designed and constructed woodworking machines for the manufacture of stocks which differed significantly from Blanchard's. Hall's series of five stocking machines never equaled the performance of Blanchard's battery largely because more handwork remained after Hall's process than with that of Blanchard. Nonetheless, the design and use of these woodworking machines clearly demonstrated Hall's creativity and his commitment to carry out all operations by machinery.

Hall's ability to manufacture interchangeable rifle parts rested not only on his skills in designing and building metal- and woodworking machinery but, perhaps more important, on his extensive use of gauges and on his rationalized design of fixtures. Hall drew upon Simeon North's experience of using the model lock as a receiver gauge to which parts were fitted and also developments in gauging at the Springfield Armory. Yet very early he extended the use of gauges for manufacturing and inspection; one of the men who worked for Hall later said that the inventor used three sets of gauges (work, inspection, and master), each comprising sixty-three different gauges. Another mechanic noted that Hall's gauges were ''more numerous and exact . . . than had ever before been used.''[88]

Production of uniform parts by machinery required skill in making the machines, tools, and gauges and a commitment to relinquish one's judgment to the ultimate authority of the gauge. Yet uniformity also demanded careful thought about how precision might be obtained and improved. John Hall elucidated what became one of the fundamental principles of precision manufacture, a principle that was the product of thought rather than action, a part of an intellectual process rather than the acquisition of skill over time. This was the principle of fixture design. A fixture is a device that ''fixes'' or secures an object in a machine tool, such as a milling machine, and holds it during the machining operation. Each time an object is fixed in a machine tool, a certain amount of inaccuracy creeps into the operation. Because fabrication of Hall's rifle parts involved a number of different machining operations, each requiring a specially designed fixture, Hall recognized that a multiplying effect would set in; the inaccuracy of each fixing would be multiplied by the number of different machine operations. To rectify this problem, Hall reasoned that if the piece were located in each fixture relative to one point on the piece, the multiplying effect could be thwarted. He called this reference point the "bearing" point and designed all fixtures for a part relative to that point. And, as Hall stressed, "this principle is applicable in all cases where uniformity is required." Indeed, this fundamental principle became universally applicable.[89]

Such clarity of thought was rarely expressed by early American mechanics, yet it seems typical of much of Hall's work. Opponents at the Harpers Ferry Armory, far more practical mechanics than he, referred pejoratively to the Yankee inventor as a "visionary theorist," a term that illustrates his approach to precision manufacture.[90] Hall's understanding of the inherent technical demands placed upon the maker of interchangeable firearms is perhaps best revealed in correspondence concerning the manufacture of his patented rifle by Simeon North under a federal contract.

After 1825 the Hall breechloader had become popular with many state militias. But because the federal arms factories could not legally produce weapons for state militias, the War Department turned to a private contract with Simeon North in 1828 as a means to supply state militias with Hall weapons. The thought of another shop producing his rifles disturbed Hall, not so much because it threatened his own operation but because he believed another maker would not understand or appreciate the demands of uniform manufacture. Any departure from Hall's rigid system of gauging and fixture design would surely imperil the uniformity of the breechloader's parts. "If the contractor should fail of full and complete success," Hall wrote Ordnance Chief George Bomford, "his arms must all be rejected and he will be ruined, as the introduction of the Rifles into the service in so defective a state as not to admit exchanging all their parts with each other, and with those made here would totally defeat the great object for which so much expense has been incurred."[91]

Despite Hall's apprehension, Bomford granted North a five-year contract for five thousand breechloaders, one thousand to be delivered annually. At once, it became evident that the War Department did not fully understand Hall's uniformity system because it gave North several production rifles to use as a pattern rather than an exact duplicate of Hall's model and an accompanying set of his gauges. North would have to prepare his own gauges to ensure uniformity with the production arm used as a pattern but not necessarily with the rifles made at Harpers Ferry. Bomford eventually provided him with the proper model and gauges, and by 1834 Harpers Ferry and North's armory were turning out rifles whose parts could be interchanged, an important milestone in the American system of arms production.

Throughout John Hall's twenty-five-year involvement with the production of his patented rifle, the Yankee mechanic never succeeded in convincing War Department officials to make the ultimate test of his system, to produce as many weapons as possible with his array of machine tools rather than to limit the number to the usual paltry (by Hall's standards) sum of a thousand per year. Once Hall had completed the first order of one thousand rifles in 1824, he began to preach not only the universal applicability of his system for all interchangeable manufacture but also the potential benefits to be won through economies of scale. When in 1828 the War Department again asked for only one thousand rifles from Hall, the inventor wrote Colonel Bomford that he had designed the Harpers Ferry Rifle Works machinery for production on a "large scale" using a "minute subdivision of Labor." Such a small order would not allow him to begin to realize the inherent economies of scale. As Hall noted, fixed costs for three thousand rifles would not be greater than for the thousand ordered.[92]

Although Bomford may have agreed with Hall, he failed to convince his superiors that in order to test fully the experimental system Hall had developed at the Rifle Works, they must see how many arms it could turn out and what effects full production exerted on the price of the rifle. Indeed, continued lack of appropriations severely restricted Hall's operations, a problem the investigating committee of 1827 noted as causing "inconvenience and embarrassment" in that Hall never had enough room at the Rifle Works to set up all of his machinery.[93] The inventor daily faced the problem of pushing one machine tool out of the way so that another could be set up in its place. This meant constant setup and adjustment of machine tools, exacerbated the problems of precision manufacture, and certainly prohibited any chance to lower costs through continuous and smoothly flowing production which the inventor envisioned—what Lieutenant Colonel James S. Tulloh described as "a kind of stream of work flowing through the manufactory in consecutive order."[94]

Continual relocation and readjustment of machine tools coupled with small demand for rifles may have been determining factors in the startling machinist-armorer ratio at Hall's Rifle Works. That with few exceptions Hall annually employed as many skilled machinists as he did armorers, according to Smith, "denotes an uncommon preoccupation with machine making."[95] Allocations within Hall's budget corroborate this conclusion. But as later experience would show, such expenditures and allocations of labor resources were characteristic of manufacture under the American system. Indeed, this was one of the primary concerns of the British Select Committee on Small Arms in 1854. Just as he recognized the universal applicability of his system of production, Hall no doubt also realized that skill within this system was confined to the construction and maintenance of machines and tools rather than to the production of the good itself. Getting into and maintaining production would require a large mechanic-operator ratio. Hall seemed to have prided himself that young boys were the best operators of his self-acting machinery.[96]

It is fitting that John Hall's Rifle Works was situated at Harpers Ferry, a point where two rivers ran together. At the Rifle Works, two important streams of development in manufacturing technology flowed together into a major stream that runs through American history. There, the idea of uniformity or interchangeability of parts was combined with the notion that machines could make things as good and as fast as man's hands, or even better. The result of this combination was the method of production usually called the "American system of manufactures" but which can perhaps be more appropriately labeled the "armory system" or "armory practice."

Smith has convincingly demonstrated the lasting influence of John Hall on American arms manufacture, particularly on the Springfield Armory and on private arms contractors of New England such as Simeon North, Lemuel Pomeroy, Eli Whitney and his nephews, and others. But the communication of information by no means flowed in one direction. Hall profited from North's prior work and that accomplished at Springfield. Nevertheless, Smith's assessment seems valid: "Hall's work [essentially carried out on the government's carte blanche] represented an important extension of the industrial revolution in America, a mechanical synthesis so different in degree as to constitute a difference in kind."[97]

From as early as 1827 until Hall's death in 1842, the Springfield Armory responded to his work at Harpers Ferry. Roswell Lee, the superintendent at Springfield (who in 1822 had expressed skepticism about carrying out complete uniformity of parts) served as the superintendent of the Harpers Ferry Armory for one year during 1827. While there he helped select men to serve on the committee that investigated Hall's operation. Judging from Lee's typical eagerness to follow any new technology that showed promise for the federal armories, he must have spent many hours at the Rifle Works contemplating Hall's machine tools, fixture design, and extensive gauging system. Hall's system must have convinced him of the possibilities for interchangeable manufacture. Hall later claimed that the Springfield Armory had adopted many of his machines and ideas without his permission, but under the contractual agreements between the inventor and the government the latter maintained full privileges to use Hall's machines as it pleased. It would have been natural for Lee to have carried technical information back to Springfield when he resumed his duties there in 1828.[98]

The decades of the 1830s and particularly the 1840s were noteworthy for the development of interchangeable manufacture of muskets at Springfield. Achievements during these years included adoption of a new model musket, extension of the gauging system, and the design of new machine tools, especially by two prolific mechanics, Thomas Warner and Cyrus Buckland. By 1850, the armory had achieved such uniformity of musket parts that it no longer fitted, assembled, and marked lock parts in the soft state before hardening them. More important, virtually all of the fabrication of the musket (except barrel welding) was carried out by machines, some of them flexible such as milling machines and other specialized such as barrel rifling machinery.

During these years, according to Felicia Deyrup, the armory experienced a decline in "efficiency" and failed to lower the cost of its product through extensive mechanization.[99] Despite pretenses to the contrary, the War Department never really expected significant cost reductions. Yet it achieved its long-sought goal of solid, easily repairable weapons constructed with uniform parts. The development of the American system of interchangeable parts manufacture must be understood above all as the result of a decision made by the United States War Department, for reasons outlined earlier, to have this kind of small arms, whatever the cost. It was willing to achieve that goal through hand labor (as borne out by a record of more than twenty years of hand filing to jigs at Springfield and Harpers Ferry) or by machines (as seen by its continued support of John Hall's experimental Rifle Works at the Ferry).

In 1838 the War Department began developing a new model musket for Springfield, which meant an extensive change in machinery, tools, filing jigs, and gauges. Apparently master armorer Thomas Warner convinced John Robb, Roswell Lee's successor, to allow him to make a "thorough reorganization" of Springfield's production system. Charles H. Fitch, the nineteenth-century commentator on the rise of interchangeable manufacture in the United States, called Warner a "projector of the movement for interchangeability." With the help of Nathaniel French, who had worked for both John Hall at Harpers Ferry and Simeon North at Middletown, and Cyrus Buckland, Warner extended the Springfield gauging system along the lines suggested by Hall. (See Figure 1.7.) Among other machines, he designed a lockplate milling machine with adjustable spindle, screw feed, and automatic disengage. Cyrus Buckland also designed a large number of machines between 1840 and 1852; some were only improvements on Blanchard's stocking machines but others were more original, such as a self-acting rifling machine. Warner also further subdivided the work processes, such as creating four different jobs in the filing of the lockplate. In 1840, the master armorer, obviously pleased with his work, wrote another armory mechanic that "what has been done was once considered impracticable and

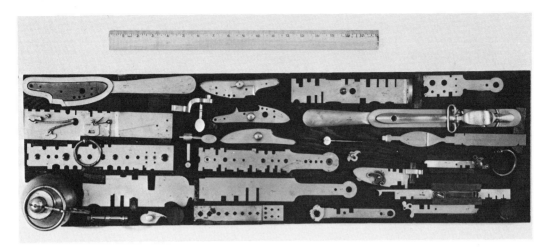

FIGURE 1.7. Inspection Gauges, United States Model 1841 Rifle. This set of gauges was one of three different types of gauges built by the Springfield Armory. Master gauge and work gauge sets were also constructed and used. (National Museum of American History. Smithsonian Institution Neg. No. 62468.)

almost impossible."[100] The Yankee mechanic helped to move the Springfield Armory a step closer to fully interchangeable firearms.

Warner left the federal service to take a job at the Whitney armory. Cyrus Buckland carried forth the long-established goal of interchangeability and the well-grounded approach to this goal. When Joseph Whitworth visited the Springfield Armory in 1853 and when the British Committee on the Machinery of the United States toured it in 1854, they saw a plant that had been partly designed and fully orchestrated by Buckland. And when the British committee purchased gunstocking machinery from the Ames Manufacturing Company for its projected Enfield Armoury, many of the machines they bought were specially designed by Buckland.[101] Moreover, he continually refined the Springfield gauges so that by 1849 or 1850, the armory's products easily interchanged.

From 1830 until the beginning of the Civil War, Springfield Armory played a critical role not only in the development of a workable, mechanized system of uniform firearms manufacture but also in the widespread diffusion of information about the system. The armory acted both as a clearing house for technical information and a training ground for mechanics who later worked for private armsmakers or for manufacturers of other goods. Felicia Deyrup noted that during the 1830s private and federal armories exchanged parts, workers, and information on production machinery. Mechanics often visited Springfield Armory, and Deyrup documents instances when the armory made "castings of valuable machines developed by the contractors."[102]

Government contracts with private armsmakers for military weapons declined during the 1840s, however, and virtually ended until the Civil War. With the demise of the contract system, old armsmakers such as Simeon North, Lemuel Pomeroy, and Asa Waters went out of existence. Yet the role played by these contractors as developers and extenders of mechanized, interchangeable firearms manufacture was taken over by a new breed of armsmakers, those of patent arms such as the Colt revolver and the Sharps rifle. Patent armsmakers capitalized on forty years of government-sponsored development in manufacturing technology, and the timing of their origin at least suggests when this development had reached a level which, although far from mature, held real possibilities for those outside the realm of government largesse.[103]

Although they contributed developments of their own, the patent arms manufacturers are perhaps best understood as crystallizers of the nascent technology developed at government armories and those of government contractors. For example, they brought drop-forging, milling, and gauging to a high degree of perfection and in doing so firmly established an armory tradition of manufacture, or more simply, armory practice. The success of patent arms manufacture seemed eventually to vindicate the costly and ambiguous production technology developed by the War Department; success allowed the proclamation of the universality of the system. Although it is tempting to treat the history of each of the patent armsmakers, only one will be discussed—Samuel Colt.

"Colonel Colt's system of manufacturing"

The opening of the new Colt armory at Hartford, Connecticut, in 1855 has been called the culmination of years of development of armsmaking technology. Colt claimed that his new factory was larger and more modern than any other private armory in the world.[104] (See Figure 1.8.) Although much of Colt's initial success depended on government purchases, he does not fit into the same category with the private arms contractors discussed earlier. Most often, the government simply purchased Colt revolvers; it did not

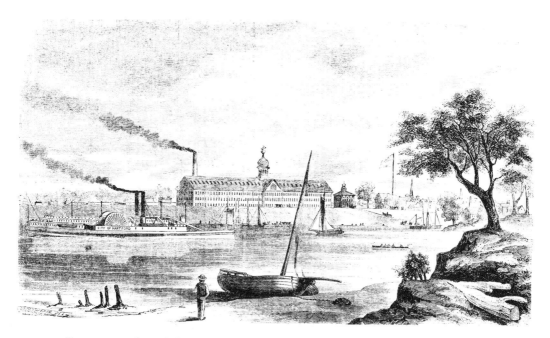

FIGURE 1.8.   Samuel Colt's Armory, Hartford, Connecticut, 1857. (*United States Magazine,* March 1857. Library of Congress Photograph.)

supply models, as it did with Simeon North, for example, nor did it make any cash advances, which had been so essential in the early days of arms contracting.

Soon after Samuel Colt received letters patent on his revolving pistol in 1835 he and a group of capitalists formed the Patent Arms Manufacturing Company of Paterson, New Jersey, Colt's Patent, to manufacture the revolver. From the beginning, Colt intended his business to be supported principally through government purchases, yet these plans never came to pass. The government purchased limited numbers of revolvers, but these weapons failed to give satisfactory service; the marines said that poor quality, not design, doomed the Colt arm. By 1842 the Patent Arms Manufacturing Company ended production.[105]

Looking back on this failure Colt could see a number of causal factors, the most important one being the poor quality of his production arm, which he believed was and would continue to be inherent in a product of hand labor. As he said in 1851, "With hand labour it was not possible to obtain that amount of uniformity, or accuracy in the several parts, which is so desirable. Nor could the quality required be produced by manual labor."[106] Even before the Paterson operation closed, Colt had decided that mechanization of production was essential; he wanted to build special-purpose machine tools at the Paterson works. William H. Miller, a Connecticut cutlery manufacturer, reported in 1890 that he had worked at the Paterson factory beginning in 1838. According to Miller, William Ball, "one of the most prolific inventors and designers of machinery [he] ever knew," developed machine tools for Colt including milling machines, "index" or "dial" milling machines, screw or "cone" machines, and drill presses.[107] When these and other tools of the Patent Arms Manufacturing Company were sold at auction in 1845, they brought in more than $6,000.

Colt received a second opportunity to manufacture his patented revolver in 1847, when General Zachary Taylor, who commanded U.S. forces in the Mexican War, requested that the War Department purchase Colt's pistols. Consequently, the secretary of war granted the inventor a contract for one thousand revolvers which specified that the lockwork was "to be made of the best cast or double sheet steel and the parts sufficiently uniform to be interchanged with slight or no refitting." Colt knew that he had neither the capital nor the time to execute the contract himself, so he subcontracted with Eli Whitney, Jr. Whitney agreed to make the patent arms and to allow Colt to own any of the special machines and tools built expressly for the production of the revolver. In addition, Colt contracted with Thomas Warner, the former master armorer at Springfield, to oversee and build tools for the Whitneyville production of his revolver. Colt paid Warner $1.25 for each weapon.[108]

Although the Whitneyville armory retained far more vestiges of hand labor than did Springfield, Colt's association with Whitney nonetheless gave the inventor an important exposure to armory practice. Colt became even more strongly convinced that complete mechanization of revolver production would be vital in achieving both quality and uniformity. In fulfilling this contract he also learned that the public would pay a higher price for revolvers that failed to meet government standards (about 50 percent of his production) than he received from the War Department for first-class weapons.[109] With a clear idea of how he wanted to make his revolver and feeling assured of new market possibilities, Samuel Colt believed he was ready to open his own armory.

In 1848 Colt moved his machinery and tools from Whitneyville to Hartford, Connecticut, and transformed an empty textile mill into a gun factory. During the next eight years the factory not only produced Colt revolvers but also built machine tools and their accoutrements for two new Colt armories, one established in London, England, the other in Hartford.[110] Colt deemed it absolutely essential that he hire an expert mechanic who not only comprehended mechanized production but who also could design new machine tools with greater precision. Initially he appealed to Albert Eames, a former employee at Springfield, who was currently working at the New Orleans mint. Eames turned the job down. Later Colt lured Elisha K. Root, a well-known mechanic and inventor, away from the Collins Company with the offer of twice his present salary or any "such compensation as you think fair and reasonable."[111] Apparently Root had declined such offers in the past, including ones from Springfield Armory and the U.S. Mint, but he could not turn down what was essentially an offer of total freedom in machine and systems building. Both the freedom and the challenge of mechanizing revolver production captivated Root.

Colt's decision to hire Root determined in large part the direction that the Colt armory would take in the production of the patent revolver. Recent scholarship suggests that Root's greatest contribution came in drop- or die-forging technology, although he made improvements in milling machinery and turret lathes. His work in drop-forging was an outgrowth of his seventeen years of experience in axmaking at Collinsville; more accurately, as Paul Uselding has pointed out, Root merely applied some of his axmaking machinery to drop-forge revolver parts.

Uselding argued that, perhaps with the exception of die-forging, "his contribution consisted, in the main, in the extensive application of special-purpose machinery to commercial uses." Root's emphasis on special-purpose machinery is abundantly clear both in his work at the Collins Company and at Colt's armory. Before he left Collinsville, Root had mechanized axmaking by designing an impressive battery of special-purpose machine tools to carry out traditional hand processes. Soon after Colt opened his new Hartford factory in 1855 a reporter noted almost 400 different machines at work; Cesari records 357 distinct machining operations, ignoring some screwmaking, for each revolver.[112] Thus between 1848 and 1856, Root, with Colt's support, continually augmented the production process by designing and adding new special-purpose machine tools.

At Collinsville, axmaking had demanded of Root almost no concern for precision manufacture. Consequently, his tenure at Colt's armory was not distinguished by an aggressive pursuit of interchangeability, with its requisite principles and practices of precision production, but rather by mechanization of work processes. When he moved to Hartford, Root had had no experience with the model-based gauging techniques used at the national armories to ensure the interchangeability of musket parts. This fact alone may explain why the Colt armory did not use as rigorous a gauging system as, say, the Springfield Armory and also why parts of the Colt revolver, despite the implications of its inventor, did not come close to being interchangeable.[113]

Despite the views expressed by James Nasmyth before the Select Committee on Small Arms, the use of large numbers of highly specialized machine tools did not result in production of perfectly interchangeable parts, even though parts were more uniform. In the nineteenth century interchangeability was achieved through careful adherence to a rational jig and fixture system (such as that created by John H. Hall) and a refined model-based system of gauging (such as that used at the Springfield Armory). The elimination of hand-fitting of parts by the production of perfectly interchangeable ones may be construed as the complete mechanization of the work process (as Nasmyth did in 1854), but Elisha K. Root and Samuel Colt were not prepared to carry mechanization this far, either because they did not know how or they judged it too expensive for practical purposes. In any case, it is naive to assume that interchangeability of parts inevitably lowered production costs.[114] No doubt for much of the nineteenth century, interchangeability raised rather than lowered the price of small arms, even in the American context of "dear labor."

The United States War Department could insist on interchangeable manufacture at its armories not only because of the tactical importance of interchangeability of parts but also because of its limited annual production and its ability to draw on the public treasury to support this costly manufacture. The goal of absolute interchangeability which had stimulated developments in gauging and in mechanization of metalworking became a secondary consideration for Colt and probably for other patent armsmakers around the middle of the nineteenth century. What is important for this study, however, is that Henry Ford would later make perfect interchangeability of parts a criterion for mass production—"In mass production," he wrote, "there are no fitters."[115]

Colt revolvers were not manufactured with interchangeable parts. Armory workmen filed and fitted machine-made parts while soft. When assembled, major components were stamped with serial numbers, the arm taken apart, and the parts hardened. After hardening, the parts with the same numbers were refitted by hand into a complete revolver.[116] When considering the establishment of a small arms factory, the British Select Committee on Small Arms heard the testimony of a number of gunmakers and mechanics who had purchased Colt's pistols to see if they were indeed constructed with interchangeable parts. None found them to be so constructed. The testimony of Colt's former superintendent, Gage Stickney, also damned the idea of interchangeability—"I have heard of it, but I defy a man to show me a case." All the British experts argued correctly that the process of hardening would throw off the fit between parts and that these hardened parts would have to be refitted with a file. Only later would mechanized grinding techniques eliminate this hand process.

The lack of interchangeability of revolver parts by no means precludes characterizing Colt's production technology as embodying the American system of arms manufacture, that is, armory practice. John Anderson and his colleagues equated Colt's techniques with the finest American practices. Certainly, Elisha Root picked up John Hall's technique of designing fixtures for his special-purpose machine tools with references to a bearing or location point.[117] And, like all New England armsmakers at midcentury, Colt employed a gauging system to control uniformity during production. But the Colt gauging system was probably not as refined nor as rigidly maintained as that used at the Springfield Armory. Both Samuel Colt and Elisha Root seemed to operate from the proposition that uniformity would be an effect, not an absolute goal, of mechanization. This conclusion is consistent with the overwhelming emphasis on mechanization rather than on the pursuit of precision manufacture at the Colt armory.[118] Mechanization of the Colt armory paralleled development outside New England armories at this time when Americans were designing machines to do anything that could be done by machines. Nevertheless, Colt's armories in Hartford and London became prime showplaces of American manufacturing technology.

Perhaps more important, Colt's Hartford armory provided a setting in which a number of Yankee mechanics worked out their rich ideas and then moved to exploit them in the outside world. These mechanics were usually sub- or inside contractors at the armory, and the inside contract system has been identified as a major stimulus to nineteenth-century manufacturing technology. Among others, Francis Pratt, Amos Whitney, George A. Fairfield, Christopher Spencer, and Charles E. Billings worked as inside contractors for Colt. Each contracted with Colt to use his shop space, power, machine tools, and materials to produce a particular part or to manage a particular operation for a set piece rate.

The contractor agreed to hire, pay, and manage his own workmen. The contractors acted as foremen but had additional administrative functions. Through efforts to lower their costs and thereby derive greater profits from the contract, the contractors sometimes made improvements in the productive processes. Charles Billings, for example, contracted with Colt as a diemaker and die forger. He became so skilled at making dies that few of his forged parts were rejected at inspection time. Pratt and Whitney sold machine tools made at Colt's armory and later founded an important machine tool firm. Billings joined with Spencer to establish a drop-forging and diemaking company. Later Spencer designed the automatic screw machine. George Fairfield applied Colt production technology to the manufacture of the Weed sewing machine, and he eventually became the president of the Weed Sewing Machine Company. The list goes on.[119]

The inside contract system had been used for many years throughout American manufacturing, particularly in New England. In some respects, inside contracting resembled the putting out system, but its particular characteristics were derived from the factory system. Although the Springfield Armory never adopted inside contracting (with the important exception of Thomas Blanchard's gunstocking operation), almost all the New England armsmakers employed it.[120] When coupled with armory gauging systems and machine tool design, inside contracting became a distinguishing characteristic of the Yankee armory practice extensively employed in all types of metal fabrication in the second half of the nineteenth century.

The contributions of other armsmakers such as Robbins & Lawrence Co. of Windsor, Vermont, and Remington of Ilion, New York, could be considered in detail but would not add greatly to our understanding of the American system of arms production in the antebellum period gained by considering the case of Colt. The inventor's Hartford armory established an archetype for the New England armory approach to manufacture. Die-forging formed the basis of this approach, and here Colt's superintendent, Elisha Root, along with others such as Charles Billings, played a significant role. Until Colt began making revolvers in the late 1840s, die-forging was just another method of shaping metal; after 1845 it was considered the one best way in most New England metalworking establishments. The Colt armory principally employed two important machine tools to machine forged parts, the milling machine and the chucking machine or turret lathe. Root designed a host of special-purpose machine tools in an effort to eliminate the need for skill in operating machines. Echoing John Hall, Colt and Root stressed the importance of the bearing or location point in the design of fixtures, templates, and jigs. This principle was now firmly established. The Colt mechanics adopted a gauging system although it appears that the Colt gauges were not as fine as those of the federal armories. Finally, the organization of labor within the armory was built around the inside contract system. This system determined the departmental structure of production. Together, these elements— die-forging; machining the forgings with milling machines, turret lathes, and numerous special-purpose machine tools (virtually all chipping or cutting tools); rational fixture design; gauging; and the inside contract system—constituted the archetypal New England armory approach to manufacture, or what John Anderson casually called the "American system" of arms production, even though some Englishmen called it "Colonel Colt's system of manufacturing."[121] Colt's Hartford armory, drawing on long-established methods used at Springfield and elsewhere, provided this archetype. Within this approach, any improvements would be in degree, not in kind.

Mechanization outside Small Arms Production

While the United States was still trying to find its political identity in the early national period, its inventors had already set out to shape the distinctive American technological character. This era witnessed the brilliant work of Oliver Evans with his automatic flour mill, Jacob Perkins with his nailmaking machinery, and Amos Whittemore with his automatic machine to make wire textile cards.[122] Countless inventors mechanized the cooperage craft, and by 1850 a wide variety of barrelmaking machinery was available for purchase.[123] Americans built pinmaking machinery; clock- and lockmaking machinery; and knife-, axe-, and swordmaking machinery. Hardly any American inventor would have disagreed with Samuel Colt that there was nothing that could not be made by machinery.

Long before the British Parliament concerned itself with American manufacturing technology, Charles Babbage compared English and American pinmaking. English manufacture was carried out strictly by hand with a high degree of labor division. The Americans, however, used machines. (See Figure 1.9.) Babbage analyzed the two methods and concluded that with the exception of heading the pin, hand labor was faster than machine. Machine-made pins also seemed inferior to those the English made by hand.[124] Such analysis did not deter Americans from their pursuits. Had the philosopher carried out a similar study on American clockmaking machinery, he might have reached the same conclusions. But Yankee inventors were on their way to highly mechanized clock manufacture. Clockmaking provides an excellent example of an industry that developed a system of manufacture that fascinated Joseph Whitworth and John Anderson almost as much as did American arms production. Yet the contrast between the system of clockmaking and that of armsmaking is striking. This contrast serves to underscore the unique way in which the armory system developed and the peculiarities of its production.

While Eli Whitney talked about mechanizing musket manufacture, another Connecticut Yankee, Eli Terry, as well as other New England clockmakers, actually set about designing and building a series of special-purpose machines to make wooden clocks. These clockmakers also made important marketing innovations, and, unlike the early arms contractors, they developed extensive private markets. Their reliance on the market rather than on the War Department created an entirely different set of expectations and responses. In developing their marketing strategy, the sewing machine and other industries of the second half of the nineteenth century that borrowed small arms production techniques owed more to the clock industry than to firearms. Moreover, the clock industry eventually demonstrated that mechanization of production could dramatically reduce costs and thereby increase sales. The Yankee machine-made clock was an embodiment of the notion of economies of scale. As Chauncey Jerome said in 1860 about the cheapness of his clock cases, "This proves and shows what can be done by system."[125]

The history of the wooden clock industry extends from about 1800 to 1837, when the Panic and the introduction of the machine-made brass clock brought about the industry's demise. (A number of wooden clock manufacturers, however, turned to brass clock production.) About 1793 Eli Terry moved to Plymouth, Connecticut, where he plied the traditional craft of clockmaking. Terry had been trained by Daniel Burnap to make brass movements and by Timothy Cheney to make wooden instruments. Like all clockmakers Terry used a hand-driven wheel-and-pinion cutting engine and a small turning lathe. But within a decade, he had embarked upon a remarkable scheme of producing a large number

FIGURE 1.9.   Howe's Pinmaking Machine, ca. 1838. John Ireland Howe first patented a pinmaking machine in 1833. Five years later, when the Howe Manufacturing Company moved from New York to Birmingham, Connecticut, Howe built a newly designed rotary pinmaking machine such as this one. Howe remained in business in Birmingham until his death in 1876. (National Museum of American History. Smithsonian Institution Neg. No. 76-15483.)

of clocks and marketing them himself rather than making them upon demand. By 1806, in a new shop, Terry began to make two hundred clocks at a time. The same year, he signed a contract with Edward and Levi Porter to make four thousand wooden movements in three years. The Porter brothers operated a company that assembled or finished clock movements purchased from other makers. This contract indicates that others besides Terry saw possibilities for the sale of large quantities of clocks. After a year of machine-building, Terry set out to fulfill the contract five hundred clocks at a time and finished the entire four thousand within the allotted period. He received $4 per movement. Terry then sold his factory to Silas Hoadley and Seth Thomas, who had worked with him to fulfill the contract.[126]

During the next two years Terry designed a new wooden clock far more compact than those previously made. About 1812 he began making plans to manufacture the "Pillar and Scroll" shelf clock, which he patented in 1816. Charles Fitch reported in 1880 that by

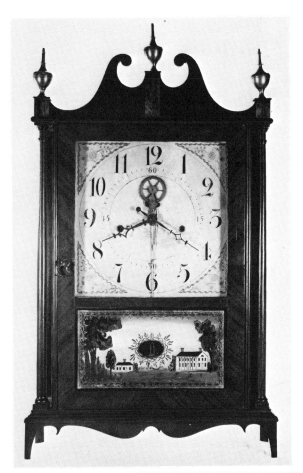

FIGURE 1.10. Eli Terry's Patented Pillar and Scroll Clock, ca. 1816. The name of Terry's clock derived from the design of the case (top). The clockwork is pictured below. (Private collection. American Clock & Watch Museum Photographs.)

1820 Terry was making twenty-five hundred clocks per year in four styles using thirty workmen. (See Figure 1.10.) How Terry actually produced his clocks is unknown, but as John Murphy points out, Eli Terry was not the only inventor/clockmaker who thought about producing clock parts with machinery. For example, on August 22, 1814, six different patents were issued for clockmaking machinery: (1) to make the time part of wooden clocks; (2) to turn and slit pinions; (3) a plate used for boring holes in parts; (4) to cut and point teeth and pinions in clock wheels; (5) to cut and point wheels; and (6) to point wire for clocks. Murphy argues that some of the patentees were or had been associated with Terry, indicating the far-ranging importance of the environment of mechanized clock production in which Terry operated. He states that Terry's "objective was not to revolutionize industrial techniques [but] . . . simply to produce clocks in quantity and cheaply." Nonetheless he helped bring about a mechanized wooden clock industry. Terry's goal should be contrasted with that of John Hall, for example. Like others associated with the Ordnance Department, Hall found inspiration in the ideal of interchangeable firearms parts rather than in mass marketing his patented firearms. The difference is at once subtle and profound. Some have argued that Terry sought and achieved interchangeable clock parts, but evidence suggests the contrary.[127] Unlike Colt, Terry did not claim interchangeability as an advertising feature of his product. Terry and other clockmakers mechanized because it was a way of producing a number of cheap clocks, not because they wanted clocks with fully interchangeable parts.

Eli Terry and other manufacturers such as Gideon Roberts and James Harrison survived in making large numbers of clocks because of the Yankee peddler system of selling clocks throughout every state and territory of the Union. Some of these peddlers sold only clocks, often extending credit for one- and two-year periods. George Mitchell, a merchant supplier from Bristol, Connecticut, not only outfitted Yankee peddlers but also financed a large number of local men in establishing factories to produce goods sold by his peddlers. Among the clockmakers he financed were Chauncey Jerome, Ephraim Downs, Samuel Terry (Eli's brother), and Elias Ingraham. The national market created or tapped by the Yankee peddler system cannot be overlooked as an important supporting factor in the mechanized, quantity production of wooden clocks.[128]

Competition in wooden clockmaking grew during the 1820s and early 1830s largely because of the ease of entry. Murphy argued that $2,000 of capital would have been adequate to set up a wooden clock factory. By mid-1830, sixteen factories operated in Bristol alone. A few miles away in Plymouth seven companies made clocks, two on an extensive scale. Other makers were located in Burlington, Goshen, Meriden, Torrington, Waterbury, Winsted, and other Connecticut towns. Before the Panic of 1837, thirty-eight thousand wooden clocks per year were produced; most companies averaged between two thousand and five thousand annually. Chauncey Jerome maintained that "no factory had made over *Ten* thousand [wooden clocks] in a year." As competition increased so did the extension of credit to buyers, and makers turned to style changes in an effort to win customers. As Murphy argued, those who prospered in wooden clockmaking were those who emphasized marketing rather than making their product.[129]

Unfortunately, almost nothing is known about the machines that Terry and others used for clock manufacturing nor does an adequate picture or engraving of the arrangement of the factories survive. There are only allusions to devices such as wheel cutters, pinion cutters, lathes, and wire pointers used in making the movements and circular saws and veneering machines employed for casemaking. Throughout the historical literature on the wooden clock industry, there are occasional references to ''gauges'' used in the sense that carpenters employ the word—a marking gauge. One reference alludes to a ''gauge'' in a way which suggests that this device might be employed to check or verify dimensions. This was a device used to indicate when a piece of wood for pinions had been turned down to the correct diameter.[130]

Recently, however, some twenty ''gauges,'' jigs, fixtures, and templates used in making late-vintage Seth Thomas wooden clocks have come to light. These devices provide more information about the level of precision in wooden clock manufacture than all the existing literary references. Because the clock on which they were used during

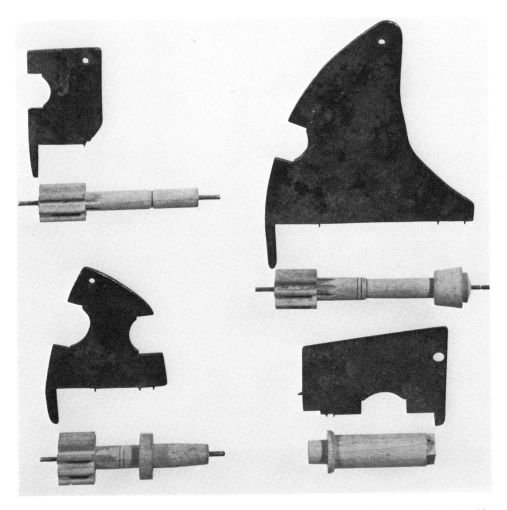

FIGURE 1.11. Seth Thomas Wooden Clock Gauges and Parts, ca. 1838. As positioned in this photograph, these gauges, which are made from sheet metal and survive from the Seth Thomas wooden clock factory, served as marking or scribing gauges. Lathe operators used them to verify dimensions of other parts. (The Connecticut Historical Society/Lewis B. Winton Collection. Connecticut Historical Society Photograph.)

production was one of the last model wooden clocks, these devices probably represent the finest specimens of then-current practice. Yet compared to metalworking gauges, they are crude. Marking gauges and verifying gauges exist. (See Figure 1.11.) Both types are roughly constructed—sheet iron hastily cut and filed—and so would give only very rough accuracy. The jigs that survive are timing jigs (used for the correct location of striking cams) and jigs for correctly bending the rods of the striker. (See Figure 1.12.) A template also survives that was used for locating the bearings on the wooden plates of the Thomas clock. (See Figure 1.13.) Although very ingenious, these devices have little in common with the jigs, fixtures, and gauges used at the Springfield Armory. The degree of precision in wooden clockmaking is so far removed from that required in small arms production that the two cannot be compared. This fundamental difference in precision also meant an important difference in the organization of production. The bureaucratic structure that evolved in American arms production to make and maintain the many precision gauges that ensured uniformity was absent in clockmaking.

The various devices used to aid the production of Seth Thomas clocks were by no means uniquely American. As British historian of design David Pye would argue, they represent an attempt to achieve the ''workmanship of certainty''—''workmanship in which the result is predetermined and unalterable once production begins.'' At the opposite end of the spectrum is the ''workmanship of risk'' in which ''the quality of the result is continually at risk during the process of making.'' Pye maintains that ''all

FIGURE 1.12.   Seth Thomas Wooden Clock Striker Bending Jig, ca. 1838. Among the surviving jigs and gauges from the Thomas factory, there are several bending jigs for mechanisms used to strike the hours, half-hours, and so on. Such jigs allowed the workman to bend the metal rods the correct way every time. (The Connecticut Historical Society/Lewis B. Winton Collection. Connecticut Historical Society Photograph.)

FIGURE 1.13. Seth Thomas Wooden Clock Plate Drilling Jig, ca. 1838. This spring-loaded jig, constructed principally of brass, was used to drill pinion bearings in the correct place in the front and back wooden plates of the Thomas clock. The small doughnut-shaped objects around each hole are steel inserts to prevent wear. The loading spring is steel. (The Connecticut Historical Society/Lewis B. Winton Collection. Connecticut Historical Society Photograph.)

workmen using the workmanship of risk are constantly devising ways to limit the risk by using such things as jigs and templates. If you want to draw a straight line with your pen, you do not go at it freehand, but use a ruler, that is to say a jig. There is still a risk of blots and kinks, but less risk. You could even do your writing with a stencil, a more exacting jig, but it would be slow.''[131] Although both the Seth Thomas devices and the jigs, fixtures, and gauges of arms production represent efforts to reduce risks, the difference in number, precision, refinement, and use is of such a degree as to be a difference in kind.

Brass clockmaking developed rapidly after 1837 and soon surpassed wooden clockmaking in the extent of mechanization and the volume of production. For centuries clocks had been made with brass wheels, yet a mechanized brass clock industry had not developed in Connecticut or elsewhere largely because of the very high cost of brass. The price declined significantly during the first three decades of the nineteenth century. Yet writers argue that even with lower cost, brass still was unfit for use in clocks because of poor homogeneity. In using brass, clockmakers had always cast their wheels and then "worked them up," a hand process involving hammering, filing, and scraping. Perhaps many a Yankee had tried to mechanize the process, but the rise of the brass clock industry hinged on the production of uniform, high-quality sheet brass. According to the historian

FIGURE 1.14.   Joseph Ives's Brass Clock, ca. 1838 Reconstruction of 1833 Patent Model. Although this particular clock was submitted to the United States Patent Office as part of Ives's application for a patent on ''striking parts of clocks,'' it illustrates the inventor's method of constructing his early brass clocks. Note the manner in which straps are riveted together to form the front and back plates and the raised bead around the circumference of the gearwheels. (National Museum of American History. Smithsonian Institution Neg. No. 81-9866.)

of the brass industry in Connecticut, brass rolling did not emerge out of its experimental phase until about 1830. The stage was then set for Joseph Ives of Bristol to initiate the manufacture of an eight-day clock with wheels and plates made from rolled brass sheets. Because the brass sheets were not as wide as his clock plates, he was forced to piece the plates together with rivets. (See Figure 1.14.) Ives nonetheless set a trend for the clock industry, which needed major revival after the destruction left by the Panic of 1837.[132]

The locus of technological development in brass clock manufacture occurred in stamping and punching. Stamping in the clock industry moved rapidly from the simple punching out of wheel or gear blanks and clock plates to production of practically finished wheels and plates. More and more operations were incorporated into the work of a single die, all carried out by one blow of the press. Joseph Whitworth reported in 1854 that thin brass was used to make wheels by raising a beading around the circumference just inside the location of the teeth. This beading provided extra lateral strength and economized the amount of brass required for the wheel. Raised beading appears on the rebuilt patent

FIGURE 1.15.   Early Chauncey Jerome Brass Clock Movement, 1839. Like all other later makers of brass clocks, Jerome adopted a single, stamped-out brass form for the plates of his clock. Note the raised beading on the gearwheels. (National Museum of American History. Smithsonian Institution Neg. No. 81-9864.)

model of Joseph Ives's eight-day shelf clock as well as in early Jerome clocks. (See Figures 1.14 and 1.15.) Thus from the beginning, beading was an important component of brass clock presswork. At the time Whitworth came to America, gear teeth were still cut in the wheel blanks rather than punched. A single machine composed of three cutters performed three sequential operations: simple cutting, rounding off the teeth, and finishing.[133] Later, many clocks were made with gears that had only been punched out.

Throughout the relatively sparse literature on brass clock manufacture in the antebellum period there are no references to a gauging system (with three different sets of the same gauges all based on a model) similar to those used in arms manufacture. This lack as well as the predominance of stamping and presswork suggests that brass clock manufacturers—like the earlier wooden clockmakers—developed their own distinctive approach to production, an approach not borrowed from or closely allied with that of small arms fabrication.[134] This raises the question of whether such British observers as John Anderson and Joseph Whitworth were aware of these differences in production of arms and other American manufactured goods.

Perhaps the difference in approach to production techniques between firearms makers and the clock industry was a result of the dissimilarity in objectives of the two industries. Whereas armsmakers, led by the United States War Department, sought to turn out a weapon constructed with uniform if not fully interchangeable parts, the clock industry—if Chauncey Jerome is a good example—desired above all to turn out vast numbers of cheap clocks. Mechanization provided the means to quantity production, not a means to interchangeability per se. Clockmakers wanted to sell a one-day shelf clock to everyone, American and European alike. By 1850, for example, Jerome operated two factories, one at New Haven and the other at Derby, Connecticut, and these factories annually turned out 130,000 and 150,000 clocks respectively. Jerome had lowered his prices to $1.50 for a complete one-day clock; later accounts reported the actual manufacturing costs of the clock at fifty cents. Jerome marketed his clocks extensively in Europe, particularly in England, where he operated his own wholesale store at Liverpool. Writing in 1860, the boastful Yankee claimed he had sold ''millions'' of his brass clocks throughout the world.[135]

When markets seemed to sag or competition pushed too hard, clock manufacturers introduced a new model. Hiram Camp complained about spontaneous changes that characterized the mature clock industry: ''The desire to make and sell great quantities has led the manufacturers to bring out new designs until dealers have become amazed *and* bewildered. . . . The expectation of something new prevents the sale of the old.''[136] From the beginning of wooden clock manufacture and the Yankee peddler system through the development of the brass clock industry in the antebellum period with its large wholesale network, the production of large numbers of inexpensive or cheap clocks was always accompanied by an impressive emphasis on marketing strategy. Emphasis on marketing by the clock industry as well as in the sewing machine, reaper, and bicycle industries suggests that marketing should be considered as an important aspect of the American system of manufactures—a component not considered by the British commentators on American technology in the antebellum period.

Among the least understood processes of the antebellum clock industry is assembly, a process which proved consistently to be problematic until the development of the assembly line in the twentieth century. With such a large output, clockmakers must have faced serious problems in assembling their products. Murphy reports that one man could assemble seventy-five movements per day, but he also notes that Hiram Camp, one of the major sources on the industry, said that brass clock assembly was "a slow process."[137] Beyond these statements, little information survives, particularly in the form of company correspondence. The historical record in clock manufacture is so inadequate—for example, the lack of any significant description of machinery or the survival of such machinery—that the historian risks peril with any generalization. Nonetheless, the annual assembly of parts for 150,000 clocks—over 3,000 weekly—should be regarded as a major feat. British observers of American technology in the 1850s were deeply impressed by the ease in assembly of muskets with interchangeable parts but failed to comment on clock assembly. This raises the question of whether the Committee on the Machinery of the United States of America led by John Anderson did indeed "learn the whole of the American system" as they were asked to do by the parliamentary Select Committee on Small Arms.[138]

The "American system" on the Eve of the Civil War

Immediately after the Select Committee on Small Arms ended its hearings, John Anderson and his fellow committee members came to the United States for the "purpose of inspecting the different gun factories in that country, and purchasing such machinery and models as may be necessary for the proposed gun factory at Enfield." Despite this charge, Anderson and his colleagues did far more in the United States than inspect arms plants. Based on the experiences of Joseph Whitworth and George Wallis in the previous year, the committee decided to tour a wide variety of American manufacturing establishments along the eastern seaboard and as far west as Pittsburgh.[139]

Whitworth, the leading machine tool builder in England, and Wallis, headmaster of the Government School of Design in Birmingham, had come to the United States in 1853, serving as commissioners to the second Exhibition of the Industries of All Nations, the New York Crystal Palace Exhibition. Originally led to believe that the exhibition would open June 1, the British commissioners arrived only to learn that even when the formal opening ceremonies were to be held—now delayed until July 14, 1853—only half of the displays would be finished. Some decided to use their time in the United States by individually touring the country to study the subject matter of the departments on which they had intended to report. The *Special Reports* of Joseph Whitworth and George Wallis were submitted as substitutes for reports on the exhibition.[140]

During June and early July, Joseph Whitworth toured manufacturing establishments in some fifteen American cities, mostly in New England but also including Washington, Baltimore, Philadelphia, Pittsburgh, New York, and Buffalo. He visited steam engine factories; railway shops; spike-, nail-, and rivetmaking factories; cutlery plants; clock and lock factories; armsmaking establishments; and woodworking factories, among many other industries. Although he occasionally pointed out certain backwardness or crudeness in American production, Whitworth's dominant theme throughout his report was "the eagerness with which they [the Americans] call in the aid of machinery in almost every department of industry. Wherever it can be introduced as a substitute for manual labor, it is universally and willingly resorted to."[141] By machinery, Whitworth meant special-purpose machinery.

The English machine tool builder declared that Americans had not matched his country's classic metalworking machine tools, which he called engineering tools.[142] America's woodworking machinery, however, greatly impressed him. "In no branch of manufacture," he wrote, "does the application of labour-saving machinery produce by simple means more important results than in the working of wood."[143]

Springfield Armory's battery of gunstock-making machinery—Blanchard's machines as redesigned by Cyrus Buckland—impressed Whitworth more than anything else about American technology. In his report, he delineated each of the sixteen machining operations and recorded their various "machine times" as well as the time required for two small hand operations. Whitworth calculated that a complete musket stock could be made in slightly more than twenty-two minutes. Curiously, he added that the "complete musket is made (by putting together the separate parts) in 3 minutes. All these parts are so exactly alike that any single part will, in its place, fit any musket." (See Figure 1.16.) Throughout Whitworth's report, this is the only mention of interchangeability, and he did not explain how it was achieved. Certainly if American manufacturing technology were viewed from Whitworth's perspective, mechanization rather than interchangeability distinguished the "American system" of manufactures.[144]

George Wallis's report echoed that of Whitworth in that when mentioning production methods for the manufactured goods he evaluated, he commented on the "large amount of highly ingenious machinery [which] is constantly and most successfully employed."[145] To be sure, Wallis also pointed out pockets or industries in America where handicrafts still dominated production methods. His emphasis on systematic and efficient production of goods in America and failure to mention the manufacture of products constructed with interchangeable parts indicates that interchangeability did not appear to Wallis in 1852 as an overriding characteristic of all American manufactures.

After Whitworth and Wallis returned to England, the subject of interchangeability of parts was of major interest during the proceedings of the Select Committee on Small Arms. It is interesting that this discussion took place in a military context. For example, the select committee chose to interview Samuel Colt and not Alfred C. Hobbs, an American who had established a lock factory in England using American manufacturing techniques. The idea of interchangeable firearms captivated the British Board of Ordnance and many members of the select committee much in the way it had the United States Ordnance Department in the 1800s. The question now was how well the Americans had achieved this goal, and opinion on this issue differed considerably. For this reason, when Anderson's committee came to the United States it took care to settle the issue of whether Springfield Armory muskets were truly interchangeable. With the permission of Springfield's superintendent, Anderson and his committee went into the main arsenal and selected at random one musket made in each year from 1844 to 1853. These ten muskets were then disassembled and corresponding parts of each put into separate boxes. These parts were mixed up. Anderson's committee then selected "at hazard" parts from each box and handed them to a workman, who, using only a screwdriver, reassembled the ten muskets "as quickly as though they had been English muskets whose parts had carefully been kept separate."[146] Colt's pistols may not have interchanged, but Anderson no longer doubted the complete interchangeability of parts of the Springfield musket.

Anderson must have wondered why the products of the Springfield Armory differed from Colt's. Earlier he had described Colt's London armory as follows:

> This manufactory is reduced to an almost perfect system; a pistol being composed of a certain number of distinct pieces, each piece is produced in proportionate quantity by machinery, and as each piece when finished is the result of a number of operations (some 20 or 30), and each operation being performed by a special machine made on purpose, many of these machines requiring hardly any skill from the attendant beyond knowing

FIGURE 1.16. Musket Assembly, Springfield Armory, 1852. This illustration may be the only antebellum representation of assembly operations at the Springfield Armory. Its lack of detail makes it unclear why the workman needed a vise; verbal descriptions such as that of Joseph Whitworth do not help clarify this question. (*Harper's New Monthly Magazine*, July 1852. Eleutherian Mills Historical Library.)

how to fasten and unfasten the article, the setting and adjusting of the machine being performed by skilled workmen; but when once the machine is properly set it will produce thousands. Hence there are more than a hundred machines employed, many of them similar in principle to each other, although differing in the form of the cutting instruments.[147]

Yet in "learning the whole of the American system" at Springfield, Anderson became aware of the fundamental importance of the "hundreds of valuable instruments [jigs and fixtures] and gauges that are employed in testing the work through all its stages, from the raw materials to the finished gun, others for holding the pieces whilst undergoing different operations such as marking, drilling, screwing, etc., the object of all being to secure thorough identity in all parts." Elsewhere Anderson argued that "it is only by means of a continual and careful application of these instruments that uniformity of work to secure interchanges can be obtained." The complexity of such a system of fixtures and gauges "cannot be appreciated by those who have been engaged on a ruder system" of manufacture.[148] Yet Anderson discussed this part of the "American system" only within the explicit context of small arms production, with Springfield Armory being the exemplar.

When Anderson and his committee described non-firearms-manufacturing establishments in the United States, they identified elements common to almost all American manufactures, including firearms. American manufactures, they concluded, were characterized by the "adaptation of special tools to minute purposes," "the ample provision of workshop room," "systematic arrangement in the manufacture," "the progress of material through the manufactory," and the "discipline and sobriety of the employed."[149]

Above all, John Anderson identified a certain universality to American manufacturing technology. Whenever an article was to be manufactured repeatedly, the "system of special machinery" could and should be used.[150] "The American machinery is so different to our own," he later wrote, "and so rich in suggestions. . . . A few hours at Enfield [newly equipped with American-made machinery and tools] will show that we shall have to contend with no mean competitors in the Americans, who display an originality and common sense in most of their arrangements which are not to be despised, but on the contrary are either to be copied or improved upon."[151] Elsewhere Anderson urged gunmakers in particular to follow the lead of American armsmakers. "If the military gunmakers of England are wise in their generation," he stressed, "they will not despise this system of manufacture, but, on the contrary, will adopt it, for it will secure for them a high vantage ground in competing with other parts of the world."[152] In concluding the report of the committee, Anderson warned that the "contriving and making of machinery has become so common in [the United States] . . . that unless the [American] example is followed at home . . . it is to be feared that American manufacturers will before long become exporters . . . to England."[153]

Anderson's warning that American manufacturers would become exporters to England came to pass in the second half of the nineteenth century. English manufacturers did not wish to adopt the American practices that Anderson, Colt, Nasmyth, and others had declared to be universally applicable. But neither did all manufacturers in the United States. Indeed, when Anderson toured Derringer's pistol factory in Philadelphia, he was "astonished" to find that traditional hand methods were still in use.[154] At the very time Anderson was issuing his warnings, an entirely new industry—that of producing sewing machines—was getting started. The history of three of the leading sewing machine firms suggests that not all adopted the "system of special machinery" so prevalent in America nor the "American system" of interchangeable manufacturing perfected at the Spring-field Armory. One firm in particular, Singer Manufacturing Company, managed for a long time to compete successfully using "ruder" European methods, even in the American economy, where "the high price of labour" was believed to prevail against the employment of such manufacturing techniques.[155]

LOSSING&BARRITT.SC.N.Y.

CHAPTER 2

The Sewing Machine
& the American System
of Manufactures

Intelligence of what was transpiring at the [Springfield] Armory was widely diffused. . . . The news reached and enchanted the sewing machine men. . . . When we look back, it appears as though they all fell into the plan tumultuously; but this was not so. It was adopted in a gradual manner. Some did not fully comprehend that the full benefit of the system would not be got unless every piece was made by it. They would leave a piece, here and there, to be finished according to the idiosyncrasies of the workmen, and these pieces made confusion until the full system was carried out.
—''The 'American' System of Manufacture,'' *American Machinist* (1890)

The gun-maker's tools were carried to the sewing machine manufactory, but as the demand grew for a better quality of work these tools were improved until we find the sewing machine now in possession of the improved milling machine, the perfected screw machine, the turret lathe, a complete system of ''jig'' working, and a system of measuring by decimals; often extending to tens of thousandths and frequently beyond the ten thousandths. . . . Gauge work is an outgrowth from a rude system that originated in the armories, but has been perfected and systematized in the sewing machine manufactory.
—''What the World Owes to the Sewing Machine Workman,'' *The Sewing Machine Advance* (1890)

The development of a workable sewing machine occurred at a propitious moment in American manufacturing history. After a half century of often intense and expensive development, small arms manufacture had reached a level of maturity. No doubt echoing many a Yankee mechanic, Samuel Colt had told a British parliamentary committee that anything could be made by machinery, not just guns. The sewing machine offered Colt's followers an excellent opportunity to demonstrate that point. Adopting and adapting small arms production technology, some sewing machine manufacturers clearly proved that Colt was correct. But at least one company demonstrated that machine production similar to Colt's was not essential for commercial success in America.

No sewing machine company could carefully choose and develop its production technology at the outset of the industry because each was embroiled in patent litigation with the others. No single inventor or company had gained a patent position sufficiently strong to dominate the industry. In fact, litigation threatened the very existence of the industry.

The Great Sewing Machine Combination, the first important patent pooling arrangement in American history, changed all of this. For a fee of $5 on every domestically sold machine and $1 on each exported one Elias Howe contributed to the pool his fundamental patent (1846) for a grooved, eye-pointed needle used in conjunction with a lock-stitch-forming shuttle. Allen B. Wilson, through the Wheeler and Wilson Manufacturing Company, placed his 1854 patent on the four-motion cloth-feeding mechanism into the pool. I. M. Singer & Co., a partnership of inventor Isaac Merrit Singer and lawyer/capitalist Edward Clark, provided a number of its patents, including Singer's monopoly on the needle bar cam (1851). In addition, the Grover and Baker Sewing Machine Company, whose president, Orlando Potter, had been the chief architect of the pool, added some of its patents. Members of the pool could use these patents freely as could any other manufacturer willing to pay a license fee of $15 per machine. In addition to Howe's fee of $4 per machine, a set amount of the $15 (initially $7 per machine) went into a litigation fund actively used to protect the patentees and licensees. The balance was then divided among pool members.[1]

Although the Wheeler and Wilson, Singer, and Grover and Baker companies had been manufacturing sewing machines since the early 1850s, the patent pool allowed them to consider expanding their manufacturing operations without fear of litigation. The pool and the rising demand for sewing machines (particularly for those of Singer and Wheeler and Wilson) provided manufacturers with the opportunity to choose their production technology. This choice was by no means obvious. The background and outlook of the heads of companies proved to be the critical factors in the decision-making process rather than any particular factor endowment of land, labor, and capital. By considering the infant sewing machine industry in the context of the production technologies selected, it is possible to sharpen our understanding of the American system of manufactures in the second half of the nineteenth century.

The production of three different sewing machines—those of Wheeler and Wilson, Willcox & Gibbs, and Singer—will be discussed in this chapter. The Wheeler and Wilson machine was initially produced by traditional hand methods, but soon its manufacturer hired mechanics who had worked in prominent American armories and who quickly introduced the techniques that had captivated the British Board of Ordnance. Brown & Sharpe, the manufacturer of the Willcox & Gibbs machine, used armory practice from the outset and made important contributions to manufacturing technology that went well beyond sewing machine production. The Singer Manufacturing Company, which became the preeminent sewing machine manufacturer in the nineteenth century, employed for a long time the "ruder" European method of manufacturing. Singer was slow to adopt the American system of interchangeable manufacture, and as late as the early 1880s still had not completely and successfully adopted this system. Singer's history provides an outstanding case that contradicts long-held assumptions about American manufacturing in the nineteenth century.

The Wheeler and Wilson Manufacturing Company

Manufacturer and capitalist Nathaniel Wheeler teamed up with inventor Allen B. Wilson to form the sewing machine company that led the industry in sales until 1867, when Singer made its runaway move toward complete domination of the industry. Wilson contributed four major inventions, and Wheeler provided the business know-how that led to early success. Early in the company's history, Wheeler guided it into manufacturing by the armory system of production, but then Wilson dropped out of the picture.[2]

Born in 1820, Nathaniel Wheeler grew up in Watertown, Connecticut, and worked with his father in carriage manufacture. Although he took over his father's business for a few years, he eventually left to undertake the manufacture of small metal articles such as buckles, buttons, and neckerchief clasps. Biographers have noted that originally Wheeler made these products by hand, but slowly he mechanized the work processes. In 1848 Wheeler combined his company with the partnership of George B. Woodruff and Alanson Warren, and the new firm, Warren, Wheeler and Woodruff, built a new factory for the manufacture of small metal articles. Operations were superintended by Wheeler. While in New York City near the end of 1850, the young manufacturer saw the Wilson sewing machine. Immediately he negotiated a contract to make five hundred of them for the company that held Allen B. Wilson's patents on the two-motion cloth-feeding mechanism. Wheeler also persuaded the inventor to move to Watertown so that he could improve the machine and oversee its manufacture.

Before Allen Wilson moved to Connecticut, he redesigned his sewing machine by replacing the shuttle with a rotary hook and bobbin mechanism that would also make a lock stitch. While in Watertown he perfected this device and patented it August 12, 1851. Three years later, though no longer involved with the company, he would patent his fourth major improvement, the four-motion cloth feed, still the basic principle for cloth feed in today's sewing machines. In 1851, Wilson joined in partnership with Warren, Wheeler, and Woodruff, and under the name Wheeler, Wilson and Company the men began to manufacture sewing machines with rotary hooks. By mid-1853, the company had made and sold three hundred machines. (See Table 2.1 for production figures of Wheeler and Wilson machines.) *Scientific American* noted that "the price of one all complete is $125; every machine is made under the eye of the inventor at the Company's machine shop, Watertown, Connecticut."[3] Each sewing machine was unique. Wheeler, Wilson and Company built them individually or in small batches with a small line of general metal-working tools.

With a glimmer of success, the partners incorporated, forming the Wheeler and Wilson Manufacturing Company with capital of $160,000. Wilson withdrew from active participation in the company but continued to make improvements on his sewing machine. The departure in 1855 of the corporation's president, Alanson Warren, and its secretary and treasurer, George B. Woodruff, both of whom had been partners with Wheeler since the 1840s, is difficult to understand. Perhaps the most plausible explanation is that both men were discouraged by the continued court battles over patent rights. Warren's subsequent history is unknown, but Woodruff joined the Singer company in 1857, eventually became a successful Singer salesman, and for many years headed Singer's European marketing force.[4]

Woodruff was succeeded by William H. Perry, a Connecticut Yankee, born in 1820. Perry's arrival determined the future of the company's production technology. Although originally a schoolteacher in Newington, Connecticut, Perry had learned the machinist's trade working for his brother, one of the many inside contractors at Samuel Colt's armory in Hartford. Eventually he assumed part of his brother's contract, but in 1855 he moved upriver to take over the bookkeeping for Wheeler and Wilson. Within a year he had been named secretary and treasurer of the company, now under the presidency of Nathaniel Wheeler. More important, he had become superintendent of its factory.[5] The stage was set for Wheeler and Wilson's adoption of New England armory practice. Moving from

TABLE 2.1. PRODUCTION OF WHEELER AND WILSON AND WILLCOX & GIBBS SEWING MACHINES, 1853–1876

Year	Wheeler and Wilson	Willcox & Gibbs
1853	799	n.a.
1854	756	n.a.
1855	1,171	n.a.
1856	2,210	n.a.
1857	4,591	n.a.
1858	7,978	n.a.
1859	21,306	n.a.
1860	25,102	n.a.
1861	18,556	n.a.
1862	28,202	n.a.
1863	29,778	n.a.
1864	40,062	n.a.
1865	39,157	n.a.
1866	50,132	n.a.
1867	38,055	14,150
1868	n.a.	15,000
1869	78,866	17,251
1870	83,208	28,890
1871	128,526	30,127
1872	174,088	33,639
1873	119,190	15,881
1874	92,827	13,710
1875	103,740	14,502
1876	108,997	12,758

Source: Frederick G. Bourne, ''American Sewing-Machines,'' in *One Hundred Years of American Commerce,* ed. Chauncey M. Depew, 2:530.

teacher in a common school to machinist/contractor at one of the best-known showplaces of American production technology, Perry had learned only one way to produce metal products—the Colt way.

The same year Perry assumed the superintendent's job, the Wheeler and Wilson Manufacturing Company moved west from Watertown to Bridgeport into a former temple of American manufacturing, the clock factory once occupied by Chauncey Jerome.[6] When it began to manufacture sewing machines at Bridgeport in 1857, production resembled that used for revolvers at Colt's armory in Hartford: major running components initially drop-forged and then machined; machining work performed by numerous specialized machine tools operated sequentially; uniformity of parts controlled by a model-based gauging system along with a rational jig and fixture system; and work executed by an inside contracting system coordinated by the factory superintendent.[7] The small Wheeler and Wilson factory seemed a microcosm of Colt's armory.

When installing the American system of manufacturing at Bridgeport, William Perry drew upon more extensive resources than his own experience at the Colt factory. At once he hired Joseph Dana Alvord, a New England mechanic who had learned his trade under Nathan P. Ames, the noted armsmaker of Chicopee, Massachusetts, and had worked at the Springfield Armory for eight years. In 1851 he had moved from Springfield to Hartford, where he worked as a contractor at the Robbins & Lawrence–Sharps Rifle Company. That year Robbins & Lawrence, a firm based in Windsor, Vermont, had attracted attention by its display of interchangeable firearms at the London Crystal Palace Exhibition. George Eames, one of the Singer Manufacturing Company's production experts, wrote that Alvord ''was an exceptionally able man, both as a manager and a progressive mechanic, and many principles of tool building that he introduced into the early history of sewing machine work have been copied and extended so they are now the regular procedure in the art of tool building.''[8]

In addition to hiring Alvord, Perry also attracted Alvord's friend and fellow Robbins & Lawrence contractor, James Wilson. By 1862 Perry, assisted by Alvord, Wilson, and others, had created what they called ''a great machine shop,'' capable of producing almost thirty thousand sewing machines annually. (See Figure 2.1.) The company claimed that with its accurate, specialized machine tools, jig and fixture design, and system of gauging, parts never needed a ''stroke of a file'' during assembly.[9] (See Figure 2.2.)

In 1855, when William Perry began as a bookkeeper for the company, Wheeler and Wilson made fewer than twelve hundred sewing machines. The following year—the year of the patent pool and the company's move to Bridgeport—sales increased to slightly over twenty-two hundred machines. Wheeler's action in hiring Perry as superintendent and

FIGURE 2.1. Machine Shop, Wheeler and Wilson Manufacturing Company, 1879. (*Scientific American*, May 3, 1879. Eleutherian Mills Historical Library.)

THE NEW No. 8

FIGURE 2.2.   Assembly Room, Wheeler and Wilson Manufacturing Company, 1879. As depicted in this illustration, workmen individually assembled the Wheeler and Wilson sewing machine at worktables. Contrary to the earlier claims of the company, files and vises are evident. The bearings of finished sewing machines lined up down the center are being broken in by running the machines. (*Scientific American,* May 3, 1879. Eleutherian Mills Historical Library.)

bringing him into the company as secretary and treasurer was, in essence, an exercise in technical decision making: Perry knew only the Colt way to make metal products. But Wheeler himself was a manufacturer who had used predominantly hand methods of working metal. He knew that he could continue to use hand methods to produce sewing machines with a reasonable profit. But through his own business as well as through observation of developments in Hartford, he realized the productive benefits of mechanization and the solid rationality of uniformity of parts in consumer goods subject to wear and mechanical failure. More than anything else, he sensed that with patent matters secure and the growing enthusiasm over the sewing machine, Wheeler and Wilson would experience strong growth. Between 1855 and 1856, its business almost doubled. A company publication of 1863 stated that Wheeler ''has never hesitated a moment in the faith that the world would appreciate a good sewing-machine sufficiently to recompense the manufacturer for an outlay of half a million dollars in facilities for manufacturing; and he has always been ready to adopt every improvement, until the perfection of workmanship and height of ornamentation, combined with usefulness, have nearly been achieved.''[10]

The company looked forward to the time "when every foot of the five acres of flooring of the Wheeler & Wilson Sewing-Machine manufactory will be fully occupied by 800 men, who could find room to work, and who, by driving the present machinery [could] turn out 100,000 sewing-machines a year." Increased production would mean that "they could be still further reduced in price, and then every family could be provided with one of these indispensable articles."[11]

Whichever technology he chose—the classic European method or the American system—Wheeler wanted to be capable of meeting his goal of a hundred thousand machines per year. William Perry (and possibly others) convinced Wheeler that the methods used in American armories could meet this challenge. Perhaps Wheeler never needed convincing. In any case, he chose to meet the growing demand for sewing machines with this special version of the American system of manufactures. With Perry as superintendent, the system was adopted in toto. Alvord provided the precision manufacturing know-how he had learned at Robbins & Lawrence, which was vital for interchangeable manufacture. The system, along with the sound design of the Wheeler and Wilson sewing machine, allowed the company to expand annual production up to its peak in 1872 of 174,088 machines. (See Figure 2.3.)

FIGURE 2.3.   Wheeler and Wilson Sewing Machine, ca. 1876. (National Museum of American History. Smithsonian Institution Neg. No. 17663-C.)

Throughout its history after 1856, Wheeler and Wilson developed an increasingly refined version of New England armory practice, which might be called "high" armory practice. For instance, the initial drop-forging procedures were extended and refined. By the mid-1860s some parts of the Wheeler and Wilson sewing machine underwent four forgings before entering the machining rooms. Under Alvord's direction, gauges, fixtures, and jigs were refined until the company claimed absolute perfection in manufacturing interchangeable parts.[12]

FIGURE 2.4.   Punching Out Needle Eyes, Wheeler and Wilson Manufacturing Company, 1879. A series of small, hand-operated machines were used to manufacture sewing machine needles. Two of the most difficult operations included punching out the eye and grooving the needle. The final step of straightening the needles was done by hand with a hammer. (*Scientific American*, May 3, 1879. Eleutherian Mills Historical Library.)

Yet one critical part of the Wheeler and Wilson sewing machine did not easily lend itself to the American system of manufacture: the needle. A description of needle manufacture at the Wheeler and Wilson factory in 1863 reads like Adam Smith's classic account of pinmaking and the division of labor, published almost a century earlier in *The Wealth of Nations*. The entire production process, from the cutting of the wire to the polishing of the finished needle, was done by highly subdivided hand labor. As the company pointed out, "It is the great amount of hand labor, in a country where mechanics' wages are so much higher than in Europe, that makes this kind of needles so expensive." Wheeler and Wilson had "several times hired English needle-makers, who had served a long apprenticeship at the business."[13] But the company found American labor—even if hand labor—superior to that of England.

Fifteen years later, needlemaking still largely eluded American know-how, but costs had been lowered substantially by the employment of poorly paid women and children. Some mechanization had also been carried out, and the number of processes had been reduced from fifty-two to thirty-three.[14] (See Figure 2.4.) Needle manufacturers such as the Excelsior Company of Torrington, Connecticut, attacked the problem of hand labor in making needles. By the 1880s much handwork had been eliminated, yet needles still were straightened one by one with a hammer on an anvil.[15]

J. R. Brown & Sharpe and the Willcox & Gibbs Sewing Machine

Whereas the Wheeler and Wilson sewing machine had originally been the product of skilled machinists in the job shop and then became a uniform article turned out by a factory operating under the armory system of manufacture, the Willcox & Gibbs sewing machine was manufactured under the armory system from the very beginning. The first few years of its commercial production offer a marked contrast to that of Wheeler and Wilson and particularly to that of the Singer sewing machine.

The Willcox & Gibbs Sewing Machine Company was a partnership of inventor James E. A. Gibbs, a Virginian, and James A. Willcox, a Philadelphia hardware merchant and perhaps also a patent model builder. Gibbs's sewing machine, covered by three patents, used a single thread to make a chain stitch rather than a shuttle or rotary hook with bobbin to form a lock stitch. Willcox recognized immediately that this machine could be made and sold far more cheaply than lock-stitch machines. But when he first saw the machine it had not been perfected. Consequently he set his seventeen-year-old son, Charles, to work with Gibbs to perfect it and at the same time began making arrangements for its manufacture.[16] Early in 1858, Willcox had decided that he wanted J. R. Brown & Sharpe of Providence, Rhode Island, to produce the Willcox & Gibbs sewing machine. (See Figure 2.5.)

Surviving records do not indicate why Willcox chose Brown & Sharpe. The Rhode Island company had had no experience producing metal articles comparable to sewing machines, but rather made custom clocks and watches; jewelers', clockmakers', and watchmakers' tools; stub files; wire gauges; rules; and drawing instruments. In addition, it had an extensive watch and clock repair business.[17] Through William G. Angell, Willcox learned about Brown & Sharpe's possible interest in manufacturing items such as sewing machines. Late in 1857 or early in 1858, Lucian Sharpe went to Philadelphia to talk with Willcox. They agreed that Brown & Sharpe would make an estimate of the cost of

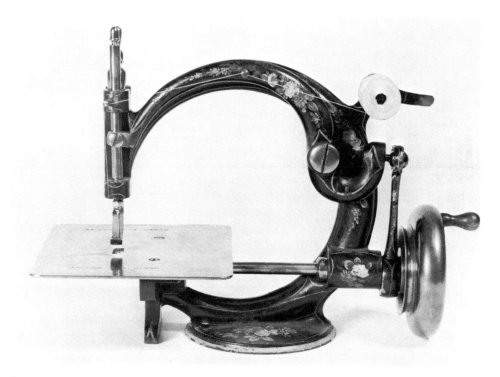

FIGURE 2.5.   Willcox & Gibbs Sewing Machine, U.S. Patent Model, 1858. (National Museum of American History. Smithsonian Institution Neg. No. P6393.)

producing the Willcox & Gibbs sewing machine once the machine itself had been satisfactorily perfected. In late February 1858 Charles Willcox brought a nearly perfected sewing machine to Providence, and within a month, Brown & Sharpe had set out to manufacture twelve sewing machines for the Willcox & Gibbs company.[18]

The historian would probably expect that a job shop such as J. R. Brown & Sharpe with its three small engine lathes, two hand lathes, an upright drill, a hand level planer, and a "donkey" planer would have fulfilled this agreement by building the twelve sewing machines one by one.[19] But like so many American job shops that blossomed into factories, Brown & Sharpe set about designing a model along with special tools, jigs, fixtures, and gauges before it ever completed a commercial sewing machine. James Willcox had apparently allowed the company plenty of leeway in what appears to have been a gentleman's agreement on manufacturing the machine. As Lucian Sharpe understood that agreement, Willcox would pay $3 per day "for all work upon a model and the first twelve machines." In addition, Willcox would be responsible for changes on patterns, dies, and tools, "provided we [Brown & Sharpe] should not build a sufficient quantity of sewing machines to compensate us for the tools."[20] How many machines constituted a sufficient number is unknown, but apparently Brown & Sharpe eventually assumed these costs.

There is no hint that Willcox demanded that his sewing machines be made with the American system of interchangeable parts manufacture or "armory practice" as New England mechanics called it; he desired only machines that were "got up right."[21] Brown

& Sharpe saw in the manufacture of Willcox & Gibbs sewing machines an opportunity to expand and diversify its business. That the shop chose to produce these sewing machines with the American system—a system with which neither partner had had any intimate experience—rather than with its accustomed job shop approach reflects the attention this production system had attracted in New England, if nowhere else.

Brown & Sharpe's initial experiences with production under armory practice compose a litany of problems repeated time and again by shops and factories adopting this unfamiliar system. Throughout this litany runs a theme of faith in the system once it is fully installed and its demands have been completely met. It was this faith in armory production technology that initially compelled Brown & Sharpe and others to adopt the system, helped them to cope with its problems, and led them to meet its demands.

Work on the Willcox & Gibbs sewing machine began about March 15, 1858. At once Brown & Sharpe made drawings of the machine from which patterns for castings were made by a local foundry, the New England Butt Company. These drawings also served as the basis for construction of tools and jigs. Three to five full-time machinists and James Willcox's son, Charles, worked on the tools, but by the end of May, they had still not been finished nor, of course, had any of the twelve sewing machines. Lucian Sharpe wrote James Willcox that setting up for production of the machine "has taken much longer than anticipated" and the "tools proved to be three to four times as expensive as was contemplated by us at the commencement though they will doubtless be cheap in the end if many machines are manufactured." Dies for forging had also been expensive.[22]

Even before the tools for the dozen sewing machines had been completed, Brown & Sharpe agreed with Willcox to make a hundred machines, fifty of them the size of the twelve machines under way and fifty smaller. The Rhode Island jobbers learned too late that "the making of tools for the small machines at the same time with the others has been a mistake, inasmuch as it has caused nearly twice the delay that would otherwise have occurred, and they could have been made cheaper after the experience acquired upon the first lot." Toward the end of July, however, Brown & Sharpe believed that it could at last see light at the end of the tunnel: "Preparation for making other lots of sewing machines can be made with advantage, and if not done now further delay will be caused hereafter." These plans included the acquisition of new machine tools. Yet Brown & Sharpe still had not been able to calculate production costs and believed this would be possible only when the hundred sewing machines had been finished.[23]

The hundred sewing machines were still unfinished at the end of August, and Brown & Sharpe's initial optimism had been tempered by technical problems. Because of excessive wear on drills, reamers, and taps, it had been forced to anneal (or soften) the castings of the sewing machines. Although annealing added trouble and expense, it helped eliminate problems that had thwarted interchangeability. Because some turning operations had consumed so much time, the company fitted its lathes with back rests and turned pieces with two cutters instead of one. It also purchased additional screw machines. These developments prompted Lucian Sharpe to conclude cautiously that "everything seems to be going on as fast as possible and we . . . hasten the time when machines can be turned out as fast as the market may demand; yet we cannot but feel nervous at the continued postponement of that much to be desired day." Sharpe stressed, however, "We are yet of the opinion that when once the tools are done and in successful operation that the machines can be turned out in as great quantities as you can desire." Caution had turned to frustration by early September, when Sharpe wrote Willcox that "we are very much disappointed and at times discouraged. Since you left more work has been done on tools than [sewing] machines, and it will be several weeks yet before the tools can be completed.''[24]

No doubt by this time James Willcox was seriously questioning the ability of Brown & Sharpe to produce the Willcox & Gibbs sewing machine. The increasing numbers of cheap sewing machines entering the market made him anxious to ''lose no time in getting them out.'' Evidently he expressed his dissatisfaction in a letter to Brown & Sharpe. The company replied that it was not surprised at Willcox's dissatisfaction. But it stressed that ''if we had turned out a quantity in an imperfect state as would have been the case had they been made without the templets and other tools it would without doubt [have] injured the reputation of the machine if it did not kill it entirely.'' As Sharpe explained, ''Our first trouble was in getting a hook to work and after that was accomplished we made tools by which we could turn out any quantity of machines.'' Fabrication of these tools had not been as simple or as cheap as Brown & Sharpe had anticipated. With the goal of ''producing perfect work,'' the company had spent ten times as much on tools as it had expected. But as Sharpe was careful to point out, ''By the experience now acquired we now hope to turn out nearly perfect machines at the outset instead of proceeding in the usual way with such things.'' Sharpe added that ''you doubtless are as well aware as ourselves of the importance of pushing the tools.''[25]

This letter must have put James Willcox at ease, for soon Brown & Sharpe undertook to produce an additional thousand Willcox & Gibbs sewing machines and planned to ''increase our tools if as many machines are wanted as contemplated.'' Lucian Sharpe pointed out to Willcox that the company was ''relying upon the sewing machines to give us business enough for these tools for two or three years at least; else it would not be good policy to increase our facilities as much as we have.''[26]

At the end of October 1858, eight months after beginning work on the Willcox & Gibbs sewing machine, Sharpe wrote Willcox that fifty of the original one hundred machines were being ''finished and put together.'' Doing so had been neither cheap nor simple, however, because ''it takes our best men to put them up. Being the first that have been made they go together with more difficulty than the others will hereafter.'' As he had in several previous letters, Sharpe noted that the special tools of the American system were demanding of skill, time, and money. Brown & Sharpe had hired a mechanic who had previously worked at the Robbins & Lawrence–Sharps Rifle Company in Hartford and was well-versed in the American system. This mechanic had told Sharpe that it was not uncommon for ''$10,000 [to be] spent upon tools for a single gun lock and $25,000 for tools for a rifle.'' Given this warning, Sharpe suggested to Willcox that work had progressed comparatively well. The mechanic, reported Sharpe, ''says we have got along with the tools as fast as he expected.''[27]

Brown & Sharpe demonstrated an abiding faith in the techniques used at small arms factories. Assembling the first sewing machine parts required great fitting by the most skilled workmen, but the next batch promised to be better until eventually near perfection would be reached. Brown & Sharpe believed that only the special tools, jigs, fixtures, and gauges—the hardware of the armory system—would provide the means to perfection in sewing machine manufacture. Lucian Sharpe also told Willcox that Brown had calculated that five thousand sewing machines could be made annually using the company's present tools, and if more lathes were added, this number could be doubled ''without materially increasing the small tools we have been so long in making.''[28]

The Willcox & Gibbs sewing machine met with instant popularity, and James Willcox continued to have faith in Brown & Sharpe's ability to manufacture it. The Rhode Island job shop soon became a factory. Still, Joseph Brown and Lucian Sharpe were unwilling to abandon fully the custom manufacturing on which they had originally thrived. Within a few years they began selling machine tools and small tools they had designed and built for sewing machine manufacture.[29] Yet Brown & Sharpe continued to make the Willcox & Gibbs machine until the 1950s because the business proved profitable.[30] (See Figures 2.6 and 2.7.) The company reaped more than financial profits from sewing machine manufacture. Because manufacture of sewing machines differed significantly from small jobbing and from machine tool building, it provided an important training ground for mechanics. Joseph R. Brown believed that a mechanic should know all three kinds of work. In addition to leading the company into machine tool manufacture, the Willcox & Gibbs business supplied Brown & Sharpe with an opportunity to test the tools it marketed and thus kept the company conscious of the real needs of manufacture in such vital elements as precision, speed, and ease of operation.[31]

FIGURE 2.6.   Brown & Sharpe Factory, Providence, Rhode Island, 1860s. Workers on the upper floor display Willcox & Gibbs sewing machines from the windows of the shop where they were made. The company moved to a new, modern factory also in Providence in the early 1870s. (Brown & Sharpe Manufacturing Company, North Kingstown, Rhode Island.)

FIGURE 2.7.   Brown & Sharpe's Shop Where Willcox & Gibbs Sewing Machines Were Made, 1879. Assembly operations are being carried out in the foreground. (*Scientific American*, November 1, 1879. Eleutherian Mills Historical Library.)

Throughout most of the remainder of the nineteenth century, Brown & Sharpe continued to refine sewing machine production processes. Because annual demand for the Willcox & Gibbs machine never exceeded thirty-four thousand, refinements centered on improvement of quality and reduction of cost. (See Table 2.1 for production figures for Willcox & Gibbs machines.) Among the well-known mechanics who contributed to this development were Frederic W. Howe and Henry M. Leland. Howe had learned the machinist's trade at Gay, Silver & Co. in North Chelmsford, Massachusetts, and worked for a number of years for Robbins & Lawrence of Windsor, Vermont, one of the wellsprings of private armory practice. There, he drew the plans for and supervised the construction of the rifling machines purchased by the British for the Enfield Armoury. From 1859 to 1861, Lucian Sharpe corresponded with Howe, who was working in Newark, New Jersey, "in relation to improvements in tools for facilitating various operations on our sewing machine work." Brown & Sharpe obtained screw machines through Howe, who also prescribed using a miller to machine the large sewing machine castings rather than planing them. A one-step milling operation enabled Brown & Sharpe to eliminate four operations on the planer.[32] During the early 1860s, while Howe worked for the Providence Tool Company, he collaborated with Joseph Brown to design a machine to mill the grooves in twist drills, which resulted in the so-called universal milling machine.[33] In 1868, Howe went to work for Brown & Sharpe. His major contributions to the company had already been made, but he superintended the design and construction of Brown & Sharpe's new factory in 1872. Howe's design impressed many, including Henry M. Leland, who in 1927 said, "I felt then and I believe now that their new plant was, and for a long period of time remained, the finest of its kind in the world."[34]

Henry Leland, the eventual creator of the Cadillac Motor Car Company and a master of precision work, also contributed to the manufacture of the Willcox & Gibbs sewing machine. At the completion of his apprenticeship to Charles Crompton, a loom builder in Worcester, nineteen-year-old Leland moved to Springfield, Massachusetts, in 1863, where he worked as a tool builder for the United States Armory. After the Civil War he built tools at Colt's Hartford armory and several other well-known machine shops.[35] Leland was hired at Brown & Sharpe in mid-1872, shortly before the company moved into its new building. Initially he ran the screw machine section of the sewing machine department. During the next few years he refined Brown & Sharpe's screw gauges and instituted a system whereby all screws and small parts were held in stock rather than being manufactured for each batch of machines. From 1878 to 1890, Leland headed the sewing machine department. He accepted this job only after Lucian Sharpe (who directed the company after Joseph Brown's death in 1876) had reluctantly agreed to end the contract system (there were four contractors in the sewing machine department), to institute a piecework system, and to allow the head of the department to determine the pay of the workmen.[36]

For the next twelve years, Henry Leland gained an education in production technology. The skilled New England tool builder later wrote: "My vision of the possibilities of manufacturing broadened. My interest became intensified. I realized that manufacturing was an art and I resolved to devote my best endeavors and my utmost ability to the Art of Manufacturing."[37]

Before Leland instituted the changes upon which he had insisted, a friend whom he called Mr. Ripley performed a cost analysis of the sewing machine department over a month-long period. A year later Ripley repeated his study and found a 47 percent reduction in labor cost, which was attributed to the ending of the contract system and initiation of piecework. Leland claimed that this reduction was matched by an equal improvement in the quality of the work. Elaborate inspection sheets, demanded by the Willcox & Gibbs Sewing Machine Company, were eventually dispensed with because of the persistently high standards of workmanship.[38]

Much of the improvement in workmanship resulted from developments in precision grinding in which Leland played an important part. Although machined soft parts were made accurately, the hardening process warped and distorted them, necessitating either hand-fitting them during assembly or grinding them to gauge after hardening. After a few years of making the Willcox & Gibbs sewing machine Brown & Sharpe chose the latter method. According to Leland, the company had adapted ordinary lathes "specially fitted with guards to protect the 'ways' of the lathe[s] from the emery grit and other abrasive materials" but had made few changes in grinding techniques before he started work there. Despite their special fittings, the grinding lathes were plagued by grit getting into the v-shaped guides—or ways—on which their carriages traveled. Not long after Leland became responsible for sewing machine production, he recognized that "in order to do accurate work on a manufacturing basis, it was absolutely essential that these [ways] should be so constructed and protected that they would remain perfectly straight and in absolute alignment with the lathe spindles and lathe centers." Leland tried to impress upon Joseph Brown that not only could such a machine be designed and built but that it "would have an almost unlimited field of usefulness." After several conversations, Leland noted, Brown finally "grasped the idea that there was a wide field for a machine of this character in the manufacturing world generally."[39]

With Leland's encouragement and advice, Joseph Brown designed what became known as the universal grinding machine. At the same time, a grinder with protected ways but that would only grind straight work was developed. This machine was particularly well suited for production work.[40] It overcame the problem of distortion in hardening and helped eliminate fitting during assembly. Brown & Sharpe's developments in grinding technology—especially the universal grinder—also improved the precision of gauges, which in turn improved the precision of sewing machine parts.

In addition to encouraging developments in grinding and eliminating the contract system, one of the hallmarks of New England armory practice, Henry Leland instituted a strict procedural approach to sewing machine manufacture. All operations on various parts were enumerated on sheets along with the necessary tools, jigs, fixtures, and gauges.[41] Although such a listing of operations had perhaps been done in the minds of hundreds of New England mechanics since the days of John Hall—and perhaps had actually been written down—Leland's insistence that these operations be recorded offers a commentary on the mechanic who had discovered the ''Art of Manufacture'' and who had taken an intense interest in process rather than the building of any particular product or tool.[42] Leland believed—and proved—that with this procedure, work could be more closely followed by foremen and materials more smoothly moved in proper sequence. Thus before he left Brown & Sharpe, Leland had already begun to systematize the American system.

In summary, although Joseph Brown and Lucian Sharpe entered the manufacture of sewing machines as jobbers, they adopted at the outset the American system of interchangeable parts manufacture as developed in the nation's armories. Only by building special tools, jigs, fixtures, and gauges, Brown & Sharpe believed, could the firm manufacture uniform, high-quality products that would succeed in the marketplace. The construction of the hardware of armory practice presented serious problems, yet throughout months of work it maintained an abiding faith that the system would ultimately work and repay the substantial investment of time, money, and effort. In the end, the company reaped great and varied benefits from its production of Willcox & Gibbs sewing machines. From sewing machine manufacture it entered the machine tool business. It is interesting to speculate, however, whether this would have happened if Brown & Sharpe had experienced the demand for sewing machines which the Singer company built up. By the time Henry Leland left Brown & Sharpe, Singer was producing more than a hundred times the number of sewing machines made annually by Brown & Sharpe. In 1890, Singer Manufacturing Company may have built as many machine tools as Brown & Sharpe, but the company consumed all of them rather than selling any outside. It is ironic that Singer, long the recognized leader of the sewing machine industry, was exceedingly slow to adopt the American system of manufacturing.

Singer Manufacturing Company

The origin of the Singer Manufacturing Company is rooted in a partnership between an inventor and a capitalist. Near the end of 1850 Isaac Merrit Singer, who worked variously as a mechanic, inventor, and actor, redesigned the Blodgett and Lerow sewing machine at the Boston shop of Orson C. Phelps, where the machine was being repaired. Singer's changes constituted a new invention, which he patented August 12, 1851. (See Figure 2.8.) Even before he obtained the patent he formed a partnership with Phelps and George B. Zieber, who had helped him with the invention. Singer, Phelps & Co. immediately began to make and sell Singer sewing machines. Soon Singer purchased Phelps's interest,

FIGURE 2.8. Patent Model, Singer Sewing Machine, 1851. Although submitted as a patent model, this machine bears a serial number, 22. Note the cam mechanism used to raise and lower the needle bar, which is the essential aspect of Isaac Singer's invention. (National Museum of American History. Smithsonian Institution Neg. No. 45572-D.)

but the latter continued to manufacture machines for Singer and Zieber at a specified price per machine. After another series of share manipulations, a young New York lawyer, Edward Clark, acquired a one-third interest in Singer's company. Clark and Singer soon bought out Zieber, and in 1851, under the name I. M. Singer & Co., they set up a permanent sewing machine business with headquarters in New York.[43] (See Figure 2.9.)

Orson C. Phelps's method of constructing sewing machines for Singer's company offers a stark contrast to that used by Wheeler and Wilson and Brown & Sharpe. Phelps was a scientific instrument maker, and naturally his shop possessed little equipment capable of making the Singer machine, a cast-iron device weighing over 125 pounds with three large spur gears.[44] Unlike Brown & Sharpe, which had made instruments as fine as his, Phelps began by building sewing machines themselves rather than first building special tools with which to manufacture them. This head-on approach is not surprising since Phelps gained his livelihood through the traditional craftsmanship of the instrument

FIGURE 2.9.   Showroom, I. M. Singer & Co.'s Central Office, 458 Broadway, New York City, 1857. The Singer company spared no expense in marketing its sewing machine. (*Frank Leslie's Illustrated Newspaper*, August 29, 1857. Smithsonian Institution Neg. No. 48091-B.)

maker. The actual process of production at Phelps's shop consisted of obtaining major components of the machine from various job shops in the Boston area and then fitting them together. Castings, including the base, the head, and initially the gears, were purchased from a local founder, Alonzo Josselyn. After a few machines were made, the company began buying gears with cut rather than cast teeth. Phelps even jobbed out lathe and planing work and bought bolts and nuts from hardware merchants. Boxes full of bills and invoices from 1851 suggest that the major processes of metalworking on the Singer machine at Phelps's shop were hand filing and hand grinding. The company did, however, purchase a drilling machine to bore holes in the castings for various shafts and screws.[45] The piecemeal approach taken at the Phelps shop seemed to work satisfactorily. Although originally uncomfortable with twenty or twenty-five workmen in his shop filing and grinding on sewing machines, by the end of March 1851 Phelps was "just beginning to feel encouraged" about the operation.[46]

Already the company had begun to set up a vigorous marketing program, which involved using trained women to demonstrate to potential customers the capabilities of the Singer machine. These women also taught buyers or their operators how to use a sewing machine. Even as early as 1851, internal correspondence reveals a profound confidence in the company's marketing strategy and in its machine being "much the best of the whole lot [on the market]."[47] Within a year after forming their partnership, Edward Clark and Isaac Singer decided to move their manufacturing operations to New York. Hence Phelps was no longer able to make the Singer machine.

Despite increasing sales of sewing machines (though still well under one thousand per year), I. M. Singer & Co. continued to make the product the same way as in Boston.[48] This job-shop approach to manufacture worked well enough, and neither Clark nor Singer was particularly interested in developing manufacturing processes. Nor did either partner know any other way of making sewing machines than that used at Orson Phelps's Boston shop. Clark spent his time and energy defending Singer's patent and the company from damage suits filed by Elias Howe. Singer continued to make improvements on and other inventions related to the sewing machine, but apparently he spent more time enjoying himself in the big city than he devoted to business.[49] In New York. the company rented space and power in a building owned by the New York and New Haven Railroad. The Singer company began to do its own lathe and planing work, which forced it to purchase a line of general machine tools. Three basic machines made up Singer's machine tool inventory: lathes, planers, and boring machines (or drill presses). The company purchased its lathes—mostly small engine lathes—primarily from Leonard & Wilson of New York, commission merchants for a number of foreign and domestic machine tool builders. Occasionally it bought a lathe from Ezra Gould of Newark, New Jersey, reputed to make "the very best" lathes. Hand and power planing machines were obtained from Leonard & Wilson. The Singer company believed in using hand planers more extensively than costly power machines. As Edward Clark wrote, "A good [hand] planer . . . we know to be worth twice as much as the steam arm for the work [we] have to do." Singer purchased other tools from Leonard & Wilson or from Schenck's Machinery Depot, another New York machinery commission merchant house. Mackrell & Richardson provided castings for the Singer machine, and James Fairbanks cut all of its gears. From W. N. Seymour & Co., Singer obtained vast quantities of British-made files and weekly sent hundreds of them to be recut by James Latham of Newark, New Jersey.[50]

An illustration of the Singer factory, printed in 1854 in *United States Magazine*, shows these various tools in operation, but more important is its depiction of dozens of workmen standing at benches along the outside walls of the factory filing away at parts held in vises.[51] (See Figure 2.10.) Nowhere—in this illustration, in published materials, or in manuscripts—does one find the special tools, jigs, fixtures, and gauges that constituted the hardware of the armory system of manufactures.[52] Nor do we find its software such as the inside contract system used at Colt, Wheeler and Wilson, Brown & Sharpe, and other New England factories. Every Singer workman was paid by the company at an hourly wage.

Singer's method of manufacturing sewing machines resembled European practice far more than it did the system developed in New England armories. The company acknowledged that the machine was "produced by hand at the bench." The significance of this fact is manifold. Such other manufacturers as Colt and Brown & Sharpe had identified their production system—armory practice—as the source of their success. But I. M. Singer & Co. stated firmly that " a large part of our own success we attribute to our numerous advertisements and publications. To insure success only two things are required: 1st to have the best machines and 2nd to let the public know it."[53] (See Figures 2.11 and 2.12.) Having the best machine implies not only excellence in design but also quality in manufacture. Edward Clark, as leader of the Singer company, did not question that the European method of manufacture provided the quality necessary for success.

FIGURE 2.10. I. M. Singer & Co.'s New York Factory, 1854. Note the scarcity of machine tools compared to the large number of hand filers and fitters. (*United States Magazine*, September 15, 1854. Smithsonian Institution Photograph.)

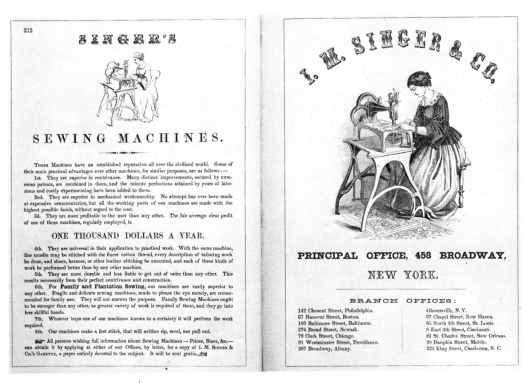

FIGURE 2.11.   I. M. Singer & Co. Advertisement, 1857. One of the means by which the Singer company "let the public know" about its sewing machine was through extensive advertising, such as this two-page spread. (David Bigelow, *History of Prominent Merchants and Manufacturing Firms in the United States . . .* , 1857. Eleutherian Mills Historical Library.)

Clark's views present a remarkable contrast to Lucian Sharpe's conviction that had Brown & Sharpe not adopted the armory system of production, the quality—and therefore success—of the Willcox & Gibbs sewing machine would have suffered greatly. Sharpe's view, as Samuel Colt's, was that quality and uniformity of parts went hand in hand. I. M. Singer & Co. took a European view of the matter, believing that hand-finishing by many skilled workmen provided the best means to achieve and maintain product quality. It chose to build its fortune by emphasizing advertising and marketing rather than by refining methods of manufacture. Clark saw no pressing need to adopt or develop a system of manufacture based on special-purpose machines nor to attempt to achieve close uniformity of Singer sewing machine parts. Using the method of manufacture the company had chosen, any uniformity that occurred was a result of many parts of the machine being cast. All of these cast-iron or cast malleable iron parts required machining and filing to be fitted into a complete machine, so the predominance of casting provided only the crudest uniformity, a uniformity worlds apart from the absolute standard sought by the United States Ordnance Department.[54]

Born in upstate New York, where his father was a small manufacturer, Edward Clark had had little or no contact with the New England approach to manufacture. He had attended college at Williams but graduated in 1831, well before the American system had attracted any notable attention, and had entered a law firm in New York City.[55] From this time until his death, he lived either in the city or at his rural estate near Cooperstown in

FIGURE 2.12.   Demonstrating the Singer Sewing Machine, 1850s. Demonstrations of the Singer sewing machine played a major role in the company's marketing program. (*Illustrated News*, June 25, 1853. Smithsonian Institution Neg. No. 48091-D.)

upstate New York. Isaac Singer, too, had remained outside the sphere of New England arms manufactures. Although both men had heard about the so-called American system of manufactures by 1852 when they moved the business to New York, neither had had any direct experience with or reason to advocate adopting this system of production. From the context of a letter written by Edward Clark in 1855, one wonders if Singer and Clark even recognized any difference between the methods they had adopted and those in New England which had attracted so much attention from the British. During 1855, the Singer company set up what proved to be an abortive factory in Paris, France, under the direction of William F. Proctor, a workman from the New York factory. In a letter to Proctor, Clark listed the machine tools the company planned to send to Paris. This list—a few lathes, hand planers, and boring machines—could have been a record of machine tools in any job shop in America or any factory in Europe, but Clark believed that ''no reasonable effort shall be wanting on our part to get you fitted out with an establishment which shall be a model one, and a credit to American skill in the mechanical arts.''[56] Perhaps Clark partook too much of the American mechanical arts chauvinism exhibited by Samuel Colt or Alfred C. Hobbs. At any rate, he was enjoying vicariously the praise heaped upon these men by the British press after the London Crystal Palace Exhibition. It would be many years before the company actually developed a production technology comparable to that used at Samuel Colt's ''model'' factory in London or the one in Hartford.

Table 2.2 shows the growth in the output of Singer sewing machines from 1853 to 1880. The tripling of figures between 1855 and 1856 relfects not only the 1856 pooling agreement but also the initiation by Edward Clark of installment purchasing—or as it was called then, the hire-purchase system. By advancing a certain percentage of the total price of the machine, a customer could ''hire'' a sewing machine, make monthly payments on it, and eventually own it.[57] In 1856 Singer also brought out its first machine intended exclusively for use in the home. At the same time, Clark overhauled the company's sales system by buying back territorial rights it had sold shortly after its formation. Most of the men who had purchased these territories had not sold sewing machines as vigorously as Clark wished. The territorial rights system had also prohibited a tight, central control, which after 1856 became one of the hallmarks of the Singer company.[58] And, of course, the jump in sales reflects the Singer factory's improvements in manufacturing capability

TABLE 2.2.   OUTPUT OF SINGER SEWING MACHINES, 1853–1880

Year	Number
1853	810
1854	879
1855	883
1856	2,564[*]
1857	3,630
1858	3,594
1859	10,953
1860	13,000
1861	16,000
1862	18,396
1863	21,000[†]
1864	23,632
1865	26,340
1866	30,960
1867	43,053
1868	59,629
1869	86,781
1870	127,833
1871	181,260
1872	219,758
1873	232,444
1874	241,679
1875	249,852
1876	262,316
1877	282,812[‡]
1878	356,432[§]
1879	431,167[§]
1880	500,000[‖]

Sources: Unless otherwise noted, all figures are from Frederick G. Bourne, ''American Sewing-Machines,'' in *One Hundred Years of American Commerce,* ed. Chauncey M. Depew, 2:530.
[*]This was the year that the patent pool was formed.
[†]John Scott, *Genius Rewarded; or the Story of the Sewing Machine,* p. 28.
[‡]*Sewing Machine Advance* 1 (August 1879):62.
[§]Scott, *Genius Rewarded,* p. 29. These figures agree with those in [‡].
[‖]My estimate, based on weekly production figures at the New Jersey and Scotland factories.

acquired through additional purchases of general machine tools and employment of an increasing number of machinists and other skilled workers. Between 1858 and 1859 Singer's sales again almost tripled. This jump reflects the introduction of an improved family model sewing machine, the Letter A, which for the first time competed successfully with the popular Wheeler and Wilson sewing machine.[59] (See Figure 2.13.) Between these two spurts of growth, the company braced for increased sales by building its own factory.

On a lot 100' × 100' on Mott Street in New York City, I. M. Singer & Co. built a six-story fireproof building in which it hoped all manufacturing operations could be carried out. It intended to cast its own parts in a foundry located in the basement, which was also occupied by an eighty-horsepower steam engine. By the time the new building was finished in July 1858, however, Clark and others believed the factory would be insufficient. Consequently, the company purchased fourteen additional city lots on Delancey Street and built a five-story building in which the foundry was erected. Here all the castings were poured and cleaned after cooling. At the Mott Street factory, the castings, along with other parts, were machined, finished, and fitted together into complete sewing machines. A separate area of the Mott Street works was set aside for patternmaking and toolbuilding.[60]

The opening of the Mott Street factory in 1858 marks the beginning of the Singer company's rise as a well-known American manufacturer. But unlike the Wheeler and Wilson Manufacturing Company when it moved into its new factory at Bridgeport in

FIGURE 2.13.   Singer Model A Family Sewing Machine, 1858. (National Museum of American History. Smithsonian Institution Neg. No. 58984.)

1856, I. M. Singer & Co. did not adopt the American system of interchangeable parts manufacture nor did it "resort," as Joseph Whitworth would have all American manufacturers, to the use of special-purpose machinery. For the next fifteen years at least, as B. F. Spalding pointed out in 1890, Singer "compromised with the European method [of manufacture] by employing many cheap workmen in finishing pieces by dubious hand work which [Spalding believed] could have been more economically made by the absolutely certain processes of machinery."[61] Basing his statements on "precise information," Henry Roland (a pseudonym of Horace L. Arnold) pointed out in *Engineering Magazine* in 1897 that the Singer Mott Street factory "had no milling machines and no gang drillers, and was doing its work on lathes and planers. The parts did not 'gage' at all; assembling was very expensive; and, after a machine was adjusted and in sewing order, all of the parts were kept by themselves while the frame was being japanned, and afterwards put back on the same frame, as they were far from interchangeable."[62]

In 1863, however, Singer began to mechanize its production processes, building more and more specialized machinery, until by 1880 its American factory—by then at Elizabethport, New Jersey—was jammed with automatic and semiautomatic machine tools. In Roland's words, the Singer works had been "brought up . . . to the armory plant standard."[63] Writers throughout the technical literature proclaimed the Singer plant as a progressive establishment. Sometime between 1880 and 1882 the factory first began producing sewing machines with interchangeable parts that did not require hand-fitting while soft, marking all critical parts with the same serial number, hardening, and refitting parts with matching numbers. Study of this process of increased mechanization and the eventual achievement of accuracy by interchangeability reveals some of the complexities and subtleties of the movement from the American system to mass production.

The first step in understanding this process must be to gain an appreciation for the leadership of the Singer company. In 1885 the company's chief production expert pointed out to the U.S. Bureau of Labor that "the President, Vice President, and all our chief and most successful agents arose from the bench."[64] He believed that their entrance as machinists and laborers rather than as executives had contributed to the company's success. George R. McKenzie, a Scotsman who served as vice-president and later president had been a case and model maker for Singer during the 1850s. William F. Proctor, the machinist whom the company had sent to set up its Paris factory in 1854–55, became the first treasurer of the Singer Manufacturing Company in 1863 and served in that capacity the remainder of his life. Throughout the 1860s and 1870s Proctor and McKenzie made all of the major administrative decisions relating to manufacturing operations. Paradoxically, this "up from the bench" phenomenon tended to retard the company's process of mechanization and upgrading of the uniformity of products. Neither McKenzie nor Proctor had had any contact with the New England system of manufacture. Their experience as skilled workers at the Singer factory during the 1850s, with its predominantly European or hand method of manufacture, determined in large part the production methods in the 1860s and 1870s—helping perpetuate European practices.

By 1863, the year I. M. Singer & Co. incorporated under the name Singer Manufacturing Company, annual sales had reached twenty-one thousand sewing machines. Throughout the preceding year the company had received an unusual number of complaints from agents and customers about the quality of the sewing machines they had received. Needles, gears, and shuttles continually caused problems. The company had set up its agency business to accommodate repair work, and many of the repairmen had actually worked in the New York factory. When parts broke, agents sent to New York for replacements, which had to be hand-fitted. The company supplied special files to fit new gears together, to file shuttle races, and to adjust certain dimensions of new shuttles. Repair problems, coupled with an inability to meet rising foreign demand, forced the company to question its present manufacturing setup. In an attempt to improve its sewing machine needles, Singer & Co. recruited Jerome Carter from the Ladd & Webster Sewing Machine Co. of Boston. Carter joined Singer in 1862, building special-purpose needle machinery similar to that he had designed for Ladd & Webster. Although needlemaking always seemed to cause the company problems, Carter evidently eliminated a large portion of them. Eventually, he became one of the few stockholders in the company.[65] His arrival helped set the stage for developments in manufacturing initiated soon after the company incorporated.

In 1863 the new Singer Manufacturing Company took its first major step away from European manufacturing methods toward those used in American armories by hiring Lebbeus B. Miller, a native New Jersey mechanic. According to Miller's obituary in the *Transactions of the American Society of Mechanical Engineers,* he was hired to "design and supervise the construction of special tools for the production of interchangeable parts" for Singer machines. Born in 1833, Miller had been an apprentice of Ezra Gould, the Newark, New Jersey, machine tool builder from whom the Singer company had purchased some of its small engine lathes during the 1850s. Yet Miller had had little experience with interchangeable manufacture. After serving his apprenticeship, he apparently continued in Gould's employ until 1861, when he went to work for the Manhattan Firearms Company. Originally located in Norwich, Connecticut, the company had moved to Newark in 1859. Within a year, Miller became superintendent of its branch factory, also in Newark. Miller's employment at the Manhattan Firearms Company put him briefly in touch with a New Jersey version of New England armory practice.[66]

Between 1855 and 1868 the Manhattan Firearms Company produced a patented, medium-priced revolver intended to capture part of Colt's extensive business. When Miller was hired by the company, Andrew R. Arnold was its general superintendent. Arnold had worked at the Colt armory before taking the New Jersey job, perhaps only a short time before Miller started there. Apparently Arnold introduced the Colt approach to pistol manufacture at Manhattan, for in 1861 he wrote Elisha K. Root, Colt's superintendent, requesting the price and delivery time of a drop forge, a piece of equipment which, he noted, the secretary of Manhattan Firearms had never seen in operation. By the time Miller left Manhattan Firearms in 1863, the company annually produced about six thousand arms with methods that resembled—in rudimentary form at least—those used at the great Colt factory.[67]

L. B. Miller immediately began to change Singer's production processes along the lines he had learned from Andrew Arnold at Manhattan Firearms. The first ten milling machines owned by Singer were made by Manhattan Firearms and purchased in 1863.[68] In addition to the use of milling machines, the adoption of drop forging and the construction of special machine tools characterized the first twenty years of his work. As at Colt's armory Miller's reliance on fine gauges played a minor role compared to the introduction of specialized automatic and semiautomatic machine tools. He introduced a few gauges for very critical dimensions, but he continued to employ a large number of fitters to correct errors that might have been eliminated by a highly refined jig, fixture, and gauging system. Gradually, however, the Singer factory refined its machine tools, jigs, and fixtures and extended and refined its gauging system until by the 1890s the company heralded its "Singer Gauge" system of more than fifteen thousand go no-go gauges.[69]

Miller's easygoing disposition well suited company executives such as president Edward Clark, vice-president George McKenzie, and treasurer and general factory superintendent William Proctor. Had Miller been like Elisha Root in pressing for the immediate and wholesale construction of hundreds of specialized machine tools—in other words, the complete mechanization of sewing machine manufacture—it is likely that he would not have survived. A radical change would have been too much for the men who had worked their way up from the bench. Miller succeeded, however, because he always seemed to know what these men wanted and generally fulfilled their wishes. He retired as superintendent of the Singer Manufacturing Company in 1907 at age seventy-four. Largely through Miller's history with the Singer company we learn about the development of its production technology.

An 1866 description of Singer's manufacturing operations reveals some of the changes initiated by Miller as well as the preponderance of methods antedating his employment. Cast iron and cast malleable iron still predominated for parts of the Singer machine, which consisted of more than one hundred pounds of cast iron and only about eight pounds of wrought iron and steel. Miller had, however, introduced drop forging for the steel shuttles used to make the lock stitch. A few other parts were drop-forged, too, as were the hinges for the wooden cabinets.[70] Planing continued to be an important method of removing large amounts of metal: the cast-iron bed of the machine was planed to obtain desired smoothness, and shuttle races within the beds also were planed rather than milled. The company no longer purchased its cut gears but made its own—at this time, bevel gears—by means of automatic gear-cutting machines. It appears that Miller built, if not designed, these machines. A gear was finished after the completion of operations on two different machine tools, the first a type of miller and the second a type of grinding machine.[71] An 1868 inventory of tools at Singer's factory in Glasgow, Scotland, listed the latter machines as "gear grinding machines." Nine of these grinders finished gears at the Glasgow plant, which turned out about one hundred sewing machines per week.[72] Finally, Miller helped introduce inside contracting, a system that characterized the New England armory approach to manufacture and would be used at Singer until the early 1880s.[73]

Although exact figures are not available, production of Singer machines at the Mott and Delancey Street factories exceeded four hundred per week during 1865 and 1866. The *New York Times* noted in 1865 that between 935 and 1,100 employees worked at the factories and that their wages "constitute[d] the principal expense in the [manufacture of sewing] machines."[74] The widening of domestic and international markets, coupled with intensification of advertising and other marketing techniques, brought about a rapid increase in sales of Singer's products. The New Family machine, introduced in 1865, had rapidly gained acceptance and would continue to be purchased by household consumers into the twentieth century. (See Figure 2.14.) In 1867, Singer sold more than forty-three thousand machines (mostly New Family but some manufacturing machines as well), and by 1869 sales reached almost eighty-seven thousand. This growth occurred in both domestic and foreign markets, so that in mid-1867 the Singer directors decided to "organize and to establish within the United Kingdom, a branch manufactory for the purpose of finishing sewing machines and articles connected with the use and sale thereof."[75]

Vice-President George McKenzie took charge of finding a suitable location for the British works, and because of the low cost and docile labor as well as the good shipping facilities, he settled on Glasgow.[76] In the beginning the directors' use of the word "finishing" accurately described the processes carried out at Singer's Scottish works. The New York factory shipped parts (other than the iron legs and head, which were cast

FIGURE 2.14.   Singer New Family Sewing Machine, 1865. The newly incorporated Singer Manufacturing Company introduced this machine in 1865 and continued its manufacture into the
twentieth century. It quickly became the staple product of the company. (National Museum of
American History. Smithsonian Institution Neg. No. 58987.)

by an American in Scotland) for one hundred New Family sewing machines every two
weeks, and the cheap, skilled workmen in Glasgow machined, filed, and fitted the sewing
machines together. As of December 1867, thirty-one men and boys could fit together only
thirty machines a week—which is to say that even after four months of experience with
the Singer machine, fitting them together still required great time and effort. Gradually,
however, the Glasgow factory began to contract outside for some of its castings rather
than obtain them from New York. It also began to work bar stock for making small sewing
machine parts. L. B. Miller's toolroom in New York supplied patterns for these castings,
jigs and fixtures for the machine tools, and most of the machine tools themselves.[77] By
mid-1869, Singer had rented another building in Scotland. Together the Glasgow factories

finished about seven hundred machines weekly. Two years later, with additional tools, these factories turned out fourteen hundred sewing machines per week, or about sixty-five thousand per year. *Engineering* called Singer's Glasgow factory "the largest in the United Kingdom."[78]

At the same time that Singer opened its second factory in Glasgow, the directors decided to move the main factory out of New York to "some convenient location on tidewater." William Proctor pressed the other directors to move quickly because, he argued, the output of the New York factories was limited to two thousand machines per week.[79] While planning a new factory, Proctor requested that the directors spend money for additional tools to raise weekly output to three thousand machines and also to plan for "increas[ing] it to 5000 sewing machines per week." Originally purchasing a site in Bridgeport, Connecticut, the directors finally settled on ten acres of waterfront at Elizabethport, New Jersey.[80] The factory they built there in 1873 was soon producing about a thousand machines daily. (See Figure 2.15.)

L. B. Miller played a critical role in the expansion of production capabilities in the decade between 1863 and 1873 both in the United States and Scotland. Although company records from this period are sparse, those that survive show a clear pattern in Miller's work. A complete 1868 inventory of the Glasgow factory lists the following special tools and machine tools that had been sent from New York: bevel gear-cutting and grinding machines, milling machines, wire-straightening machines, a needle bar edging machine, and shaft-cutting lathes. The New York factory had also sent sets of forging dies, milling cutters, and lathe tools for working individual parts of Singer machines. Jigs included a collar jig, turning jigs, a pitman jig, a pinhole jig, a camwheel jig, a jig for marking treadle rods, and lifter jigs. Glasgow's gauges consisted of an arm gauge, a cloth

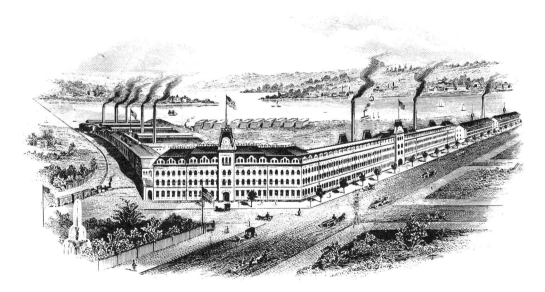

FIGURE 2.15.   Singer Manufacturing Company's Elizabethport Factory, 1880. Built in 1873, this factory was reported to be the largest factory in the United States making a single product under one roof. (John Scott, *Genius Rewarded; or the Story of the Sewing Machine*, 1880. Singer Company Photograph.)

feed gauge, balance wheel gauges, needle bar gauges, and screw gauges. The factory either checked its gauges or gauged its parts on six surface plates. The tool that dominated the Glasgow inventory, however, was the file; thousands of files in various shapes and sizes are listed among the factory's tools and fixtures.[81] Files were probably predominant in the New York factories as well.

Since labor cost more than twice as much in the United States as in Scotland, one would expect the Scottish Singer factory to use more hand labor and less machinery for its production than in America. Yet throughout the nineteenth century, the Scottish factory closely followed the American factory's production processes. New machines or processes adopted by the American factory were taken on by the Scottish factory. As gauges were refined in New York and New Jersey, duplicates were sent to Scotland.[82] The company's reasoning for this close control lay in its perception of its market in Europe. When Singer initially began to sell its European-made machines, an outcry erupted over their supposed inferior quality compared to those made in the United States, which had begun to dominate European markets.[83] Standardization of manufacturing methods and of the end product between the two factories therefore seemed essential. Singer initially spent a considerable amount of its advertising dollars to convince Europeans that the Glasgow product was identical in quality to the New York Singer. The French government was so skeptical that it commissioned William J. M. Rankine, the famous theoretical engineer of the University of Glasgow, to inspect Singer's Glasgow factory. Rankine concluded that machines made there were equal to those made in the United States and that ''I am convinced that all the parts of all the machines sent out by the Company from their Glasgow establishment, are wholly manufactured, as well as put together, in the Factory which I have inspected.''[84] Emphasis on standardization perhaps worked to move the company toward greater uniformity in the parts of its sewing machines. Yet increasing mechanization of production, resorted to in order to fill rapidly increasing orders, also played a role.

At the end of 1868, the directors of the Singer Manufacturing Company appointed L. B. Miller general superintendent of the American factories. Between 1869, Miller's first year as superintendent, and 1873, the year Singer opened its Elizabethport works, sales almost tripled, from almost 87,000 to more than 232,000. And in this period the Singer factories could not supply their sales agents with enough sewing machines. It is not surprising that the directors approved Miller's request for money to build and purchase machine tools in large quantities. In 1871, for example, when Singer increased its output by about 54,000 sewing machines, the company spent $58,000 on new tools in addition to $42,000 for repair and maintenance of existing tools. These expenditures, however, represented less than 5 percent of total expenses of manufacture; the cost of packing (in wooden crates) the sewing machines made at the New York factories alone that year exceeded total tool expenditures by more than $10,000.[85] Pushed by Vice-President McKenzie, Miller sought for the factory not only to produce as many machines as the company's sales agents could sell but also to bring down production costs.

About the time the directors promoted Miller to factory superintendent, George McKenzie began to show a strong interest in the manufacturing operations. By the early 1880s he wanted to know exactly what had occurred each week in the company's factories, especially the production costs and where they could be cut. In the Singer directors' discussions in 1868–70 about a new American factory, McKenzie expressed his desire to check the efficiency of Singer's manufacturing operations against that of the Wheeler and Wilson factory and the Springfield Armory, two plants that had received widespread attention for their production technology. He suggested that the company commission an "expert" to carry out such a study.[86] No evidence exists that an expert was ever hired, but before the Elizabethport factory opened in 1873, McKenzie, William Proctor, or L. B. Miller may have gone to New England to see and study "high" armory practice at Springfield or the Wheeler and Wilson plant. In any case, they witnessed the New England armory system of manufacture at the Providence Tool Company between 1870 and 1873.

The Providence Tool Company had gained an international reputation for its ability to produce large numbers of high-quality firearms and similar consumer durables. In 1870 Providence Tool contracted to manufacture twenty-seven thousand Domestic brand sewing machines for Singer, delivered in equal monthly installments between March 1871 and January 1, 1873.[87] Singer apparently wanted to capitalize on the cheaper sewing machine market. But rather than jeopardize its current sales by making a cheaper machine, bearing the Singer name, the company set up the Domestic Sewing Machine Company. Though separate, the Domestic Sewing Machine Company was nonetheless controlled by Singer, which operated as a middleman between manufacturer and marketer, taking $3 per machine off the top. Near the end of 1872 this contract was extended and the number raised to one hundred thousand sewing machines to be delivered in 1873 at a rate of "300 daily or more." On December 31, 1873, Providence Tool stopped making the Domestic machine because, according to John Anthony, the company's president, "we entered into large contracts for rifles, which called for all our force and room, and large additions to both."[88]

Unfortunately, Anthony's explanation for the termination of Providence Tool Company's contract with Singer and Domestic will not stand historical scrutiny. Although it may seem ancillary to the manufacture of Singer sewing machines, this episode bears directly on the Singer company's thinking about production technology. It also creates an anomaly not easily dealt with by the historian. The Singer company, rather than Providence Tool Company, terminated the contract because of "differences [which] have arisen between the . . . Domestic Sewing Machine Company and the . . . Providence Tool Company in regard to the manufacture and sufficiency of [the sewing] machine under said contract."[89] Simply, the Providence Tool Company failed to produce sewing machines of sufficient quality to meet the Singer officials' approval. Contractually both parties agreed only that the sewing machines were to "first class." But throughout the tenure of the contract, the companies clashed over the workmanship and uniformity of the Providence product and finally hired an impartial referee to pass judgment on sewing machines made by the tool company according to agreed-upon "standard" machines.

Not long after the companies reached the agreement on the production of one hundred thousand Domestic sewing machines for 1873, conflict arose about the quality of the machines. Both sides believed that with the Providence Tool Company's installation of a completely new gauging system, "no serious trouble in the inspection is likely to occur." Yet by May the gauge system had not eliminated the elements of contention. Therefore, the two sides agreed that an impartial third party would mediate and would judge production machines by two approved machines that would serve as "suitable standards" of the "character of work" required. The Providence Tool Company warned Domestic representatives that "at present we shall find it difficult to get out many machines as perfectly fitting in some of these parts as these [two standard] samples are." Two weeks later Providence advised Domestic that, "under our new [mediation] arrangement, and under the new 'Standard' we have, we may not be able to turn out the full hundred thousand machines this year.'' But since the Domestic Sewing Machine Company had indicated that one hundred thousand machines probably would not be sold in 1873, Providence Tool believed that it could turn out as many as Domestic would require. Informed guesses indicated that it actually produced less than sixty thousand Domestic sewing machines.[90]

When in May or June 1873 Singer officials decided to terminate their contract with Providence Tool Company, together with the Domestic Sewing Machine Company, they set up the Domestic Manufacturing Company to make the Domestic sewing machine after January 1, 1874. The articles of agreement forming this new company, in conjunction with the state of manufacturing art at Singer, are difficult to interpret. In the agreement, Singer specified that ''the said Domestic sewing machines manufactured under this contract shall be manufactured in a workmanlike manner, and shall be done on the principle of making the parts interchangeable.''[91] Singer's contracts with the Providence Tool Company had never specified interchangeability. Providence had, however, manufactured the machines under the armory system because it constructed special tools and special machines. Providence also used a gauging system but never claimed to Singer officials that its sewing machines were completely interchangeable. The contrary is true. For example, when Singer officials suggested that the Singer factory, rather than Providence, make the hemmer attachment for the Domestic because Singer could make it five cents cheaper, John Anthony had protested, ''It seems to me the hemmer should be made where the machine is made. It will hardly answer to tie up a hemmer in each parcel of apendages [sic], as each one needs more or less fitting to its special machine.''[92]

The Domestic Manufacturing Company began its production in 1874. It used the special machines, special tools, and gauges purchased from the Providence Tool Company at the end of 1873. Whether Domestic actually produced sewing machines with interchangeable parts cannot be ascertained. What is certain is that the Singer Manufacturing Company had not produced sewing machines with interchangeable parts and would not for at least another eight years.[93] Singer, however, had terminated its contract with Providence because of the poor quality of the machines Providence manufactured—or perhaps because once Providence raised its product to standard it could no longer turn out the desired number of machines. In the new agreement creating the Domestic Manufacturing Company, did Singer demand a product more nearly perfect than the one Singer itself manufactured? Had it asked more of the Providence Tool Company than of itself? It should be recalled that the Domestic sewing machine was sold as a much cheaper grade machine than the expensive Singer. And did Singer construe ''the principle of making the parts interchangeable'' to mean only production of goods by specialized machines, tools, jigs, fixtures, and gauges or, in other words, only in the most relative sense of meaning? These questions may never be answered definitively because the necessary documentation is missing. Nonetheless, it seems that there is a plausible explanation for this episode in the Singer company's history.

According to B. F. Spalding, the Singer Manufacturing Company fully adopted the American system of making interchangeable parts by special tools and machinery when it moved into its new factory at Elizabethport, New Jersey, in 1873. No longer, Spalding says, did Singer ''compromise with the European method by employing many cheap workman in finishing pieces by dubious hand work.''[94] The company had hired L. B. Miller in 1863 to introduce ''interchangeable parts manufacture.'' Miller became Singer's superintendent in 1868. A year later Vice-President McKenzie wanted to hire an expert to compare manufacturing methods at the Springfield Armory and the Wheeler and Wilson sewing machine factory with those used at Singer. These two Yankee establishments claimed to produce products with absolutely interchangeable parts. And, finally, in 1870 Singer contracted with the Providence Tool Company, one of the better-known New England manufacturers which had made many arms during the Civil War, for the production of the Domestic sewing machine.

All of these developments point to an intensification of interest in the American system by Singer officials between 1869 and 1873. At the Providence Tool Company, Singer officials probably had their first intimate experience with full-blown Yankee armory practice. The Singer men who had worked their way up from the bench found this system unsatisfactory. The fits achieved by armory practice, as conducted at the Providence Tool Company, were too sloppy by Old World standards. Joseph Whitworth had stressed this point in his testimony before the parliamentary Select Committee on Small Arms in 1854. When held to these standards, the Providence company found it could not produce the quantity it had originally envisioned. This is the crucial dilemma of the American system so poignantly described by Eugene Ferguson and later by Daniel Boorstin.[95] Singer could not accept the imperfectibility inherent in the American system and in armory practice, an imperfectibility it had been able to eliminate through its unique blend of Old World and New World manufacturing methods. Only hand-filing and hand-fitting yielded the close fits demanded by the Singer officials' instinct for workmanship.

An intriguing paradox arises here. The Singer Manufacturing Company could neither live with nor without the American system. Because of rising sewing machine sales—in 1873, almost a quarter of a million—Singer was forced to find ways to produce more sewing machines. The impending end to the sewing machine combination in 1876, when the company could no longer have the protection of a tariff on other sewing machine manufacturers, would compel the company to lower its production costs. By 1873, when Singer officials found the Providence Tool Company's product unsatisfactory, they had been thoroughly captivated by the idea of interchangeability. During the next eight years the company moved closer and closer to this elusive goal. By 1881 L. B. Miller and his assistants were almost in a frenzy to attain it, and finally in April, the superintendent declared Singer parts "absolutely" interchangeable.[96]

Looking back on their decision to terminate their contract with the Providence Tool Company, Singer officials would probably have concluded that their action was too hasty, that their demands for perfection from armory practice were too rigid. Once they adopted it themselves, they experienced the same difficulties in turning out close-fitting parts as had the Providence company. Only through a long and often painful process of refinement did Singer eliminate these problems and at the same time achieve high-volume output. Armory practice, particularly the principles and practices of precision manufacture, placed terrific demands on the manufacturer who adopted it. Singer officials believed, however, that if these demands were met they could turn out a half million machines annually.

Although almost no company records pertaining to Singer manufacturing operations survive from the 1870s, there are two good descriptions of the factory. One apparently was written in 1874, a year after the new Elizabethport works opened; the other dates from 1880, just when company officials began feeling overwhelmed by production problems.[97] The latter publication included a series of excellent steel engravings which provide valuable visual information. (See Figures 2.16–2.24.) For brief periods during the 1880s, manuscripts exist which show the intense efforts of Singer personnel to overcome manufacturing problems.

According to a historical article in the *Elizabeth Daily Journal*, George R. McKenzie

FIGURE 2.16.   Singer Foundry, Elizabethport, 1880. (John Scott, *Genius Rewarded; or the Story of the Sewing Machine*, 1880. Eleutherian Mills Historical Library.)

FIGURE 2.17.   Singer Forging Shop, Elizabethport, 1880. (John Scott, *Genius Rewarded; or the Story of the Sewing Machine*, 1880. Singer Company Photograph.)

FIGURE 2.18. Singer Screw Department, Elizabethport, 1880. (John Scott, *Genius Rewarded; or the Story of the Sewing Machine*, 1880. Eleutherian Mills Historical Library.)

FIGURE 2.19. Singer Needle Department, Elizabethport, 1880. (John Scott, *Genius Rewarded; or the Story of the Sewing Machine*, 1880. Singer Company Photograph.)

FIGURE 2.20.   Singer Polishing Room, Elizabethport, 1880. (John Scott, *Genius Rewarded; or the Story of the Sewing Machine*, 1880. Eleutherian Mills Historical Library.)

FIGURE 2.21.   Singer Japanning (Painting) Operations, Elizabethport, 1880. (John Scott, *Genius Rewarded; or the Story of the Sewing Machine*, 1880. Eleutherian Mills Historical Library.)

FIGURE 2.22. Singer Assembling Room, Elizabethport, 1880. Note the presence of files and the machine tools. (John Scott, *Genius Rewarded; or the Story of the Sewing Machine*, 1880. Eleutherian Mills Historical Library.)

FIGURE 2.23. Testing Singer Machines, Elizabethport, 1880. Women at the left check the assembled machines under a variety of sewing conditions (thread count, material, and so on). (John Scott, *Genius Rewarded; or the Story of the Sewing Machine*, 1880. Eleutherian Mills Historical Library.)

FIGURE 2.24.   Setting up Singer Machines for Shipment, Elizabethport, 1880. (John Scott, *Genius Rewarded; or the Story of the Sewing Machine*, 1880. Eleutherian Mills Historical Library.)

alone decided that the new factory would be located in Elizabethport, New Jersey, rather than in Bridgeport, Connecticut. Although the directors had purchased the Connecticut site while McKenzie was in Europe, they deferred to his desire to build in New Jersey. After that, superintendent L. B. Miller was largely instrumental in picking the exact location and arranging for construction. In October 1870 the company obtained a ten-acre lot with good access to railroad facilities and to upper New York Bay at Elizabethport. The new factory opened in 1873. It had facilities to produce every part of the Singer machine except the wooden cabinet, which was made at the Singer Case Factory in South Bend, Indiana, established in 1869. Both in size and quantity of production the red brick Elizabethport works easily qualified as the largest sewing machine manufactory in the world. As Charles Chapman, the author of the first published account of these works, wrote in 1875, "You could scarcely believe in your own mind that such works were necessary for so small a machine." Chapman added: "You had often heard the name of Singer and had seen his machine all over the United Kingdom, and, in fact, all over the world, and for some reason or other . . . you had an idea that Singer was a myth, but now that you saw the gigantic works, you soon altered your mind." When Chapman visited the Singer factory in 1873 or 1874, L. B. Miller showed him through the works, "describ[ed] the use of the various machines, and also . . . explained the working of the different departments." Rarely did the company allow anyone but employees into the factory, and even more rarely was the works described in print. For these reasons, Chapman's narrative is invaluable in describing Singer practice.[98]

As late as 1874, most parts of the Singer New Family machine and its equivalent commercial model were made of cast iron, thus requiring a large foundry. The Elizabethport foundry covered about sixty thousand square feet—or well over the size of a standard football field. On its floor, molds could be laid out for more than thirty tons of melted pig iron. Singer took pride in having adopted a foundry system that dispensed with skilled molders. It had a dual basis—a special press or molding machine which packed the sand around the pattern in one operation, plus intensive division of labor. After being broken out of the mold, the cooled cast-iron pieces were rumbled (cleaned through a tumbling process) and then taken to the filing and drilling department. Here, workmen filed away rough edges that would impede proper placement in the various fixtures used during machining operations. Other workmen used drill presses to clean out cast holes as well as to drill new holes. These holes were also used as bearing points in fixing parts to machine tools.[99] With good casting work, only a little filing was necessary to secure the pieces correctly in the fixtures. Bad castings often required "very much fitting to prepare them for the fixtures."[100] The two principal castings were the base and the arm; joined together these made up the head of the machine. Most of the other parts fitted into the head, so the machining work on it was critical. Chapman described at length the machining operations on the arm and base of the New Family machine.

The contractor of the machining department or one of his inspectors checked each casting as it came from the filing department. If acceptable, he issued a receipt for the casting and sent it to the first machining station. For arms, the first operation was reaming the bottom hole for the vertical shaft. Ten milling operations followed. Chapman reported that about forty milling machines carried out the arm work. Many holes had to be drilled—five for oiling bearings, four for screws to hold the arm and base together, four for screws to secure the needle bar presser foot faceplate, three for a spring adjustment, and one each for the thread spindle and thread take-up lever. After the necessary screw holes had been tapped, the bottom of the arm was faced where it joined with the base. At this point the base, which had undergone numerous machining operations, and the arm were screwed together. Then came the arduous task of putting in the two shafts along with the two bevel gears and one thrust bearing.[101] Probably cast in the arm, the bearing holes for the horizontal shaft were reamed to correct size. Then the holes for the vertical shaft were reamed.

Because the size of the bearings varied significantly from machine to machine, workmen either selectively chose shafts whose journals fit closely into the bearings or else fit the journals to the bearings. Custom fitting was most likely the case. Workmen also custom fit the bevel gears onto the shafts as well as the thrust bearing located on the vertical shaft. They did this by tightening down a set screw, securing the gear or bearing at the correct location on the shaft. When all were properly set, tapered holes were drilled through the gears and shafts and then the set screw was removed permanently.[102] Tapered steel pins rather than screws held the gears and bearings in place. Once the machine was together with gears, shafts, and bearings correctly fitted, a workman stamped a serial number on the base and gears. The gears and shafts were then taken out of the machine head, and the head was sent to the japanning department, where it received its jet-black finish. After being ornamented, the head was matched up again with the parts of the same number and the machine put back together. Before being joined to the arm, the base underwent this same process of machining parts, fitting them, marking them, and disassembling them.[103] No doubt many of these parts were fitted soft and then sent to the case-hardening room while the machine heads were being japanned and ornamented.

Why were Singer parts stamped with the same serial number? Chapman related that L. B. Miller told him that each part was numbered "not because one part of a machine will not fit any other machine of the same size" but "in order that the company may prevent any unscrupulous persons from selling bad machines with the Singer name on them."[104] Such an explanation is suspect. I have attempted to interchange parts on New Family sewing machines made in approximately the same year Chapman visited the Singer factory. Generally it is impossible.[105] Shafts and the parts attached to them are particularly unswitchable. Nevertheless, parts of the Singer machine were uniform or standard, and they were made with armory production techniques, that is, by special tools and machine tools used in conjunction with rationally designed fixtures and checked in some manner by a gauging system.

Still, Miller deceived Charles Chapman. Parts were stamped with the serial number not to thwart "unscrupulous persons" but because they required filing and fitting to put them together and because they had to be taken apart for painting and ornamentation. Only by matching up the numbers could the workmen correctly reassemble a machine. Had Miller's excuse for marking been true, the company would have—as it did some years later—stamped the model and individual part numbers on all the pieces along with "Si Man Co" rather than the serial number. That parts had to be fitted to the sewing machine head before the head was painted (and thus before final assembly) also makes Miller's statement suspect. The historian faces a danger in placing too much emphasis on interchangeability per se. Yet since the Singer company itself deemed it important—as indicated not only by Miller's claim to his British visitor but also by Singer advertising—some attention must be given to this question.

Why were Singer's New Family model sewing machines not constructed with absolutely interchangeable parts? Was the Singer factory not technically capable of interchangeabilty? If it was, what were the economic factors that led it to produce parts with large tolerances and to have workmen fit them together by using files? Or were Miller and his men working at the limits of precision manufacture in America of the 1870s? In other words, was interchangeability in sewing machine manufacture (with its demand for very close fits between shafts and bearings, for example) simply beyond reach? Because of the dearth of Singer manuscripts from the 1870s and because of the general ignorance among historians about precision manufacture at this time, these questions cannot be answered with any certainty. Yet they should be raised. The question of noninterchangeability can be addressed conjecturally.

As pointed out earlier, in the 1850s and early 1860s Singer made its sewing machines by using what B. F. Spalding termed European methods, that is, with general machine tools and with much handwork rather than with special tools, jigs, and fixtures. Singer's skilled machinists literally built sewing machines one by one. Gradually during the 1860s, after Singer hired L. B. Miller, the company introduced special machines and special tools. Yet tradition weighed heavily, particularly among Singer executives who had worked their way up from the bench. The production process became a mixture of European methods and the American system of manufacturing. The company's directors believed that Singer produced superior sewing machines by mixing the two different methods. With this blend of technology, Singer probably manufactured a sewing machine better in workmanship, that is, with closer fits, than those produced wholly with armory practice, such as the Domestic sewing machine manufactured by the Providence Tool Company. Among others, Joseph Whitworth, in his remarks of 1854, had stressed the importance of hand finishing in obtaining quality workmanship. Economic arguments, based on a paucity of data, could be made in a number of ways. It might be argued, for example, that it was cheaper to produce sewing machines by the Singer blend than by either wholly American or European methods. On the other hand, it may be fairly questioned whether during the 1850s and 1860s Singer people paid close attention to production costs. Because of the patent pool and because of its extensive use of advertising and other promotional techniques,[106] Singer was able to sell sewing machines at five to ten times its cost of production.

The net result of all of these questions, even if unanswered, is another question: What was the state of progress of the armory system of manufacture by 1874? If Singer had found armory techniques unable to meet the standards of quality it desired, the technical development of the system since the days when Whitworth and Anderson had visited the Springfield Armory had perhaps not gone very far. If Singer had found highly refined armory practice to be too expensive, this raises the question of its economic efficiency in the context of 1870s America. And finally, if Singer was simply a technologically backward—albeit large and important—manufacturer, one must question the general diffusion of the American system of manufacturing in the post-1850s period. (This question becomes extremely important with the McCormick Reaper Company, as discussed in Chapter 4.)

B. F. Spalding offered an explanation for the noninterchangeability of Singer sewing machine parts in 1890, but it defies commonly held beliefs about why the American system was widely diffused in the second half of the nineteenth century. Spalding claimed that the "unparalleled demand for their [Singer's] sewing machines had drawn them into this way [that is, "employing many cheap workmen in finishing pieces by dubious hand work which could have been more economically made by the absolutely certain processes of machinery"], leaving no spare time for making tools."[107]

This assertion is flawed because, however Singer's production technology during the 1870s is viewed, it cannot be seen as remaining static. Over the years, Miller and his contractors continually made changes in production methods and refined the production system. In response to a complaint from the New York offices about the manner in which the cloth feed wheels were fitted to the bottom of the vertical shaft, for example, Miller wrote in 1875 that the "Feed Wheels have been fitted by the same process for years. . . . At the commencement of 1874 [however] we made a radical change in *this* particular, as well as in some others, and machines made during that time are much better fitted than they had been previously."[108]

Symbolic of the changes made and the greater precision obtained in production, sometime between 1874 and 1880 Singer changed the name of the department where the sewing machines were put together from the fitting department to the assembling department. In its public relations literature, the company began putting the word "assembled" in quotation marks to describe the process by which the machines were put together. This charge in nomenclature seems to signal a consciousness that the machines were no longer "fitted" together. Literature of a later date would play up "The Singer Assembly System," enabled by the production of parts that were "perfectly interchangeable." As late as 1885, however, Singer called the employees who worked in the assembling department "fitters" and paid them, on average, forty cents per day more than ordinary machine tenders.[109] One is reminded again of the stricture Henry Ford laid down in his article on mass production in the *Encyclopaedia Britannica:* "In mass production there are no fitters."

Between 1874 and 1880 Singer doubled its annual production of sewing machines, from a quarter of a million to over half a million. Within another six years, yearly production again doubled. Since Singer operated a factory in Scotland, not all these machines were made at the Elizabethport works. But well over half of them were.

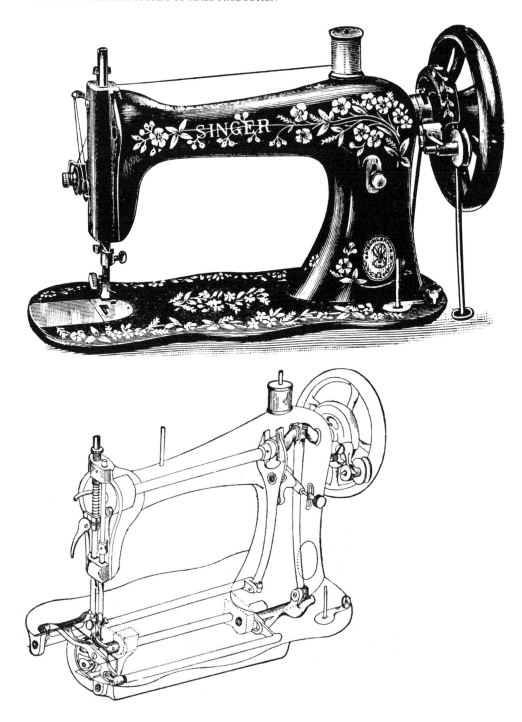

FIGURE 2.25.   Singer Improved Family Sewing Machine. The Improved Family machine was first introduced in 1881. (*Catalogue of Singer Sewing Machines for Family Use*, 1893. Eleutherian Mills Historical Library.)

Expanding production fourfold in twelve years to meet rapidly rising sales prospects, Singer perhaps experienced a "technological imperative" to mechanize fully, reduce tolerances, and thereby eliminate the cumbersome and expensive process of hand-fitting. Although there is evidence that files were not totally eliminated from Singer's assembly room, it is fair to say with John Scott in 1880 that "each of these [Singer machine] parts has been . . . so accurately worked by the machine which made it, when the numerous and varying pieces come together in the assembling process, it requires little and often no adjustment whatever, and each fits in the place made for it, resulting in a complete and harmonious whole."[110] Unfortunately, the gap in manuscript and other documentary material between 1874 and 1880 does not provide an understanding of how—or even if—this dramatic change at Singer came about. Substantial manuscript material from the early 1880s suggests, however, that the battle for quality and quantity manufacture had by no means been won.

When John Scott, George Ross McKenzie's attorney and a well-known biographer of the late nineteenth century, published *Genius Rewarded; or the Story of the Sewing Machine* (1880), he noted that Singer's Elizabethport factory "is believed to be the most complete, systematic, and best-equipped in the world [and] . . . is believed to be the largest establishment in the world devoted to the manufacture of a single article."[111] Although it would be difficult to challenge the latter assertion, an intimate look at the inner workings of the factory betrays a level of chaos that runs counter to the meaning of "systematic." Much of this chaos resulted from the company's efforts to produce a new model sewing machine called the Improved Family sewing machine (Improved Manufacturing for the commercial model of the same machine).[112] (See Figure 2.25.) While trying to get the new machine into production, the factory simultaneously faced increasing demands for the New Family machines which it had been making since 1865, as well as for several commercial models. Miller often found his factory unable to supply the company with enough machines; unfilled orders reached as high as forty thousand machines and usually ran about twenty thousand. Added to these problems, Elizabethport was building most of the special tools, special machine tools, fixtures, and gauges used to equip the rapidly expanding factory in Scotland, which in 1884 moved into a new works larger than the American factory. The company also started a small factory in Montreal, Canada, in 1883, which drew additionally upon the resources and energies of the Elizabethport works. In light of these pressures, it is no surprise that the Singer factory in New Jersey pursued sewing machine production and the refinement of manufacturing processes with great intensity and with a certain disorder.

Throughout late 1880 and early 1881, a team of Singer mechanics tried to design a production system for the new oscillating shuttle model. As was common with any new attachment the company planned to introduce, Singer had given great advanced billing about the new model to its army of sales agents throughout the United States. The new machine possessed such attractive features as increased sewing speed, quieter operation, smoother cloth feed, better stitching, and bobbins that held more thread. Consequently, sales agents had ordered large numbers of the machines even before the factory was ready to produce them. When the factory could not get the machine into production, the executive offices in New York became increasingly concerned.

Faced with other manufacturing problems, L. B. Miller had relied almost exclusively on the various inside contractors to design the required tools for the oscillating shuttle machine. At the end of March 1881, for instance, he reported to McKenzie that he had "just been up to [W. H.] Jackson['s shuttle department] and looked over the new Shuttles finished this week, and if they come out all right, he will be ready to go right on with them in a regular way."[113]

Concerned that Miller could not handle all the details of factory operations, Vice-President McKenzie hired Albert D. Pentz to supervise some aspects of the production of the new model and particularly to set inspection procedures. Pentz had worked in the Mott Street factory since 1870 and then had been sent to Chicago to head Singer's extensive repair shop there. While in Chicago he had invented several new sewing machine mechanisms, some of which the company adopted for the new oscillating shuttle model. Why McKenzie chose Pentz to deal with the production of the new model is unclear except that McKenzie believed strongly in bringing men who had exhibited faithful service up through the ranks of the company. When Pentz arrived at the New Jersey factory in March of 1881, he began to institute changes. The time was ripe; Singer president Edward Clark had told McKenzie, "I cannot help feeling impatient to increase the production of the small O.[scillating] S.[huttle] machines as there is a loud call for them from our agents, but I suppose it is best not to be in a great hurry lest some mistake should be made."[114]

McKenzie had asked Pentz to give him his general "impressions of the status in the factory" and advice on how its operations might be improved. Both McKenzie and Pentz clearly saw the inside contractors as the root of both success and failure. Pentz told McKenzie that he was "of [the] opinion that you have here, on the whole, a most excelent [sic] and capable lot of men in the important positions, *but* there is . . . need of considerable more of the harmony which, while not being 'mutual admiration,' is a mutual endeavor to forward the company's interests." Hans Reiss, contractor for both the machining and assembly departments, was, according to Pentz, "doing all in his ability—and his power is great—to forward the work." Yet Pentz believed that Reiss had too much work to do and should be relieved of one of his departments. Despite "stand[ing] head and shoulders above all the rest of [the Singer contractors]," Reiss was "irritated continually at the high standard demanded of him, [and] considers all imperfect work which will not pass inspectors to be *thrown out* by spite." William Inslee, contractor for the adjustment department, although "a good close workman," was guilty of "much fault finding with the work received from Reiss, much of which fault is merely theoretical and having no practical value." The contractor for shuttles, W. H. Jackson, impressed Pentz "as being the Yankee prototype of Mr. Reiss, but his ideas lack in the practical, while being very ingenious indeed.," [115] (See Figure 2.26.)

Superintendent L. B. Miller reported weekly to McKenzie about the output and progress of the Elizabethport works. On April 5, for example, he wrote that the japanning or painting department had improved its weekly output to almost 7,000 machines. Reiss's machining department had exceeded this by 250 machines, but his assembly department had managed to put together only some 6,000 sewing machines. Other departments had fallen slightly behind this figure. Of the 6,000 finished machines only 300 were the Improved Family model.[116]

Miller's letter indicated little concern about the small number of Improved Family machines produced during the previous week, largely because he had been preoccupied with the erection of new foundry additions and because he relied upon Pentz to deal with the new model. Yet faced with increasing pressure from Singer executives to increase rapidly the production of the new model, Miller began to pay more attention to the machine and to Pentz. William F. Proctor, the secretary of the company, who had

FIGURE 2.26.   Caricature of a Yankee Inside Contractor, 1880. The Yankee inside contractor was often seen as a cunning, wealthy, well-dressed mechanic who put his own interests before his company's. ([James W. See], *Extracts from Chordal's Letters*, 1880. Eleutherian Mills Historical Library.)

formerly been responsible for the production end of the business, started making regular visits to the factory to impress upon Miller and others the importance of getting out the Improved Family.[117]

The factory's problems in producing the new machine stemmed largely from its design. Except for its basic shape, none of its parts or mechanisms resembled the old New Family machine. The Singer factory had eased into large-scale production of the New Family over a fifteen-year period, but it was being asked to produce overnight a far more complex machine that required closer machining work, most of which was different than that on the New Family. Consequently, much of the problem centered around obtaining close uniformity. Pentz concentrated his efforts on this problem. Miller reported to McKenzie that "Mr. Pentz has just showed me an I[mproved] F[amily] Machine which he has followed through Reiss' Dept., watching both the machining and the gauging, and which on the final test proves correct, (absolutely). He says that now the gauges agree with each other, and the tools are correct, and if we not get good work it is the workmanship." Two days later Pentz proudly announced, "I have been able to reduce the gauges to such a system that the machines are practically interchangeable, especially the I[mproved] F[amily]." Miller tempered Pentz's optimism, however, by pointing out the troubles in starting production of the intricate shuttle for the new machine. Only a few had been made thus far, though he thought that "we can simplify the work and reduce the cost when we get fairly into it."[118]

By the middle of April, optimism pervaded the factory. More than seven thousand machines had been made the week of April 12 to 17. With plenty of castings and japanning kilns in working order, everyone believed that "there is no reason why this total should not be increased in the next two weeks. . . . Pentz is getting to work in good style and his work begins to tell. Altogether, prospects are improving." Miller cautioned, however, that "it now depends on Reiss['s assembling department] and we are doing all we can to enable him to get up to 8000 per week, and we *think* we shall succeed with him, though it may take a few weeks to accomplish it."[119]

For two weeks, everything at Elizabethport looked rosy to Miller and Pentz. Miller wrote McKenzie that he had nothing "to report from Factory, except 'progress.' The Building is going on favorably. . . . We are getting out Machines quite well we think, and with prospects of doing still better." Pentz told the vice-president that "we are trying to produce 8,000 machines as a week's work and we can do it without doubt if the japanshop can handle their share of it." He added that through his efforts tension between inside contractors had lessened to such a degree that the weekly meeting of the contractors with Miller and executive officials had taken on "some of the characteristics of a *bear* [*beer*] *garden*."[120]

Although factory officials saw progress, the men from the New York office saw problems. Sydney Bennett, who handled many of George McKenzie's affairs, found the weekly production of six thousand New Family machines and a thousand machines of other makes "most unsatisfactory" and "miserably small." William Proctor's words were more balanced. "The most unsatisfactory part of our business here at present is at the factory," he wrote McKenzie. He found it almost impossible to account for the factory's "inability to make the new machine, or even to get the tools right for making them." Proctor admitted, "I go over there fully determined to blow every body sky high, and am met with the most plausible excuses possible. They certainly are doing a great deal, and there is a good deal to do." Although he could not pinpoint the factory's weakness, Proctor had satisfied himself that "Reiss is not now the great bugbear as formerly."[121]

When George McKenzie called Pentz and Miller to account for the low production figures at Elizabethport, both gave good reasons. Pentz maintained, as Proctor suggested, that machining and assembly (Reiss's departments) were no longer the problems. Rather, "the Jappanery is the weak brother in this establishment, we are continually being checkmated by this department in our efforts to get a respectable weeks work. . . . When the work does not come out of that department, and an investigation is attempted, one is met by such a flood of excuses, figures, and promises." The production expert promised the vice-president that the factory would turn out weekly between five hundred and eight hundred Improved Family machines "if the present indications are correct." Miller supported Pentz's "good service" and noted that "he seems to be the right man for the place." But rather than cast doubts on the contractor of the japanning department he recognized the need to construct more japan kilns, "which we now need more than any other one thing to facilitate us in getting out machines, and getting them out so that they look well." Not until the new foundry was completed could new kilns be built. Until that time, even the New York officials realized that "we may expect trouble from the kilns until the new foundry is completed and the kilns changed."[122]

Despite Pentz's and Miller's opinions about japanning, production problems at Elizabethport did not reside in this department alone. As Pentz acknowledged when he wrote McKenzie on June 1 that "We have succeeded in transferring part of the trouble from the jappannery to the assembling department . . . [which] does not seem able to put the shafts and gears into them [the heads] as fast as needed."[123]

Even as late at 1881, some fitting and filing were required to put sewing machines together. Rather than question the precision of work produced at the Singer plant in general, Pentz blamed the assembly problem on the contractor of the machining and assembling departments. Pentz regarded Hans Reiss as the "most valuable" contractor at the Elizabethport factory yet thought he had "too much to do, and inferior assistance to help him do it."[124] As events unfolded, McKenzie discovered more fundamental problems than simply an overworked contractor and faulty japanning kilns. Eventually, he called into question the entire inspection system used at Elizabethport.

Sydney Bennett, McKenzie's assistant, warned his boss about production problems at the American factory. "The need for [New] Family machines is clearly understood at the factory," he wrote McKenzie, and "it is perfectly clear to me that before there is a new machine [Improved Family], you will have to go and live down at the factory for a few months." Bennett explained why only six Improved Family machines had been made the previous week. The machining department had miscentered the shuttle race on the entire week's production. As Bennett bemoaned, "The factory is full of machines which won't pass inspection, and Smith [the inspection contractor] found them doctoring in the old fashion way to knock bad work into shape—in some cases opening, and in others closing the race."[125]

Pentz saw the matter differently. "I am much bothered," he wrote McKenzie, "by having my instructions to inspectors changed." The week's production of Improved Family machines had been ruined, Pentz argued, "for the reason that my instructions had been so changed by a gentleman who is not a mechanic [and] who frequently strains at an unworkmanlike gnat and swallows a mechanical camel." The nonmechanic to whom Pentz alluded was probably Smith, the contractor for the inspection department, who was responsible for the final testing of the sewing ability of the machines and for a visual inspection of the workmanship and ornamentation of Singer products. The problem always seemed to be the criteria or standards for these inspections. Earlier, Pentz had appealed to George McKenzie to set the standard for Singer's portable hand sewing machine: "No one appears to know just how perfect they should be. I found them inspected up to the I[mproved] F[amily] standard, and took the liberty to lower that standard to a small extent, as the construction of it will not permit of it being so closely scrutinized."[126]

Yet despite Pentz's best intentions, the inspection standards on the Improved Family had not been determined. In lengthy letters written during June 1881, Pentz painted a vivid picture of the chaos at the Elizabethport factory resulting from efforts to produce eight thousand machines weekly including the new and more sophisticated model. Pentz described how his standards for inspection and testing had been changed or ignored by several of the inside contractors. Even Miller's orders, made after discussion with Pentz, had not been precisely followed. As Pentz concluded in one letter to McKenzie, "The inspectors say that they get [so] many diferent [sic] kinds of instructions that they don't know what to do." Pentz later wrote a letter of grievance against the inspection contractor Smith for, Pentz charged, consciously deceiving both him and Miller.[127]

When George McKenzie asked L. B. Miller about the problems with Smith and his inspection department, the superintendent assured him that "there is no cause for uneasiness on your part so far as his [Smith's] interference here is concerned. I learned some time ago that he was not very practical and I have been obliged to bring him up with a short turn, and since [then] there has been no annoyance from him." Satisfactory production reports for the last two weeks in June, coupled with Miller's assurances, must have allayed McKenzie's fears, at least temporarily. Nonetheless, Pentz's belief that "it is suicide to make too many changes at once, which tend to rattle the man, or men, who do the work" must have distressed McKenzie and put him on the lookout for future problems.[128]

L. B. Miller had insisted that he and the Singer contractors were "pushing to get out machines in every way we can think of," and finally on June 28 he proudly reported to the vice-president that the factory had "finished more machines than ever in one week before, and the last 3 weeks a total of 24,347 Machines."[129] This dramatic improvement had occurred, Sydney Bennett believed, because "Miller has gone into the works more & done more real pushing than I have ever seen him do before." Pentz, however, pointed out that "there is [still] that weakness in the japannery; if we get quantity we lack quality and *vice versa*. When the new kilns are done we will then know if the lack of capacity is in the tools or the man [the japan contractor]." The new kilns verified Miller's belief that japanning problems resulted from a lack of adequate facilities and put aside the suspicion that the contractor was at fault. Miller reported to McKenzie that with the new kilns, finished in early August, "the results are entirely satisfactory. We get the work out in half the time we ever did before . . . and it looks better than anything we ever did in the way of baking japan. . . . We are no longer weak in japanning."[130]

Despite these production successes, Singer executives still believed that the factory's performance in manufacturing the new Improved Family and Improved Manufacturing machines was a "succession of errors & setbacks." Delay had become normal to those in the New York office. One week in July, Elizabethport made only two Improved Family machines because two parts were lacking and "Pentz's inspection [was] unusually severe." Only recently Pentz had assured McKenzie that he was having the new machines "closely inspected" and checking them "through every process and operation." Pentz took pains to explain, however, that "I expect perfection in no one" and demanded only what was practically right.[131] Yet the issues of quality control and inspection standards continued unresolved even when McKenzie returned to the United States and attended weekly production meetings at the Elizabethport factory.

Upon his return to New York, McKenzie confessed to President Clark his fear that "there is something wrong at the Factory," which needed his utmost attention. The president, who could take much of the responsibility for Singer's success, admitted to McKenzie that "sometimes I think we have been trying to do too much business, and that it is getting to be impossible to manage it." Yet Clark assured McKenzie that "it is best to assume that the long delay in producing the new machines in adequate quantity has been for the best. Such delays have occurred before in the history of our business. But it is certain that this delay has been a disappointment to us and to all our agents." Assured by McKenzie, he, too, recognized that "there is no other system by which a large Sewing Machine business can be carried on with success except our own."[132]

The problems caused by trying to expand output of present models as well as beginning production of a new model continued to plague the Singer Manufacturing Company. George McKenzie's intercession at the New Jersey factory apparently produced results with the Improved Family but none as encouraging as anyone hoped for. There is no evidence that McKenzie or Miller ever questioned the basic production processes that had been adopted or developed at the factory. Certainly McKenzie seems to have accepted Albert Pentz's view that production problems arose from inside contractors' performance rather than from the hardware or the production system. Despite all efforts by Miller and imprecations by Singer executives, Elizabethport's output continued on a plateau of about six or seven thousand machines per week until production managers began to institute major changes in their system of manufacture rather than trading accusations. Some of these changes were made between 1882, when McKenzie succeeded Clark as president, and 1885, but most occurred after McKenzie retired in 1886. Unfortunately, no historical record of the changes made after the McKenzie regime exists, and we may only speculate about their nature. The events before 1886, however, inform these speculations.

Operations at Singer's Elizabethport factory in 1882 mirror those of 1881 except that George McKenzie found little time to be concerned with them. Rapidly expanding markets in Europe had forced the Singer directors to initiate the construction of a bigger factory in Scotland, and responsibility for locating a site and beginning the building fell to McKenzie. McKenzie also worked at getting increased production from the hard-pressed Scottish works, a need he found greater than expanding Elizabethport's output. At least for part of the year, Albert Pentz helped McKenzie at the Scottish factory. L. B. Miller, as usual, reported to McKenzie on the week-to-week events at Elizabethport. The American superintendent found that in pushing production workmen had occasionally forgotten to gauge their work. But he assured McKenzie that, once this was discovered, "we have insisted on the machines being made to guage [sic] . . . and [shall] see that they are kept to the guages [sic]."[133]

Vice-president McKenzie reported back to Miller that because he and Pentz had pushed the Scottish factory so hard to raise its output, some of the contractors "have been using too soft material. They will have to change to harder material and consequently [will] not be able to turn out so many machs. altho' they are working night & day." Consequently, he warned, Elizabethport would have to assume the burden of supplying the increased European demand.[134] As he might have anticipated, the American factory failed to meet the challenge. McKenzie lost his patience. "I am at a loss," he wrote Sydney Bennett, "to see how those in charge at [the] Elizabethport factory do not find out why the japan should be chipped off the balance wheels, & that band covers should be [too] short. There is always something." McKenzie continued, "If it is not one thing it is another, and as I told you before it is of the greatest importance that the production should be increased." Inasmuch as Elizabethport had "all the facilities at their command,"[135] McKenzie could not understand its failure to meet the demand for increased production. Because he was then busy locating and purchasing a suitable location for the new European factory, he deferred a thorough investigation of the problems at Elizabethport. But early in 1883 he began to instigate major changes at the great American factory.

Perhaps the most important change wrought by McKenzie was in appointing Edwin Howard Bennett as assistant superintendent to L. B. Miller. For a long time McKenzie and other Singer executives had realized that Miller was overworked, and Bennett's appointment made a decided difference in certain aspects of the factory's history. Bennett had worked for Singer since 1861, when he began as an assistant to his father, a mechanic in the factory. Until 1883, he had worked chiefly as Singer's steam engineer and as its millwright; in addition, after 1879 he had been in charge of boiler construction for Babcock & Wilcox, a company owned by Singer.[136] As had Pentz's, Bennett's efforts showed up principally in matters of precision, inspection, and standards. Either because of his personality or his ability, however, he made a greater contribution than Pentz, who was still working at the factory. McKenzie met on several occasions in March 1883 with Miller, Bennett, Pentz, and Philip Diehl to attack some of the deep-seated problems of the Singer production system.[137]

On March 27, 1883, the Singer factory experts and President McKenzie made a resolution, which, if fulfilled, would effect a radical change in production methods. For about ten years the factory had been moving toward this change, and a careful observer might have predicted it. But the manner in which the resolution was adopted betrays elements of discontinuity—of a break with the past: "Decided that each piece commenced in a department shall be finished there to guage [sic] ready for assembling and no part shall be made in the department where it is assembled into the machine."[138]

As we have seen, Singer's production process had emerged as a compromise, or blend, between European custom-building techniques and the American system of manufacturing. As late as 1883, some parts of Singer sewing machines apparently were being made and custom-fitted in the rooms where they were supposed to be "assembled." Chapter 1 stressed the importance for the development of the American system of the War Department's decision to have arms constructed with interchangeable parts. To be sure, certain technical constraints existed, but the key decision was an administrative one. With Singer in 1883, when the company made perhaps six hundred thousand sewing machines, an administrative decision shaped the company's production system so that it could, in the sense that Henry Ford used the expression, "mass-produce" sewing machines. No longer would McKenzie and Miller allow parts to be made and fitted in the assembling department; now every part would be made to gauge in the machining department and then assembled in an entirely different location.

McKenzie and his experts subsequently developed other policies to raise and maintain the quality of product. The president ordered Pentz "to examine some needles in every lot in every respect as to size, straightness, quality, temper, etc." The production committee also decided that "the shafts and boring shall be examined by the Company's Inspector . . . after they leave the machining room and before they enter the assembling

[room]." This policy in particular would ease assembly problems. In order to ensure uniformity in the quality of the materials, the company resolved that immediately after each steel delivery, the stock was "to be thoroughly tested . . . by finishing a few pieces from each lot." To document decisions or problems at the factory, Singer executives demanded that minutes be taken of the weekly contractors' meetings with the superintendent. And, near the end of 1883, Miller and McKenzie decided to create a gauge department, which would be responsible for the production and maintenance of gauges used throughout the factory. Philip Diehl was assigned responsibility for the department's organization. With the creation of this department, Singer moved toward a rigid gauging system.[139]

The changes instituted by the Singer company occurred at a propitious moment in its history. In 1883 Singer opened a small factory in Montreal, Canada (to make three hundred machines per week), and prepared for the opening of the new works in Scotland (planned to produce eight thousand sewing machines per week). Events during the years between 1883 and 1886, when most of the problems were being worked out at both factories, reveal how seriously the production experts in Elizabethport had taken their resolutions of 1883 and how intensely they pursued precision and systematization. They also reveal the perennial problem of deciding what was good enough and what was too good, which is perhaps best seen in the company's attempt to set up the Montreal factory. Planning began early in 1883. At the end of March, McKenzie expressed to Miller his desire that "the fitting of this factory [be] pushed more." Not surprisingly, he decided to send Albert Pentz to Montreal to help set up the factory and initiate production. McKenzie also discussed with Miller possible candidates to superintend the Canadian factory. Although Miller recommended Frederick Lander, McKenzie hand-picked George Leach for the job.[140]

After several weeks in Montreal, Pentz reported his progress to McKenzie. He stressed that although the power system had been put in working order, much work remained to be done. Most important, he said, "E[lizabeth]port must hurry the remainder of the tools as many of them are connecting links; to lack one is to want all." For some unknown reason, Pentz returned to New Jersey before the Montreal factory had begun production, having stayed in Canada perhaps two months. On July 25, 1883, a month after Pentz had returned, Miller reported to McKenzie that he had received a message from Montreal: "Ten (10) N[ew] F[amily] machines ready to sew off, find many small points to correct." Miller added that this was not "a very encouraging report either as to quantity or quality, and is very indefinite as to when we shall have finished Machines ready to send out." With McKenzie's approval, Frederick Lander, Miller's original choice, replaced Leach as superintendent of the Montreal works. According to Miller, the main trouble at the Canadian factory was "imperfectly machined" parts, difficult to understand since Elizabethport had supplied the machine tools, special tools, fixtures, jigs, and gauges.[141] As the new superintendent, Lander neither improved the machining on the New Family nor significantly increased production. The company gave him almost a year to straighten out the factory before asking for his resignation.

Lander's work at Montreal proved to be a comedy of errors which illustrates the complexity of production of armory practice and underscores some of the demands on those who used it. By February 1884, McKenzie and Miller, concerned about the small and poor-quality output of the Canadian factory, visited it. Maintaining some faith in Lander's abilities, they allowed him to return to the Elizabethport factory "and spend a few days in posting himself as to the details of the I.F." When he left the great American factory he believed that "he had it all, and went back with entire confidence in himself." Yet when E. H. Bennett investigated Lander's performance in mid-May 1884, he found his work "anything but flattering." L. B. Miller reported to McKenzie some of the problems Bennett had discovered in Canada. Although Lander had complained to McKenzie that the Montreal factory had not been sent gauges from Elizabethport, Bennett "found that some of the gauges he had received he had never used, and some that he admitted having received were lost, and could not be found." To both Bennett's and Miller's amazement, "one of the very important fixtures for setting the arm and table correctly together, and which we would not think of trying to get along without, he had put aside as "too slow." " Miller hastened to point out to McKenzie that he had "labored to impress [Lander] with the idea that it was all important to have the machining done accurately, and so make a large saving in the finishing, but it all seems to no avail."[142]

Though acknowledging that Bennett had given him "a severe raking over," Lander defended his action concerning the gauges. He claimed that the jigs, fixtures, and gauges sent from Elizabethport had not been made at the same time and were "never properly tested." He found them "so far out that no two of them correspond." Lander blamed Elizabethport for never having sent a model with which to check these devices. E. H. Bennett held a different view: "The ignorance of the people there as to the requirements of machines and their uses, is something to be wondered at." Even though he thought production and quality would pick up only after Lander was relieved of his duties, Bennett gave him three weeks to shape up. By July, Bennett's opinion had changed little. He wrote the president that Lander was "of no earthly use to us . . . [and had] no ability either as a mechanic or manager."[143]

One of Philip Diehl's brothers replaced Lander at Montreal. Bennett promised McKenzie that the factory was "now getting into much better shape and after it is once revised you will have no further trouble." While Bennett was in Montreal in July, he initiated a piecework system throughout the factory. Evidently, McKenzie feared that such a system would destroy the quality of the Singer product. Bennett assured him that with other changes he had made and with piecework, "I expect it to excell the work made at E[lizabeth]port—that is that it shall be machined closer to gauge—making less hand work."[144] Although not all the problems of manufacture in Montreal were eliminated by Bennett and Diehl, the former was able in mid-1885 to report "matters at the factory progressing favorably in all respects."[145] Few problems would crop up which demanded the president's attention. The Montreal factory's performance satisfied the Singer executive's objectives.

This seemingly trivial episode illustrates the profound changes that had taken place in manufacturing technology at the Elizabethport works. Perhaps as late as 1880 or 1881, Singer executives, particularly McKenzie and Miller, would probably have tolerated the situation in Montreal. But the changes to which the Singer production experts had committed themselves in 1883 made it impossible for them to accept such cut-and-try methods. Accuracy, system, and efficiency had become important watchwords, words which had gained currency by the nature of the technology they had chosen to use and develop at the Elizabethport factory. Bennett—and no doubt others—clearly recognized the inherent demands of this technology, and he scorned those who would not, or could not, recognize and meet those demands.

A paradox remains about the changes in production technology made at Singer between 1883 and 1886. Problems of quality and quantity persisted; the standard of Singer products vacillated; and output of the Elizabethport factory dropped well below 1881 and 1882 figures.[146] Maintenance of quality and quantity was part of the "manufacturer's condition," a result of the nature of the armory system of manufacture and its inherent demands. The manufacturer was obliged to maintain a constant vigil over tools and workmen. Success depended upon how well the manufacturer had systematized or managed this vigil. As a writer in *American Machinist* pointed out in 1884, "Those very appliances, which are supposed by some to insure perfect work *must be used just right*, and to adopt their use is not to relieve superintendents and foremen from care, but to impose upon them new burdens in constantly keeping all hands up to their duties."[147] Singer production experts, pushed by McKenzie, focused their efforts on organizing or on systematizing the plant and the men in it. It is important to recognize their new but abiding faith in the hardware of the armory system—in their automatic and semiautomatic machine tools, their jigs, fixtures, and gauges—and their belief that the success of their manufacturing operations depended almost wholly on the human dimensions of this system.

The Montreal factory experience had clearly demonstrated this change in their thinking. E. H. Bennett precisely articulated this view in his reflections on the status of the Elizabethport factory in mid-1884. "I am more convinced than ever," he wrote McKenzie, "that the best results from this [Improved Family] machine can only be obtained by close workmanship and the application of tools with the least possible amount of hand labor in their production. To accomplish this result, I believe will require the services of a thorough practacle [sic] mechanic to superintend the inspection of all operations, parts, and gauges." Bennett added that the position he had in mind "would be very similar to the one now held by Mr. Pentz, only we want less science and more practice." Moreover, insisted Bennett, "No changes of model or gauge points should be allowed except in committee of the whole, where a complete record should be kept, all changes and reason for such and a free exchange had between both [Elizabethport and Scotland] Factories."[148] Machines, tools, and gauges were essential, but it was people—the right people—who through skill, knowledge, and the exchange of critical information bound these together into a workable, productive system of manufacture.

Bennett seemed to ignore the question of how good the Improved Family sewing machine had to be and who would determine this standard. Perhaps he felt that the matter had been settled, but A. D. Pentz thought that they were "no further ahead than we were two years ago" in determining standards of quality. He wrote McKenzie that he had "been able to hold successfully the middle course, so frequently urged upon me by yourself, between exacting superiors who demanded impractibillities [sic] and hurried subordinates who offered inferiority."[149] The Singer Manufacturing Company executives and production experts battled annually, if not weekly, to place their product somewhere on the vast scale between perfection and inferiority. McKenzie could look back at Edward Clark's advice to William Proctor in 1855 always to "have the best machine," but this counsel appeared ambiguous at best. The success or failure of the product in the marketplace always seemed the final test. Singer's obvious success there throughout the latter part of the nineteenth century indicates that, even if perennially unsure about "how good is good enough," Singer usually made the correct judgment (or guess). In this sense, Singer had mastered the finer, nontechnical aspects of the American system of manufactures.

The company also succeeded in some of the technical aspects of this system. What it often lacked in precision and uniformity, it gained in productivity—in sheer quantity of output. Despite a drop in the weekly output of sewing machines between 1881–82 and 1885, L. B. Miller proudly recognized that his factory had made important strides in real productivity. He reported in 1885 that "by the addition of machines [at Elizabethport] the same number of employees will produce double the number of mchs. they would ten years ago."[150] Even matters of precision improved at Singer. President McKenzie relied more and more on Bennett's judgment rather than that of Pentz. He became the "thorough practical" mechanic he had prescribed in 1884. The year that Bennett's role becomes apparent—1883—also marks the end of an old and the beginning of a new era at Singer. Its production experts, Miller and Bennett in particular, ended their isolation from the American technical community by joining the American Society of Mechanical Engineers.[151] Since its creation, the company had been secretive about its internal workings, so this participation in professionalism seems noteworthy.

More important, Singer ended the inside contracting system at its Elizabethport works this same year. Whenever quantity or quality of production sagged, Singer officials and some of its production experts questioned this system of manufacturing management. In all of their reforms during this period Singer officials sought increased control over the production processes. Elimination of inside contracting and the establishment of "fixed day pay foremen" provided this desired control, especially in the realm of gaining information on real costs of operations and maintaining the quality of the Singer product.[152]

Edwin H. Bennett's contributions to Singer production technology after 1883 had become apparent by 1886. During that year he traveled to Scotland to investigate Singer's new Kilbowie factory (opened in 1884) and to encourage greater procedural uniformity between the Kilbowie and Elizabethport factories. Bennett took the models and gauges, as well as parts, from the Elizabethport works and carefully compared them to those made at Kilbowie. With few exceptions he found close agreement. The production expert had also written a blue book for the Elizabethport factory which delineated all of the machining operations and work-flow routes for the Improved Family sewing machine. Bennett's blue book also specified inspection procedures and limits of precision. In sum, it codified Singer factory production operations for the first time in the company's history, and it made clear—implicitly, at least—that control over these operations had been assumed by Singer factory managers.

The Kilbowie factory received a copy of the Singer blue book. Bennett told the Kilbowie managers that he did not expect the Scottish factory to follow the Elizabethport book absolutely, but deviation could be justified only if "the work can be done cheaper and more efficiently." Elizabethport officials expected these changes to be noted in the book. In procedures of inspection Bennett found variation between the two factories. At Elizabethport, company employees inspected parts of the Improved Family between the numerous operations, whereas in Scotland departmental inspectors (working for the contractors) checked them when finished. Bennett insisted that Kilbowie follow Elizabethport's practice. Finally, Bennett's long-sought improvements were reflected in his demand to Kilbowie managers "that the supply of reamers & files to workmen are [sic] to be kept short as much as possible as in the case of presser [foot] bar brackets, these were being filed which ought not to have been done."[153] Bennett summed up his opinion of the Kilbowie works, which had been "arranged somewhat in the same way as the Company's chief Factory at Elizabethport," and which George McKenzie had helped organize: "The

Factory and its situation is simply grand & I have no fear of contradiction when I say it is the largest, best equipped, & most complete in the world. This factory—with proper tools, its advantages in Labor & Material [—] will produce the I. F. Machine at a price which will be a greater percentage of difference from E'port than has even been the case in the N. Family.''[154]

Yet, perhaps because of Bennett's wish for more practicality and less science, manufacturing technology at Singer remained more of an art than a science. Bennett, Miller, Pentz, McKenzie, and others had created a rigidly bureaucratic procedure for producing sewing machines (a departmental structure with oversight mechanisms such as the gauge and inspection departments, all of which were governed by the writ of a blue book), but they failed to develop a science of manufacturing. The best indication of this lack is in the design of their gauging system. The system neither eliminated human judgment (and therefore human error) nor adequately defined and controlled quality. As late as 1886 Singer's gauges were absolute gauges; they defined the dimensions of the model sewing machine rather than the limits of acceptable variation from the model's dimensions. With such a system, Singer experts realized, ''limits of size must be given to the inspectors as it is impossible to be absolutely correct to gauges.''[155] Therefore, the work of the inspectors consisted of measuring rather than gauging. Likewise, workmen on machine tools must have been forced to measure rather than rely on the indisputable judgment of a go no-go gauge. Had the company used a rational gauging system, some of the questions of quality that plagued the factory might have been settled. From the company's viewpoint, however, it was easier to tell inspectors about changes in acceptable limits of variation than to scrap a set of gauges with one set of limits and replace them by a new set with different limits. One can speculate that the greatest change or improvement in manufacturing made at Singer after 1886 was in the system of gauging, more specifically in the adoption of limit gauges.

As we have seen, the sewing machine industry was born in the context of technical choice. Three sewing machine manufacturers, Wheeler and Wilson, Brown & Sharpe (for the Willcox & Gibbs sewing machine), and Singer responded to the demands of sewing machine manufacture in different ways. Both Singer and Wheeler and Wilson began ''building'' rather than ''manufacturing'' machines, whereas Brown & Sharpe immediately constructed special tools and other devices for the manufacture of the Willcox & Gibbs machine. After a few years of production, Wheeler and Wilson completely adopted the armory system of production and soon claimed to manufacture sewing machines with interchangeable parts. Brown & Sharpe's experience with sewing machine production led the firm to the manufacture of machine tools, a business that soon outweighed the making of the Willcox & Gibbs sewing machine. For a long time, Singer's production technology—basically a European, skilled machinist approach—remained unchanged. Although perhaps surprising in the American economic context, Singer survived with this mode of production. In the early years, Wheeler and Wilson outsold Singer. While Wheeler and Wilson was busy perfecting a manufacturing plant, Singer laid out a worldwide marketing strategy, which when finely honed ensured that consumers would want its products. Singer executives, many of whom had begun their work at the bench, paid little attention to the manufacturing end of the business until its skilled workmen, under pressure to produce increasing quantities of machines, could no longer turn out enough or good enough sewing machines. During the 1860s the company changed its production technology, but not completely. L. B. Miller gradually introduced the most important element of the American system: special-purpose machinery. The Singer factory was a compromise between two worlds of production techniques, those of Europe and those of America. The company continued to thrive.

When the Singer factory at Elizabethport, New Jersey, opened in 1873, according to one account, the company fully adopted the American system of manufacturing all parts by special machinery, thereby achieving uniformity if not interchangeability. The company still relied extensively on an army of fitters to file parts so that they might go together to form a workable sewing machine. With this system, as with the purely European and the mixed approaches, the company continued to turn out a high-grade, albeit expensive sewing machine, which dominated the market and whose name became a generic name for any sewing machine.

In 1876, the last year of the sewing machine combination, Singer sold more than 262,000 machines compared to Wheeler and Wilson's 109,000 and Willcox & Gibbs's 13,000. Singer's army of agents continued "peacefully working to conquer the world" with the Singer machine.[156] By 1881 sales had reached well over half a million sewing machines, and it had become apparent that some important changes were needed at the factory. To be sure, the company had built special machine tools, special tools, jigs, fixtures, and gauges,and had purchased other machine tools, including automatic screw machines, from New England machine tool builders such as Brown & Sharpe and Pratt & Whitney. But the company was feeling the pressures of mass consumption to an extent unknown to most American manufacturers of that time. A technological imperative arose which demanded that Singer bureaucratize its factory, laying out strict procedures and creating a task force to maintain a vigil over precision. Factory experts strove in earnest to achieve that elusive goal of absolute interchangeability, a goal that had suddenly become a criterion of mass production. In the period covered by this chapter, Singer never achieved this goal, yet it reached a point where it could keep the "supply of reamers & files to workmen . . . short as much as possible."

The historian may quibble with definitions of mass production—whether it is a doctrine, a business philosophy, a large production output, or a technological system—but by 1881 or 1882, the American system as practiced at Singer had reached its limits. Perfect interchangeability of parts, whose pursuit had shaped the approach and much of the hardware in small arms factories in the antebellum period (and which had been sought for entirely different reasons), had now become critical for mass production. A doctrine, a business philosophy, a large production output, and a technological system; mass production is all of these bound together. Not until the early 1880s for the sewing machine industry and Singer in particular does the concept take on any real meaning. Henry Ford's criterion that "in mass production there are no fitters" haunts the Singer company's history between 1850 and 1885 and leaves that history very much unfinished. This is because Singer itself, despite its gradual adoption of armory practice, left mass production unfinished. The bicycle and automobile industries, with the help of the machine tool industry, would complete what the firearms and sewing machine makers had begun.

Yet in another important respect, the sewing machine industry—particularly the Singer Manufacturing Company—pointed the way to innovation in mass production. Unlike the other cases examined in this study, this involved the manipulation of wood, not metal. The Singer company's manufacture of cabinets and tables for its sewing machine demonstrated that such furniture could be produced on a large scale, but the furniture industry did not follow. British technical tourists, including Joseph Whitworth and John Anderson, had especially hailed American woodworking technology, but the American furniture industry had remained composed of comparatively small production units. Whatever the constraints on companies in the furniture industry, they were not constraints inherent in the production technology, as the case of Singer's woodworking operation clearly shows.

CHAPTER 3

Mass Production in American Woodworking Industries: A Case Study

In no branch of manufacture does the application of labour-saving machinery produce by simple means more important results than in the working of wood.
—Joseph Whitworth, *Special Report* (1854)

In those districts of the United States of America that the Committee have visited the working of wood by machinery in almost every branch of industry, is all but universal.
—*Report of the Committee on the Machinery of the United States of America* (1855)

When Joseph Whitworth toured the industrial areas of the United States in 1853, he was deeply impressed by America's bent toward all types of labor-saving machinery. He found many ingenious metalworking machines in American shops, yet he noted carefully that on the whole the American machines were flimsy compared to those made in England. Woodworking machinery was another matter. When in 1854 the parliamentary Select Committee on Small Arms called Whitworth to testify about American manufacturing, one member asked him specifically about woodworking: "Altogether in America, you were more struck with the mode of working the wood than their mode of working the iron?" Whitworth replied, "Much more; they are not equal to us in the working of iron."[1] Visiting many of the same American establishments Whitworth had toured, John Anderson, British ordnance inspector of machinery, was equally impressed with American woodworking technology. Anderson included lengthy descriptions of woodworking machines and their operations in his *Report of the Committee on the Machinery of the United States of America.*[2] Americans could pride themselves on the knowledge that Anderson's committee had decided American gunstocking machinery was indispensable for its new small arms plant at Enfield. Not only did the committee buy a complete battery of gunstocking machinery, it hired an American mechanic to set up and manage it in England. Once in operation, Anderson told his countrymen that this woodworking machinery—above all the American machinery in England—was "a positive addition to the mechanical resources of the nation."[3] (See Figure 3.1.)

Yet less than seventy years later, when Henry Ford was demonstrating to the world the full effect of mass production, a segment of the American mechanical engineering community began to declare the woodworking industries out of date. Writing in *Mechanical*

125

FIGURE 3.1.   American Woodworking Machinery, Enfield Arsenal, 1857. Acting for the British Ordnance Department, John Anderson contracted in 1854 with Cyrus Buckland of the Springfield Armory for Buckland to design and build gunstocking machinery for the new Enfield Arsenal, which began operations in 1857. A Blanchard-type copying lathe is illustrated here. (*Illustrated London News*, September 21, 1861. Eleutherian Mills Historical Library.)

Engineering in 1920, Thomas Perry, manager of a Michigan veneer works, noted that "woodworking, one of the oldest civilized trades, is now one of the largest industries in the United States. It is doubtful whether any group of modern manufactures gives evidence of less scientific knowledge of its products."[4] Another mechanical engineer, B. A. Parks, echoed this sentiment a year later when he wrote, "The woodworking industry is one of the oldest industries extant, and yet it has shown the least development and has been the slowest to adopt modern principles of manufacturing of any industry of which the writer has knowledge."[5] Other engineers reiterated Perry's and Parks's belief that American woodworking was devoid of scientific engineering knowledge. To rectify this problem, the American Society of Mechanical Engineers established in 1925 a Wood Industries Division, which was intended to focus mechanical engineers' attention on all aspects of woodworking technology.[6]

These sharply contrasting views of American woodworking technology demand explanation. Were the assessments of the British visitors of the 1850s accurate, and if so, how can the criticisms of the mechanical engineers be explained? Could one argue that the

British were too sanguine or that the mechanical engineers of the twentieth century were too caught up in the rhetoric and ideology of efficiency and the movement to eliminate waste?

Unfortunately, despite its importance in the history of American technology, indeed in the history of American material culture, mechanized woodworking has received little attention. Relying upon the Whitworth and Anderson reports, Nathan Rosenberg argued that America's rich endowment of timber helps to explain its "rise to woodworking leadership" in the period from 1800 to 1850. Polly Earl has suggested that much American woodworking machinery was not as capital-intensive as has been commonly presumed. Often machines were hand-powered, simple, and inexpensive, if not completely jerry-built. Edward Duggan has suggested that in Cincinnati, manufacturers adopted woodworking and other "labor-saving" machinery to minimize production time (or maximize output) rather than strictly dispense with skilled labor.[7] Other historians have written on specific woodworking companies or locales, but none has dealt extensively with production itself, and certainly none has sensed the urgency of problems in woodworking expressed by mechanical engineers in the 1920s.

Alfred D. Chandler's Pulitzer Prize-winning *Visible Hand* perhaps offers the most broadly based explanation for why woodworking, despite its antebellum development, may not have kept pace with other American industries in the second half of the nineteenth century and the first two decades of the twentieth. Chandler argues that by the Civil War most woodworking industries had substituted machines for hand operations, and given the nature of this "non-heat-using" industry, throughput thereafter could be increased only, for the most part, by adding more men and machines but without increasing productivity. These characteristics, therefore, imposed limits on the size of woodworking firms and on their individual output. There were no advantages to having a massive woodworking factory because its unit costs would not be significantly lower than those of a much smaller firm.[8]

Given Chandler's interpretation of the woodworking industry, it is difficult to speak of the "mass production of furniture" or the "mass production" of most wooden consumer durables in the nineteenth century because the firms were smaller and their output severely limited relative to metalworking firms. Because of the very nature of working wood, Chandler implies, there could be no Henry Ford of furniture. By taking another look at the woodworking industries in the period 1850–1930, particularly in light of new manuscript evidence, however, more can be learned about mass production and the material culture of America.

This chapter examines woodworking in the second half of the nineteenth century principally by means of a case study, the manufacture of sewing machine cabinets. To connoisseurs of fine Victorian furniture, it may seem presumptuous if not preposterous to regard sewing machine cabinets as furniture. But historians interested in pursuing a broadly conceived study of American material culture should consider sewing machine cabinets as part of the Victorian furnishings of American homes, if for no other reason than that the manufacturers of these cabinets regarded their products as furniture and sold them as such.[9] (See Figure 3.2.)

An important recent study by Michael J. Ettema, "Technological Innovation and Design Economics in American Furniture Manufacture of the Nineteenth Century," suggests that the extent of mechanization in woodworking was directly proportional to the price of furniture. High-priced Victorian furniture, or what Ettema calls "high-end furniture," possessed ornamentation too elaborate for machinery to produce. Yet this high-

FIGURE 3.2.   Wooden Sewing Machine Cabinet Made by the Wilson Sewing Machine Company, 1876. This is a notable example of a sewing machine cabinet marketed as fine Victorian parlor furniture. (*Treasures of Art, Industry and Manufacture Represented at the International Exhibition, 1876*, 1877. Smithsonian Institution Neg. No. 76-1241.)

end, high-style furniture provided models for middle-range and low-end pieces. Since machinery could not exactly reproduce these high-style objects, producers simplified designs while trying to preserve an impression of high style. As Ettema argues, "A desire to emulate makes high style artifacts models for the entire scale. The less expensive objects are not naive imitations of inferior quality, they are less labor and material intensive, and therefore are simplified objects designed in the same style."[10] The use of machinery increased as producers moved from the middle range to the low end.

In price and style, sewing machine cabinets can be regarded mainly as "low-end" furniture. It is therefore not surprising that producers of these cabinets such as Wheeler and Wilson and especially Singer Manufacturing Company resorted to the extensive mechanization of manufacturing processes. Furthermore, these firms came to employ "heat-using" methods of production. The extent of this mechanization and the way it was carried out present a history that raises a number of questions about how representative the woodworking operations of the sewing machine industry were, why furniture manufacturers did not follow its lead, and what role the market and consumer preferences have played in the size, organization, and operation of the furniture industry. This chapter does not pretend to answer all of these questions, but it identifies some of them in the hope that future scholarship will address them more fully.

Woodworking at the Beginning of the Sewing Machine Industry

By the time of the New York Crystal Palace Exhibition of 1853 a wide variety of woodworking machinery was available in the United States. (See Figure 3.3.) At least two of the exhibitors at the Crystal Palace would become major manufacturers of woodworking machinery throughout the second half of the nineteenth century.[11] Those who toured Crystal Palace had a chance to see what can be termed "standard" woodworking machines including circular cutoff saws, scrollsaws, simple turner's lathes, planing machines, tenoning machines, mortising machines, boring machines, dovetailing machines, and a variety of related woodworking machines. (The history, operation, and capability of these standard woodworking machines have been treated elsewhere and need not be discussed here.[12]) In addition to standard woodworking machines, visitors received the very slightest glimpse of special-purpose woodworking machines. Machines were exhibited, for example, for turning, boring, and mortising wagon hubs; for dressing barrel staves; and for making felloes of wagon wheels.[13] Yet as Joseph Whitworth learned while he was a British commissioner to the New York Crystal Palace Exhibition, it was impossible to gather from the exhibition itself the extent to which American manufacturers had mechanized woodworking.

Whitworth toured many of the manufacturing districts of the United States and witnessed the application of woodworking machinery "to every possible purpose."[14] Everywhere, he and the Anderson Committee on Machinery of the United States that followed his footsteps noted specialized woodworking machines in operation.[15] Nowhere was this tendency better exemplified than in the gunstocking operations at the Springfield Armory, which were discussed in Chapter 1. These stocking machines were not manufactured by the woodworking machinery industry. Like most of the special-purpose machinery studied by the British, the Springfield machinery had been made "in house." Consequently, although adequate information survives on standard woodworking machinery of the period, little or no substantive evidence remains about the "universal[ly]" used special-purpose woodworking machinery. The historian who pays attention only to the standard machinery so colorfully described and elegantly illustrated in works such as John Richards's *Treatise on the Construction and Operation of Wood-Working Machines* (1872) will have an incomplete view or understanding of woodworking in nineteenth-century America.[16] Unquestionably, much of this special-purpose machinery operated on the same principles as standard machinery, but its single- or special-purpose features set it apart from standard machinery, as the case of the sewing machine industry's woodworking operations makes clear.

FIGURE 3.3.   Interior of the New York Crystal Palace Exhibition, 1853. (*Scientific American*, August 13, 1853. Eleutherian Mills Historical Library.)

Cabinetmaking at Wheeler and Wilson Manufacturing Company

With its initial strategy of capturing the home sewing machine market (as opposed to the commercial market of the boot and shoe and garment industries), Wheeler and Wilson quickly moved the entire sewing machine industry into the furniture business. Although it sold machines with plain wooden tables, the company also offered, for an extra $12, a nicely made (and soon highly ornamented) wooden cover to enclose the head when not in use. For $25 in 1857, customers could purchase a mahogany or a walnut case that completely enclosed the machine. Wheeler and Wilson stated that "these elegant cases . . . are ornamental parlor furniture.''[17] Throughout much of the nineteenth century, the company maintained the basic outlines of these sewing machine cases but sought consciously to ornament them in accordance with current furniture industry styles. The physical features of the sewing machine, however, with its head, treadle, flywheel, and belting, clearly imposed severe limits on the design.

By 1863, when it made almost thirty thousand machines, Wheeler and Wilson had begun to advertise its product by emphasizing the manner in which it was made. The company published a small booklet on the sewing machine which described the Wheeler and Wilson factory and was later reprinted in several American periodicals and newspapers. The journalist who wrote the booklet included a brief description of the woodworking facilities where the cases were manufactured. Of the approximately four hundred workers in the Bridgeport factory, one-fourth made cases and tables. Since Wheeler and Wilson stressed that its production system was modeled on that used at United States armories, it was no surprise to find machinery substituted for manual labor in the woodworking department, "so that one man, on the average, does as much as ten men could without machinery."[18]

Without any surviving manuscript material it is difficult to gain an adequate picture of precisely how or even how many sewing machine cases, as opposed to tables and simple covers, Wheeler and Wilson made. Yet, given the firm's location and the period in which it flourished, some general perspectives on the company's woodworking operations can be drawn.

Wheeler and Wilson's move into the same factory where Chauncey Jerome had once made more than 150,000 brass clocks and wooden clock cases annually could have been of great significance for the production technology the sewing machine company used in making its wooden covers and cases. This is not to suggest that the special machinery designed to produce clock cases was converted to sewing machine cabinet manufacture. Rather, it seems probable that mechanics who had worked in the Jerome factory later worked for Wheeler and Wilson, bringing the specialized techniques of clock case production to sewing machine cabinetmaking. Jerome's clock cases were constructed of twelve-ply veneer pieces, cut out and fitted together with highly specialized machines and then form-sanded with another special machine. "With these great facilities," claimed the Yankee clockmaker, "the labor costs less then twenty cents apiece, and with the stock, they cost less than fifty cents. A cabinet maker could not make one for less than five dollars. This proves and shows what can be done by system."[19]

The previous experience of Jerome clock case manufacture may not have been neces-

FIGURE 3.4.   Woodworking Shop at the Wheeler and Wilson Factory, 1879. Several sewing machine cabinets and cases are evident in the illustration. This shop worked from material cut to dimensions at the company's mill in Indianapolis. (*Scientific American*, May 3, 1879. Eleutherian Mills Historical Library.)

sary, however, because, as noted in Chapter 2, Wheeler and Wilson's principal production expert had a long career in both public and private armories. Therefore, it would be surprising if Wheeler and Wilson had not built special machines, tools, and fixtures for its woodworking operations. Certainly it did so for the production of the sewing machine itself. Without doubt, Wheeler and Wilson bought and employed standard woodworking machinery, including cut-out saws, planers, and the like, but it also built special machines to produce its large runs of cabinets. Because it manufactured more and more of these cabinets annually, there must have been a movement toward greater specialization of tools and division of labor.

An excellent example of such specialization is the company's practice, begun in the late 1860s, of having its rough cut-out work done outside of Bridgeport, near supplies of timber. The company built a mill in Indianapolis, where raw material was cut to dimensions and then shipped to the Bridgeport factory, where subsequent machining, assembling, and finishing were carried out.[20] (See Figure 3.4.) At roughly the same time, the Singer company began manufacturing some of its cabinetwork in the Midwest and, with growing sales of sewing machines, moved toward the quantity production of furniture in the last third of the nineteenth century.

Singer Manufacturing Company's Cabinetmaking Operations

Much slower to get into the fancy case and cabinet business than Wheeler and Wilson, the Singer company began in the late 1850s, when it introduced home or, as the company called them, "family" sewing machines.[21] For at least a decade Singer contracted with New York cabinetmakers to manufacture its cases, but in 1868 the company built a case factory in South Bend, Indiana, and in 1881 a supplementary woodworks in Cairo, Illinois. By the close of the century, Singer claimed that "the Cabinet Factory at South Bend, Indiana, and its adjunct at Cairo, Illinois, compose one of the largest and most complete woodworking establishments in a country that is celebrated for the supreme excellence of its woods, and for the highest attainments in cabinet making."[22] In 1920, when output had been sharply curtailed because of the loss of Singer's Russian factory to the Bolsheviks (a factory that demanded woodwork for about eight hundred thousand machines a year),[23] the South Bend factory turned out more than two million sets of cabinets. Employing about three thousand workers, the factory had a total floor space of 1.35 million square feet and its own electric power plant of eighty-five hundred horsepower.[24] Before the war, almost four thousand people had worked at South Bend. The development of this remarkable plant provides an excellent perspective on the mass production of furniture in the United States from around 1870 to 1920.

Before Singer built its case factory in South Bend, Francis A. Ross, among other New York cabinetmakers, manufactured its cabinets, tables, and box covers. Unfortunately, nothing is known about Ross or his operation except that he employed a young cabinetmaker named Leighton Pine. Singer's decision to establish a midwestern case factory and its growth and that of the one at Cairo came about largely through Pine's work. When he died in 1905, Pine was still general manager of all Singer's woodworking operations.[25]

In 1868, Pine, at age twenty-four, convinced Singer company officials of the desirability of building its own sewing machine cabinets in the center of hardwood stands in the Midwest, rather than in New York City. (There is evidence that as early as 1864, Singer had tried without success to contract with various midwestern furniture factories for the manufacture of its cabinets.[26] The company sent Pine to the Midwest to find a suitable location for a woodworking factory, and Pine settled on South Bend, whose town fathers offered financial inducements and concessions. Pine supervised the construction of the factory, a three-story brick building, 40 feet × 150 feet, which employed 160 workers when opened in mid-1868.[27] By this time Singer was making almost sixty thousand machines a year; within two years this figure would more than double and would double again two years later.

Singer management in New York City contracted with another of its cabinetmakers, John A. Liebert, a German immigrant, to work with Pine for a year in South Bend to get the factory into smooth operation. Singer also hired W. B. Russell, another former cabinetmaker at Ross's New York shop, to help with the operations. This arrangement caused a great deal of friction between Liebert, Pine, and Russell, which, happily for the historian, resulted in much correspondence to headquarters detailing overall operations in the factory.[28] Although there is no precise information about the machinery and methods used at the new South Bend works, which initially produced twenty-one varieties of tables, tops, and cases, this early correspondence makes clear that in November 1869 the company installed the same inside contracting system which it had recently adopted in its machine works and which was highly characteristic of the armory system of production.[29] With the three New York cabinetmakers nominally in charge of these contractors and of the general management of the woodworks, it is also clear that the factory was plagued with management problems and was not considered successful by Singer officials either in South Bend or New York.[30] Liebert left the company and returned to New York when his year's contract ended, but Pine and Russell remained. Over the next twenty-five years, Pine prevailed, yet controversy often arose between the two, causing ill-feelings and general morale problems.[31]

Two other, unrelated bits of information emerge in this early period which are important mainly because of future events in Singer's woodworking operations. There is clear evidence that the company was making some of its products with veneer, yet rather than making the veneer or obtaining it in the Midwest, the South Bend factory relied upon New York suppliers from 1868 to 1871.[32] Later, the South Bend works contracted with midwestern companies for its veneers and with the opening of the Cairo works in 1881 began both peeling and slicing its own. (See Figure 3.5.)

More important than veneer supply is the evidence that from the very beginning, Singer's woodworking operations relied on the company's central factory, first in New York and then in Elizabethport, New Jersey, for machines and tools and even for information about new woodworking processes.[33] Although not exclusive or total, this reliance on the home factory meant that the company was less dependent on the woodworking machinery industry than were other woodworking concerns. It also meant that because the same machinists built both Singer metalworking and woodworking tools and machines, there was a clear diffusion of ideas from one branch of manufacture to another. In addition, the home factory provided the case factory with all of its hardware for tables, covers, and cases, including screws, hinges, pulls, and latches.

At no point in its early history did the South Bend factory supply all of the woodwork for Singer's American and foreign operations. Yet within months of commencing operations, the Indiana works began to ship tables and cabinets, initially unfinished, to the company's Glasgow, Scotland, factory.[34] But because Singer's woodworking operation could not initially supply all domestic needs, the company continued to rely upon con-

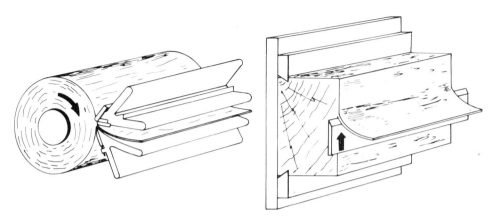

FIGURE 3.5.   Diagram of Peeled and Sliced Veneer Production. Peeled or rotary veneer (left) is wood peeled or shaved off a round, rotating timber, and it has entirely different grain characteristics from sawed timbers. Sliced veneer (right) is wood cut off a sawed timber, and it has the same grain characteristics as the sawed timber.

tracts with cabinetmakers and woodworking concerns in New York and elsewhere. These contracts not only allowed Singer to meet its needs for woodwork but also served as an important check on unit prices of South Bend woodwork. The New York officials often reminded Pine that his factory had to compete with Singer's other suppliers and therefore created a continual incentive for Pine to cut costs. Certainly Pine and the company used the inside contracting system to reduce costs; it annually renegotiated contract prices, always cutting prices when contractors showed a profit at year's end.

As with all of the Singer company's records, there is a lacuna in manuscripts dealing with casemaking operations between 1872 and 1880. During those years, the company's annual output grew from about 220,000 machines to more than 500,000. Operations at South Bend grew accordingly but precisely how and when is unknown. Leighton Pine left Singer's employ during this decade, from 1875 to 1879, to pursue his interests in the Oliver Chilled Plow Company but returned to make spectacular changes in Singer's woodworking technology during the 1880s.[35] These changes center primarily in the company's adoption of cheaper woods; its design of veneered, ''built-up'' (plywood) tables and cabinets; and its move toward bentwood or formed wood cabinets. The result was a more highly mechanized and more capital-intensive operation, which impelled the company toward the mass production of its wood products.

Manuscripts dating from 1881 offer considerable evidence that the South Bend factory, in operation for more than twelve years, was still unable to supply Singer with all of its case and cabinetwork. Among others, the Indianapolis Cabinet Company (which until recently had been owned by the Wheeler and Wilson Manufacturing Company), the Indiana Manufacturing Company, and another firm in Cincinnati held contracts for large numbers of tables and cases for Singer sewing machines.[36] (See Figures 3.6 and 3.7.) Singer even entertained bids from the Sligh Furniture Company of Grand Rapids, Michigan, but rejected them because the price per unit was too high.[37] In addition to problems of quantity production, South Bend received continual complaints about the quality of its cabinets and the problems poor cabinetwork created for agents in selling machines.[38] These problems would be heightened over the next decade as Pine and the South Bend

FIGURE 3.6. Plain Walnut Cabinet Made for Singer Sewing Machines, 1876. (*Asher & Adams' Pictorial Album of American Industry*, 1876. Eleutherian Mills Historical Library.)

FIGURE 3.7. Fancy Cabinet (Walnut, Mahogany, or Rosewood) Made for Singer Sewing Machines, 1876. (*Asher & Adams' Pictorial Album of American Industry*, 1876. Eleutherian Mills Historical Library.)

factory undertook a general program, ordered by Singer's New York officers, to use veneer when possible on all of the company's woodwork.[39] It remains unclear why Singer insisted on the use of veneer, but it is certain that this decision created a host of problems and opportunities which demanded the full financial and technical resources of the Singer Manufacturing Company.

Throughout the nineteenth century, the Singer company remained highly sensitive to consumer tastes and to its competitors' stylistic or design changes. The company held annual meetings of its principal sales agents, who supplied information about what its customers really wanted and why such-and-such a competitor was cutting into Singer's market. Singer's responsiveness to these suggestions may help to account for the company's success in the nineteenth century. Singer's move to adopt all-veneer and built-up cases may have received its impetus from the company's system of checking the pulse of consumer taste, for there is evidence that the company saw itself as following the lead of other companies with respect to woodwork.[40]

Wheeler and Wilson seems to have pushed Singer into the business of all-veneer, built-up woodwork. Not long after Wheeler and Wilson opened its Indianapolis mill (sometime during the 1870s), the company began building up five-layer tables and panels with peeled veneers and the top layer of sliced veneer.[41] Singer soon wanted to copy this technique. Leighton Pine saw advantages in doing built-up work. He had found that he could reduce blistering problems when veneer (usually walnut) was laid on solid pine by introducing another veneer whose grain crossed both the pine and the top veneer. As he explained, "This holds the fibre of the pine and top veneer perfectly and makes a fine table that can be relied on to stand cold exposures." Anticipating the future, Pine noted that this approach would work "well as the surface of built up work."[42]

At a time when South Bend was turning out more than ten thousand tables[43] (both solid and veneered) weekly as well as a large (but unknown) number of covers and cabinets, Pine and his men were adopting new machines and techniques for working wood. (See Figure 3.8.) Two deserve mention. Faced with stepped-up veneer work, South Bend designed and Elizabethport built steam-heated veneer-drying plates. The factory began experimenting with this technique in February of 1881 and soon adopted it on a large scale by building plates and presses for all sizes of work. To manufacture box covers more readily, South Bend designed a machine, adopted early in 1881, which Pine said would "punch out" the necessary panels. Unfortunately, Pine gave no further description of this machine, but from the context of the letter, it clearly was a special-purpose woodworking machine.[44]

Singer demonstrated an intense desire to secure information about what other manufacturers (not just those making sewing machines) were doing in metalworking production technology. The same was true with woodworking, particularly since the company was changing its woodworking techniques because of its decision to manufacture built-up woodwork.[45] An outstanding example of such information-gathering occurred at the factory Wheeler and Wilson had sold, which, under the name of Indianapolis Cabinet Company, began making five-layer built-up drop-leaf tables on contract for Singer. This contract provided Singer's woodworking experts with the opportunity to see how the Indianapolis firm produced its built-up work. Singer executives may have instigated this contract because they had learned that the superintendent of the Indianapolis Cabinet Company had been the cabinetmaker who had originally started Wheeler and Wilson's Indianapolis operations and had directed the manufacture of all of its cabinets there.[46]

FIGURE 3.8.   Plain Walnut Table with Paneled Cover Made for Singer Sewing Machines, 1876. The design of both the solid and veneer tables was the same. (*Asher and Adams' Pictorial Album of American Industry*, 1876. Eleutherian Mills Historical Library.)

To their amazement, Singer officials found that the Indianapolis operation "had a very poor organization for manufacturing, all their tools and fixtures are poor, their processes for making tables are not economical, and bear no comparison to the advantages that are made the most of at every possible point, to cheapen production at South Bend." James Van Dyke, the company's western technical expert, pointed up the critical differences between the two factories. At Indianapolis glue was spread by hand with a brush, whereas "at South Bend they pass the stuff between two rollers, which does the work far more perfectly and economically in materials as well as labor." With special clamping devices, workmen at South Bend could mount walnut edge strips four times as fast as with the hand clamps used in Indianapolis. The cabinet company sawed the holes for sewing machines in its cabinetwork one at a time rather than in multiple units as in the Singer factory, and Van Dyke noted that "they bore holes for belts and brass plates by hand with brace and bit instead of with power boring machines."[47] He also informed the New York office that the Indianapolis company was sure to renege on its contract because the superintendent had told him that labor and materials costs alone totaled nearly as much as Singer was paying for this woodwork. Nevertheless, Singer learned something important at Indianapolis. Van Dyke wrote George Ross McKenzie, "Those folks were cutting their own body stuff for tables (that is, shaving around the log the $\frac{3}{16}$ [inch] white wood). This gives them quite an advantage, and offsets their disadvantages as compared with South Bend."[48]

Armed with this discovery, Van Dyke called Pine to Indianapolis, where the two technicians thoroughly inspected the old Wheeler and Wilson veneer-peeling (or rotary veneer) machines. They learned from the Indianapolis superintendent the cost per thousand square feet of peeled veneer and upon comparing that figure with what Singer paid for its "body stuff" found that it could be cut for less than half the price. Immediately Pine and Van Dyke pressed for the establishment of a supplementary plant, "at some point where the timber is very cheap,'' to shave 3/16 inch veneer, to dry it there, and ''to glue it up, at the same place, to all the table sizes we need, ready for shipment to either South Bend or Elizabethport.''[49] Both men envisioned this new plant as the center for built-up construction of newly designed box covers and full cabinets.

Pine and Van Dyke sketched out in detail the plant necessary for manufacturing built-up work and figured the cost:

—A two story frame building . . . 70 ft. × 175 ft with a shed at one end to cover eight tanks for boiling the logs. . . .	$12,000
—4 Machines for shaving logs (three for the cotton [or white] wood, and one for the Walnut veneer)	7,000
—1 Grinding machine for knives	500
—25 Sets plate dryers, of 16 plates each ($500.00)	12,500
—8 Boiling tanks	2,500
—Boiler 200 H.P. & Engine 125 H.P.	10,000
—Shafting, Pullies, Belting, etc.	1,500
—Power trimming knives	500
—Steam condensor pump	450
—Cranes for handling logs, car, tracks, etc.	800
—Equipment for glueing up complete	2,000
	$49,500[sic][50]

In addition, ten acres would be required to store logs before they were shaved. It is clear from these projected requirements that the business of built-up woodworking, when pursued on the scale intended by Singer, was a capital-intensive, heat-using undertaking and certainly beyond the means of the small-time cabinet shop.

Given a favorable response from New York headquarters, Pine and Van Dyke were directed to find a suitable location for this projected veneering and built-up plant. Van Dyke emphasized that such an operation must be in an area of cheap timber. He alluded briefly to cottonwood as the timber that he and Pine had in mind to slice and glue together to provide the body for the walnut veneer. Yet by the time the wood production experts toured major timber areas in the Midwest, Pine had other ideas.

Notably energetic and always seeking to find cheaper ways to do things, Leighton Pine began experimenting in early 1881 with the use of gumwoods, or trees of the genus *nyssa*.[51] Gumwood is a sorry wood. As most contemporary and modern guides to woods and forests point out, it is a medium-weight wood, strong, hard, and moderately elastic.[52] With 5 to 6 percent shrinking, it warps and checks badly and is difficult to work. Light yellow or often white, the gumwood was used mainly for hubs of wooden wheels, for wharf piles, fruit baskets, and rollers in glass factories. Yet Pine wanted to use the vast, mostly unwanted gumwood supplies not only for the ''body stuff'' of Singer sewing machine woodwork but also for the finished wood of the products. Shortly before Pine toured forest areas with Van Dyke, he had built several sample gumwood tables, which he sent to New York for inspection. As he wrote in late May, they were ''pronounced failures, as the wood will not take a deep stain and shows light when marred.''[53] Initial rejection did not deter Pine, and on his subsequent site search he clearly wanted to be near gumwood supplies so that eventually he could use this cheap material as a means of lowering costs of Singer woodwork.

Thus at a time when South Bend was manufacturing ''batches'' of ten thousand sewing machine tables (solid walnut and walnut veneer laid on solid pine) weekly, Pine had concluded that ''built up tables are now a decided necessity'' and that with the new factory and possibly the use of gumwood, Singer would be able ''to make *better* and *cheaper* built-up tables, than is done at any other factory.''[54]

Van Dyke and Pine chose Cairo, Illinois, as the site for the veneer mill after ruling out St. Louis, Kansas City, and Glasgow, Missouri (the latter where Singer obtained the white wood—cottonwood—veneer as well as the walnut veneer it already used). Cairo offered several advantages including an abundance of gumwoods (particularly sweet gum) and also cottonwood, excellent river and rail transportation, a ready-built three-story brick factory building (65 feet × 80 feet) complete with a new 250-horsepower steam engine, and twenty-two acres of land, the land and factory available for $46,000. Headquarters approved the purchase, which Pine completed in early June 1881.[55]

As soon as he had received approval, Pine wrote the company's chief production executive, then in Scotland, that around Cairo there were ''millions'' of feet of a ''dark, fine grain, handsome figure gum wood . . . which Van and I think will cut into veneers, and much of it can be used in place of walnut, and needs no staining. It is very handsome and some we saw was as good a *French Walnut* as any one could desire. We have strong hopes of making a good thing of this wood, and I can now see no chance for disappointment.'' To allay any doubts that may have arisen in New York from its initial experience with gumwood, Pine pointed out that this wood was ''rapidly coming into favor as a substitute for Walnut, and beautiful sets of furniture are being made of it. . . . A new hotel at Cairo had considerable wainscoating [sic], partitions and trimmings made of it, and it is very handsome in color and figure, and so close grained as to require but little filling under the shellac finish.''[56]

Pine's enthusiasm for gum stemmed from the contracts he and Van Dyke had secured with timbermen that called for a price of $4 per thousand feet. Singer's timber buyer had recently bought supplies of walnut at $50 and $55 per thousand. Clearly there were strong incentives to view nyssa or gumwood as being as good as French walnut.

Traveling to New York after the Cairo property was purchased, Van Dyke arranged with the Singer Elizabethport factory to construct the ''necessary tools for cutting the logs, the plate drying apparatus, the process for glueing up, and the balance of the equipment in all its detail,'' which was scheduled for completion and installation by January 1, 1882.[57] In New York, Van Dyke questioned Singer officials about if and when the Cairo works would begin manufacturing built-up covers and drawer cases in addition to the already scheduled tables. This question would not be immediately answered.

Within a month, Pine was able to report rapid progress on the Cairo machinery. He wrote to the company's vice president, ''All of it has passed the experimental point. Our cutting machines are simply immense, both in quantity and quality of work and exceed anything I ever imagined in that line. Our drying presses are unequalled, as we know by their use here since last February.'' Pine noted that Edwin Bennett, the production expert who was instrumental in making reforms in production of Singer sewing machines in the 1880s, had designed the Cairo equipment. Elizabethport's costs for it had not been more than $6,000, and Pine claimed that it could not be purchased elsewhere for $50,000. Most important, he believed the Singer-made veneering machinery was ''the best in the country.'' After detailing the cost and performance of each piece of the Cairo machinery, Pine concluded that the ''whole plant will be so far ahead of all other like institutions as you can well imagine.''[58]

Pine's letter makes clear that when it began operation in 1882, the Cairo woodworking factory consisted of machinery designed and constructed by Singer. Its details, given the closed nature of the company, were not accessible to the woodworking community or to the public.

The institution by the New York officials of what can only be described as an absurd and chaos-producing managerial structure at Singer's woodworking factories (reminiscent of South Bend's first years) made 1882 a year of trial by fire for these operations. Total output actually decreased from 1881 production by some ninety-one thousand pieces, and cost of production increased. Details of these troubles need not detain us, but some general observations are in order.[59] First, when Pine returned to manage the Singer case factory in 1879, he ended the inside contract system. Yet because woodworking costs increased during 1880 and 1881 under a foreman system of production management, Singer executives in New York ordered Pine in January 1882 to reinstate the inside contract system. Pine protested vigorously, but he carried out the company's wishes. Singer executives also pushed Pine to find other measures to lower costs and reminded him that the cost of woodwork which the company contracted for to meet all of its demands compared very favorably with, if it was not lower than, that produced by South Bend.

Pine objected to this tactic and argued that "your comparison of the cost of outside work is not a fair one; the outside makers accept orders for such work and quantities as they can handle to advantage to themselves; we have no choice, but are compelled to make all we can, and as best we can." He suggested that if the outside contractors devoted all of their resources to production for Singer, their work "would cost much more than we now get for it. I know how easy it is for outsiders to talk, when they do not understand what *quantity* means." Pine concluded his protest by saying that he was tired of having the work of outside cabinetmakers held up to him "as a model for this factory to pattern after." If it had done so, Pine maintained, he and everyone else at South Bend would have been fired long before. The executives in New York had to recognize that the techniques used by outside cabinetmakers were "not at all adapted to the requirements of this business. . . . I know that we are far ahead of them, and . . . I am aiming to get the best system for running this business."[60]

Pine's argument seemed to border on countereconomies of scale, which along with the implication that Singer was obtaining cabinets from a number of smaller shops at cheaper prices might support Alfred Chandler's argument about the limits to the size of woodworking factories. In subsequent letters, however, Pine made clear his logic.[61] Pressure to increase output, changes in the design of Singer woodwork, and the introduction of a new sewing machine model (whose production was currently plaguing the Elizabethport factory) had all worked together to drive up unit production costs and to frustrate output. (See Figure 3.9.) Pine assured New York that this situation was aberrant.

In addition to these problems, it is clear that the first year of production at Cairo was fraught with technical problems. By midyear, Cairo's production amounted to only seven or eight thousand tables, which were reported to have loose joints where bands were butted together, broken veneer fibers, and veneer layers that were as easily separated with a knife "as mica." Plagued with personnel problems and challenged for authority by W. B. Russell, who had been given new responsibilities by the home office, Pine sought desperately to straighten out the production problems at Cairo. Yet in October he reported that the plant was able to turn out only one thousand built-up tables per day at a time when total demand exceeded twelve thousand weekly.[62]

From Pine's perspective some strides were clearly made by the end of the year. He had

FIGURE 3.9.   Singer "Drop" Cabinet, 1893. This cabinet was introduced in conjunction with the new Improved Family sewing machine in the early 1880s. (*Catalogue of Singer Sewing Machines for Family Use*, 1893. Eleutherian Mills Historical Library.)

succeeded in getting Singer into the production of gum or nyssa wood tables, covers, and cases. Pleased with the appearance of new nyssa cabinets which Pine had sent to New York, Singer executives decreed in December 1882 that "all woodwork for Europe will be made of nyssa wood."[63] Thus Singer began what two later writers called the "period of indiscrimination" in which gumwood was used as an imitation of walnut.[64] American consumers would soon be seeing nyssa wood cabinets in Singer retail stores. With new techniques of working wood, with clear incentives to use cheap gum timber, and with an output of more than eight hundred thousand units in 1882,[65] Singer exemplified the most extreme form of a large-scale woodworking industry.

In 1883 Pine and his production men ironed out many of the problems at the Singer woodworking factories, adding new machines at Cairo and learning much more about the critical factor of moisture content in the successful working of wood, especially of veneer.[66] By May, Pine wrote that the operations at Cairo were satisfactory and that the factory would produce some four hundred thousand table bodies. So successful were the South Bend and Cairo factories that by midyear their ability to produce woodwork outran demand, and they were forced to curtail output. Pine used the relative lull in factory operations to explore all aspects of production costs at Cairo. He found that for the first time Cairo produced built-up table bodies (both of walnut and gum veneer) which cost "a trifle less" than the solid walnut tables made at South Bend.[67] These successes provided Pine with the opportunity in the following year to expand the Cairo facility and its share in the total output of Singer woodwork despite cutbacks in overall production.

Singer executives approved expansion plans because of a major change in the style of portable covers (and later of much of its woodwork) which the company initiated in 1884.[68] Raised and carved panel covers were soon to be replaced by ones made of bent or formed plywood. Because of the self-fulfilling promise of nyssa or gumwood, Singer executives envisioned making these covers "exclusively" of nyssa but left open the possibility of some being made of walnut. Pine priced out both types for three-ply work and estimated that walnut covers would cost some 33 percent more than those veneered with nyssa. Within a month, Cairo started turning out bentwood covers with tools and forms built at the Elizabethport machine works. Pine noted that "it was quite a job to get the forms right to bend the body of these covers" but that the Singer team had succeeded. Beyond the mere tools, what remained for Pine to do at Cairo was to get the bentwood work "running with a system." In August he wrote that "this bent work was entirely new to all of us, and its production has been a matter for experiment, but is so no longer. They can be turned out with certainty and uniformity, and we want only enough of them to make [so that it is possible] to reduce their cost to a low figure."[69] (See Figure 3.10.)

Having successfully developed a process to make bentwood covers, the Singer company moved over the next two decades toward the adoption of this technique for the manufacture of all of its woodwork. By the twentieth century, when Singer output was nearing three million units, the layered (or plywood), peeled-veneered, bentwood cabinet prevailed. The steps taken during these years were not always sure ones, but Pine and his colleagues proceeded from the technical basis which they had established between 1868 and 1885.

Abundant evidence documents the adoption at Cairo and South Bend of increasingly specialized machines and tools for working wood, many of which were built in Singer's Elizabethport works. As he learned more about the effects of moisture content in working wood, Pine developed a rolling process to squeeze out water from peeled veneers, an idea he took from a paper mill in South Bend and from a laundry wringer.[70] Singer patented

FIGURE 3.10.   Singer Bent Plywood Sewing Machine Cover. Leighton Pine and others at Singer's woodworking factories at South Bend and Cairo developed this bentwood cover in 1884. (*Catalogue of Singer Sewing Machines for Family Use*, 1893. Eleutherian Mills Historical Library.)

this process and prosecuted companies that infringed on the patent. Pine worked it out in 1885 because the company had grown increasingly concerned about blistering of veneer on the surfaces of its products. With the process, the company claimed, veneer could be made more uniform in thickness and could therefore be laid more uniformly, thus eliminating blisters. Singer's continued use of the process for veneer and built-up work gives evidence of its beneficial effects.[71] In addition to increasing output and trying to maintain and improve the quality of the company's products, Pine was never allowed to ignore unit costs. After 1885, one important way Pine met the cost challenge was by adopting oak for Singer machine furniture.[72]

Within the sphere of cost reduction, however, Pine was forced to follow overall company management policy. Despite ending the inside contracting system at the Elizabethport factory around 1883, Singer maintained the system at South Bend until 1887, when Pine argued that inside contracting, "while . . . invaluable in a new business, and has served us well, . . . has injured us of late years."[73] This was particularly true in the finishing of cabinets, Pine reasoned, but the same ill-effects were felt in such other areas as the cut-out department. With approval from New York, Pine once again directed foremen and department heads rather than coordinating inside contractors. As at Elizabethport, Singer's woodworks management practices were tightened, new work rules were established, and greater control was vested in the hands of the company.

Although often experiencing periods of storm and trouble, Singer's woodworking operations played a vital role in the company's history and comprise an important chapter in the development of the mass production of furniture in the second half of the nineteenth century and the first decades of the twentieth. Some comparative analysis of the overall state of woodworking in this period and of present-day scholarship in this area helps round out the picture.

In the early 1880s, when Singer's annual woodworking output ranged from eight hundred thousand to almost a million units (which it valued at roughly a dollar per unit), the company employed more than 1000 workers at South Bend and Cairo. By comparison, the average American furniture factory employed 11 workers and turned out products valued annually at about $15,000.[74] In Grand Rapids in 1880, when this city was becoming one of the furniture capitals of the United States, fifteen firms together employed about 2,250 workers and produced goods valued at slightly over $2 million.[75] By 1900, Singer's South Bend cabinet works alone employed 3,000 workers and manufactured as many as seventy-five hundred units daily.[76] Even then, Singer could not produce all the cabinets it needed because in 1901 it contracted with John Widdicomb Furniture Company of Grand Rapids to manufacture two hundred thousand five-drawer oak sewing machine cabinets for almost $4 apiece. One of the largest of the Grand Rapids furniture manufacturers, Widdicomb shipped these cabinets to Elizabethport in weekly lots of two thousand. The previous year, thirty-four factories made furniture in Grand Rapids. These factories hired some 6,200 workers and turned out products valued at $7.5 million.[77] Thus Singer's contract with Widdicomb represented roughly 10 percent of Grand Rapids' total dollar-value output. Singer's own output equaled about half that of Grand Rapids. The American furniture industry as a whole in 1900 consisted of about eighteen hundred establishments employing more than eighty-seven thousand workers (an average of fewer than 50 employees per firm) and producing about $125,300,000 worth of work (an average of less than $70,000 per firm).[78] These data suggest that Singer's woodworking operations were highly atypical—even extraordinary—in the furniture manufacture of the period.

Singer's woodworking experience demonstrated that it was technically possible—if not always economically desirable—to concentrate the manufacture of sewing machine cabinets in one or two very large factories. The company's commitment to this mode of production was reaffirmed in 1901 when it built a much-enlarged cabinet works in South Bend, which, as noted above, annually produced as many as two million cabinets in the first two decades of this century. No furniture maker in this period ever matched that output (certainly not in numbers and probably not in dollar value) or the scale of employment Singer maintained. Even the manufacturers who produced ''low-end'' furniture for Sears, Roebuck never approached Singer's output, nor is it likely that they mechanized their production to the extent and in the manner in which Singer did. Few woodworking establishments had the technical resources of an Elizabethport factory on which to draw. This fact may help to explain Singer's performance, but there are other considerations.

One could argue, as did John Richards in his *Treatise on the Construction and Operation of Wood-Working Machines,* that the furniture and woodworking industry was bound to be highly segmented and, therefore, would always consist of small factories. In particular, Richards stressed that woodworking technology was widely diffused throughout the United States and that it was relatively easy for new competitors to enter the industry. He also believed that the very nature of working wood prohibited the system of production that prevailed in American metalworking establishments.[79] The general diffusion of woodworking technology may have contributed to a decentralized industry, but Singer's cabinet operations clearly show that there was nothing in the nature of wood that prevented its use in mass production. Singer ''mass-produced'' wooden sewing machine furniture. The Singer case factories adopted and developed special-purpose machinery and with peeled veneer entered the realm of a heat-using industry.[80]

The furniture industry did not to any great extent follow Singer's lead. There was some change, however, in the woodworking industries; statistics on furniture in the 1905 census, for example, demonstrate a clear trend toward greater capitalization and increased output during the previous fifty years.[81] But there is no evidence of concentration in furniture manufacturing or large-scale production comparable to Singer's.

The key to the history of Singer's woodworking operations may lie in the nature of the product and the market rather than in the material. A ready-made market for sewing machine cabinets existed in the United States and Europe. People bought a Singer cabinet when they purchased a Singer sewing machine; they did not buy a cabinet for its own sake. The Singer sewing machine, in effect, sold the woodwork of South Bend and Cairo. If Singer's woodworking factories had been forced to compete in the regular furniture market, there is no reason to believe that they would have attained the scale of operations they did in the late nineteenth and early twentieth centuries. The furniture market itself probably constrained the application of mass production technology within the industry in a way that did not occur with Singer.

This is precisely the point that the mechanical engineers of the 1920s bemoaned. In his plea "Engineering in Furniture Factories," B. A. Parks, engineer for a Grand Rapids furniture manufacturer, argued that "lack of engineering ability in the furniture manufacturing organization shows its effect throughout the entire plant; in fact, the writer is convinced that *the average manager of a furniture plant is more interested in marketing his product than in manufacturing it.*" Why was this so? "Furniture is constantly changing in style," Parks complained, and "also most plants manufacture quite an extended line, and consequently a large variety of product must be handled in any given plant." But Parks saw a solution: "A point which most managers overlook and which is primarily due to lack of engineers in the organization, is the possibility of reducing the variety of parts to be manufactured through standardization of design, interchangeability of parts, and greater limitation of line [which] would not only directly reduce manufacturing costs, but would also tend toward the development of automatic machinery, better utilization of raw product, economies in handling parts in process of manufacture, etc."[82] Obviously, Parks and his fellow members of the American Society of Mechanical Engineers believed that the character of the furniture industry could be changed by the introduction of proven mechanical engineering principles, especially standardization. They viewed furniture as a consumer durable that could be produced in the same way and with the same benefits as firearms, sewing machines, and automobiles. Neither the introduction of mechanical engineering into furniture manufacturing nor a major change in the character of the industry has occurred to any great extent in this century. Furniture may not be, after all, like other consumer durables. Yet there are those who believe that eventually furniture production will reach parity of mechanization with metalworking. The ideas of the ASME Wood Industries Division have been restated by the author of the article on the furniture industry in the latest edition of the *Encyclopedia Britannica*. He argued that recent changes in the industry had "resulted in the trend of woodwork towards engineering, a trend likely to accelerate."[83] Fulfillment of this prophecy remains to be seen.

Perhaps Singer's cabinetmaking operations can be more appropriately compared to other woodworking industries. In this century, there have been several innovations—for example, radio and television—for which wooden cabinets became important. Like sewing machine cases, radio and television cabinets were "low-end" furniture, their manufacturers had an already established mass market, and the furniture was secondary to the machine it housed. Consequently, in these industries woodworking establishments arose similar to those of Singer. Archer W. Richards, an engineer for the Grigsby-Grunow Company of Chicago, described one such example in a paper delivered to the American Society of Mechanical Engineers, "The Mass Production of Radio Cabinets." In 1928, Grigsby-Grunow established in a building owned by General Motors a woodworking factory to produce "a complete piece of furniture to house . . . the radio receiver." Richards detailed the means by which his company had within two years brought production to an output of five thousand units per day: "The furniture industry is one of the oldest in existence, and while low-priced furniture has been made, the lesson learned in mass production of other products did not seem to be accepted by the industry." As a result, the Grigsby-Grunow Company "found it necessary to build a complete cabinet plant which would combine the most modern production methods and machinery." Furthermore, Richards emphasized, "the development was made, not by experienced woodworkers of the old school, but by ingenious and resourceful production engineers, who were inspired by their chiefs to undertake and creditably to master the best way of producing attractive, efficient, and salable radio outfits. It is one of the instances that demonstrates that trade knowledge is much less essential to success than is the thorough production experience of technical engineers combined with *a tremendous market* demand for the product." Although celebrating the wonders of engineering expertise, Richards acknowledged that the "great demand for radio cabinets made it necessary for manufacturers to arrange their methods and machinery on a highly productive basis."[84] No doubt there were dozens of other companies that departed from techniques used by the furniture industry in a way comparable to the Singer Manufacturing Company. The product and the nature of the market seem to have been the crucial determinants.

One other industry—a nineteenth-century one—bears useful comparison with Singer's woodworking technology. Though not having a product whose sales were tied to demand for the "real" product, the carriage and wagon industry rapidly moved toward the use of special-purpose machinery in its operations. As discussed above, John Anderson and his British colleagues were particularly struck with the degree to which this industry used special machines and with the fact that much of this machinery was marketed on a routine basis. The Studebaker Brothers Manufacturing Company provides a particularly interesting case not only because it became in the 1880s the largest manufacturer of wheeled vehicles in the United States but because its factory was located in South Bend, where Singer's operations flourished.

The Studebaker Brothers Manufacturing Company

When the partnership of H. & C. Studebaker was formed in South Bend, Indiana, in 1852, the wagon and carriage trade in the United States was undergoing what contemporaries regarded as a revolution. Increasingly, as one writer pointed out, small-town wheelwrights found that they could "no longer make an entire vehicle as formerly with any success, but purchase wheels, axles, top frames, etc., of any and every pattern, to put together and finish. All these parts are produced in great quantities, by machine."[85] In New York alone there were sixteen spoke factories, a similar number of felloe factories and hub factories, and numerous factories that made all the parts of wagons and carriages by machinery. At its founding, however, the Studebaker company hardly resembled these

New York factories. It began as a small-town wagon shop manufacturing three wagons the first year but grew within a generation into a large integrated factory described in 1880 as *le plus grand des grandes.*[86]

Both Henry and Clement Studebaker had learned the trade of blacksmithing from their father. Although they were regarded as "practical mechanics,"[87] the success of their enterprise rested on their prime location near a good supply of timber and the availability of a "standard" line of special-purpose wagonmaking machinery, particularly wheelmaking machinery.[88] Thus the Studebakers did not have to devote their energies to developing machinery because they could buy it ready-made. (See Figure 3.11.) Joined in 1858 by another brother, John Mohler, the Studebaker brothers directed new capital into their wagon works. They were well prepared to reap benefits from wagon and ambulance contracts made by the United States during the Civil War. The company historian wrote, however, that "the Studebakers were unable to satisfy fully the demands made upon them" by the Civil War.[89] Nevertheless, the company expanded in an attempt to meet these demands. Its founders firmly believed that the war served to spread the name "Studebaker" throughout the United States. With annual sales of about $350,000 and tangible assets of almost a quarter of a million dollars, the brothers incorporated in 1868 as the Studebaker Brothers Manufacturing Company. The new company employed 190 workers and produced almost 4,000 units per year. By 1872 the corporation hired 325 employees who turned out 6,950 wagons and carriages. Two years later, immediately before a fire destroyed the factory, 550 workers manufactured 11,050 vehicles.[90]

Perhaps the fire of 1874 was a boon for it forced the Studebaker company to purchase new machinery and to build a new factory. The brothers apparently contemplated moving their company to Chicago or to Cincinnati but finally decided to remain in South Bend. They built an impressive T-shaped, three-story factory with detached lumber sheds. By the centennial year, this factory turned out a wagon every seven minutes.[91] Increasing demand led to the addition of a new blacksmith shop 100×200 feet with a ceiling expanse of three stories, another woodworking area, two stories, 80×200 feet, and another engine room and lumber shed. A reporter in 1880 described the Studebaker Brothers factory as a place marked by "order, system, intelligent supervision, the best of material, and all the mechanical helps that genius can contrive and capital produce."[92] Although impressive in size, workings, and output, the Studebaker factory seems to have developed little new production technology. It had acquired some of the latest items available from woodworking machinery companies such as J. A. Fay & Co. and Egan Company of Cincinnati and the Defiance Machine Works of Defiance, Ohio.[93] As the reporter noted in 1880, "The Studebaker works, while a world within themselves, draw upon all the mechanical world for its improvements in their line." He listed four examples of the Studebakers' "constantly drawing upon invention [from] without": "the skein-setter, the apparatus for putting on tires, the hydraulic press for forcing in boxing, and O'Brian's priming."[94]

The Studebakers saw themselves as "first-class manufacturers," who through extensive mechanization had driven down the cost of their wagons from $140 to $70 while maintaining or improving quality. Uniformity provided work "impossible with the unaided human hand. Every piece of wood and iron is marked by rule, and shaped by machinery so fixed that there can be no variation. This, with the excellent material used, insures the perfect fit of every part, and the consequent perfection of the whole."[95] The company had adopted the rhetoric of other "American system" manufacturers as well as their special-purpose machinery.

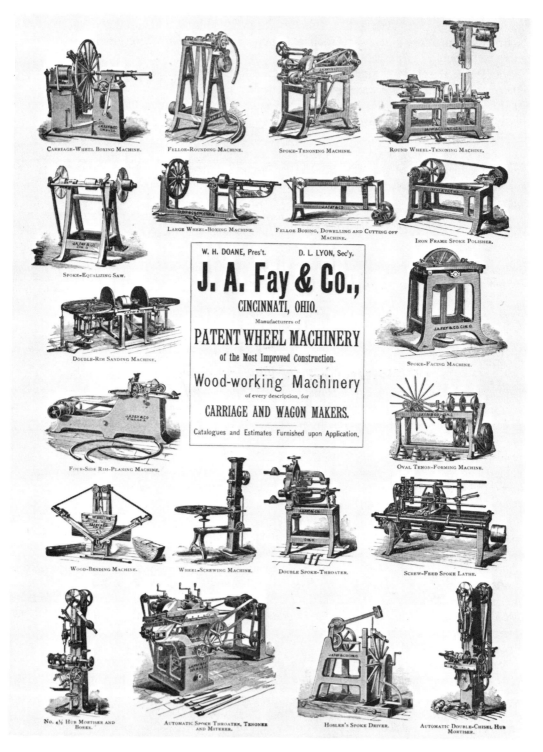

FIGURE 3.11.   Special-Purpose Wheel Machinery, J. A. Fay & Co., 1888. As early as the 1850s, companies such as J. A. Fay & Co. marketed a wide variety of highly specialized wagon and carriage wheelmaking machinery. This illustration provides a vivid example of such machinery. (*The Hub*, January 1888. Eleutherian Mills Historical Library.)

Throughout the 1880s and 1890s, Studebaker Brothers expanded to become the largest maker of wagons and carriages in the United States. It manufactured some seventy-five thousand vehicles in 1895 with a labor force of about nineteen hundred. By this time Studebaker operated at the cutting edge of production technology, at least in certain areas.[96] Two examples are suggestive. At the same time that leading bicycle makers were developing similar processes (see Chapter 5), Studebaker Brothers adopted resistance welding and sheet steel stamping techniques. Electric resistance welding was developed between 1886 and 1888 by the prolific inventor Elihu Thomson. This process automated what has always been one of the most difficult tasks of blacksmithing.[97] As early as 1889, *Iron Age* had predicted a revolution in metalworking because electric welding made automation possible.[98] Although not the first manufacturer to adopt resistance welding, Studebaker Brothers became by 1891 the owner of "one of the largest and most complete plants" for resistance welding in the United States.[99] The company butt-welded steel wagon axles of up to an inch and a half in diameter. In addition, the iron tires that were placed around the circumference of the wheels were welded with resistance techniques, as were the hub bands and a variety of small parts of spring work. About the same time, Studebaker Brothers adopted a technique to make wagon skeins (the metal sleeves that fitted over the ends of the wooden axle arms for protection from wear) out of sheet steel,[100] which required large stamping machines or punch presses and several operations with different sets of dies.

Although not in the area of woodworking, these developments in welding and sheet steel working linked Studebaker Brothers to the metalworking industries in much the same way the Singer case factory was tied to the home factory in Elizabethport. Electric resistance welding in particular radically changed a major part of Studebaker production technology because it replaced large numbers of skilled blacksmiths with machines. Its successful adoption also suggested that other changes—even in woodworking—were possible.

Nevertheless, Studebaker adhered to some methods that might be considered antiquated. For example, photographs dating from around the turn of the century show the process of hubmaking.[101] One photograph suggests that Studebaker used manually indexed machines to chip out hub mortises even though automatically indexing machinery had been sold for some years.[102] (See Figure 3.12.) Moreover, it reveals that even after they came out of the machines, the hubs required hand chiseling to finish them correctly. Another photograph shows a row of workers hand planing felloes after they had been fitted to the spokes. (See Figure 3.13.) It must have been such methods that bothered the mechanical engineers in the early decades of this century. In general, however, the Studebaker factory in South Bend, like the Singer case factory, offered testimony to the engineer's notions that large output could be achieved through extensive mechanization, standardization, and other principles dear to the hearts of mechnical engineers.

Studebaker Brothers moved successfully from wagon and carriage manufacture to automobile production, first using its manufacturing plant to produce wooden bodies for another manufacturer. In 1904, the company began to sell its own gasoline engine automobiles, but for seven years it purchased chassis from another company and added its own bodies at the South Bend works. During these years it made about twenty-five hundred gasoline-powered cars and trucks (as well as nearly nineteen hundred "electrics"). The company expanded radically in 1911 and changed its name to the Studebaker Corporation when it purchased the plant and assets of the Everitt-Metzger-Flanders Company. (The Flanders of the Everitt-Metzger-Flanders Company was the same Walter Flanders who provided rich suggestions for Ford Motor Company's production technology; see Chapter

FIGURE 3.12.   Manually Indexed Hub Chipping-out Machines and Handwork, Studebaker Brothers Factory, ca. 1890. (Studebaker Collection, Discovery Hall Museum, South Bend, Indiana.)

6.) The Studebaker Brothers Manufacturing Company produced more than a million horse-drawn vehicles in its forty-three years of business, making its factory and proprietors well known throughout the United States. The Studebaker brothers moved beyond South Bend to invest substantially in Chicago real estate and society and to establish a branch carriage factory and showroom on Michigan Avenue. From these activities came a friendship with another celebrated manufacturer, Cyrus H. McCormick, whose reaper factory was as celebrated as that of the Studebakers' in South Bend.[103] As will be seen in the following chapter, however, unlike the Studebaker factory, the McCormick works could not purchase ready-made special-purpose machinery to produce reapers and consequently relied for a long time upon general-purpose machinery and skilled machinists for its operations. Constant changes in the model of reaping machines also mediated against the McCormick factory's adopting special-purpose machinery to any large extent.

Model changes or style changes bothered the mechanical engineers who in the early part of this century wanted to introduce mass production methods in the furniture industry. The rhetoric of these engineers remarkably paralleled that of Samuel Colt, who had told the parliamentary Select Committee on Small Arms in 1854 that anything could be made by machinery. Colt and the engineers notwithstanding, however, the American furniture industry did not adopt the production technology that proved so successful in such areas of

FIGURE 3.13.   Hand Planing Felloes, Studebaker Brothers Factory, ca. 1890. (Studebaker Collection, Discovery Hall Museum, South Bend, Indiana.)

metalworking as firearms, sewing machines, bicycles, and automobiles. There was no Henry Ford of the furniture industry.

Explanations for this failure have almost always focused on the nature of the material wood. The working of wood has inherent limitations: the size of the production unit is constrained, and the choice of production machinery is narrowed. But the case of Singer Manufacturing Company's woodworking operations in the nineteenth and early twentieth centuries seriously calls this ''nature of the material'' argument into question. Singer built cabinets in such huge factories and on such a large scale that one is tempted to use the expression *mass production* to describe these operations. Though not achieving the volume of Singer's output, Studebaker also developed large-scale woodworking operations largely because of the nature of its product.

The character of the furniture industry probably has been determined more by the market than by the material. This fact greatly distressed the mechanical engineers. Products could not be standardized, and product lines could not be maintained long enough to justify the construction of special-purpose machinery and other ''efficient'' production techniques. The American system of manufactures developed in areas where producers could sell large numbers of nearly identical goods. The furniture market did not have this characteristic. Instead, fashion, style, and taste prevailed. Why Americans were willing to buy millions of the same model of Singer sewing machines and millions of all-black Ford Model T's but not something equivalent in furniture is an important question, one that requires study beyond the scope of this work.

The McCormick Reaper Works & American Manufacturing Technology in the Nineteenth Century

There is not enough consistency [in the agricultural implement industry] to admit of exact treatment, and there is great variation in methods pursued, work covered, and system employed. . . . Agricultural machines have, as a rule, comparatively few fitting points, partly because the action of these implements is not mainly between the parts of the implement itself, but upon the soil or crops, so that interchangeability is not as important a feature. This is really a prime distinction between agricultural implements and the other forms of interchangeable mechanism which have been previously treated.
—Charles Fitch, ''Report on the Manufactures of Interchangeable Mechanism'' (1880)

Father [Cyrus H. McCormick] and Mr. Spring went to the Works today with Mr. Wilkinson, who is to be our new Superintendent at the Works. He comes to us from Shumway Burgess & Co., bolt Mfrs. He has had considerable experience as a foreman of shops and as a Superintendent of Manufacturing Works. He has been with the Colt Firearms Co., the Connecticut Firearms Co., the Wilson Sewing Machine Co., and many other concerns. He will take the place of L[eander] J. McCormick as Superintendent at the Works.
—Cyrus McCormick, Jr., Diary, May 6, 1880

Charles Fitch's inclusion of a section on the manufacture of agricultural implements in his ''Report on the Manufactures of Interchangeable Mechanism'' in the *Tenth Census* is puzzling. The section seems entirely out of step with the rest of the report, and Fitch strains to find words and phrases to tie it to his sections on arms production, sewing machine manufacture, and clock- and watchmaking. Students of American technological history often read accounts of Cyrus McCormick's reaper demonstrations at the 1851 London Crystal Palace Exhibition and statements that McCormick's reapers were constructed with interchangeable parts. The obvious success of McCormick's reaper and his company suggests that the McCormick reaper works must have been one of the more technologically progressive establishments in the United States. This view does not conform to that of Fitch, who intimates that the entire agricultural implement industry was backward when compared to sewing machines, for instance.

153

Through a study of the manufacturing operations of the McCormick reaper company from its creation in 1841 until about 1885, this anomaly can be resolved. This study confirms Fitch's misgivings about the state of the agricultural implement industry in 1880; in fact, it blackens the picture. Until about 1880, the McCormick works in Chicago depended primarily on skilled blacksmiths, skilled machinists, and skilled woodworkers to build its reapers. From the beginning of the business, Cyrus McCormick's youngest brother, Leander, was responsible for the production of the reaper. Leander, whose only experience had been as a country blacksmith from Rockbridge County, Virginia, operated the reaper works as though it were a large country blacksmith shop. Rarely did he draw upon the stock of knowledge about large-scale manufacture that had developed in New England, and even more rarely did he use the special tools devised there or the New England concepts of specialization in machine tools.

Had Leander and Cyrus not had an irreparable fight in 1879–80, the reaper works might not have undergone any notable changes until Cyrus's or Leander's death. But Cyrus replaced his blacksmith brother as superintendent with a New England mechanic—the Mr. Wilkinson mentioned in Cyrus, Jr.'s, diary quoted in the epigraph to this chapter. This mechanic, who had once worked at Colt's armory, stayed with McCormick for only about a year. Yet during his tenure he convinced Cyrus McCormick, Jr., that basic changes were necessary and passed on to him that old New England quest for interchangeability. He also taught Cyrus, Jr., the rudiments of armory practice, especially the desirability of a model-based gauge and jig system. By 1881 the younger McCormick had become superintendent of the reaper works and soon would direct the entire company because of his father's failing health. Not until the early 1880s, therefore, did McCormick's reaper works adopt New England armory techniques of production.

This chapter explores the long period of reaper manufacture under a tradition of skilled blacksmithing and the changes that took place at McCormick's factory during the 1880s. This study calls into question long-held presumptions about McCormick in particular and about the extent of the diffusion of the American system of manufactures in general.

Almost from the beginning, participants in the reaper's history as well as those who chronicled its development fought vigorously over who really invented the reaper. As with the sewing machine, men had long dreamed of and tried to devise a mechanical means to reap grain, yet the basic element of such a machine eluded would-be inventors until the early 1830s. In the United States both Obed Hussey and Cyrus Hall McCormick, one from Ohio, the other from Virginia, hit upon the idea of using a vibrating knife or blade to cut the stalks of grain. Hussey made his reaper work effectively by vibrating the blade in slots cut in guide teeth or fingers. He patented this machine in 1833 and began to sell it in 1834. McCormick claimed that he had anticipated Hussey in the essentials of the reaper, but he did not patent his machine until mid-1834. McCormick's blade lacked the effective slotted finger bars of the Hussey machine. Yet, as is often the case in the history of American invention, McCormick moved through the courts to become eventually the famed hero who invented the reaper.[1]

The history of the reaper and its manufacture does not lend itself to heroic treatment. Throughout most of the nineteenth century it is a history notable for constant change of mechanism and design. As will be seen later, changes made in the name of progress militated against the rise of a mass production psychology among reaper manufacturers and retarded the adoption of techniques aimed at large-scale production.

Thirty-one-year-old Cyrus Hall McCormick began building reapers to market commercially in 1840. They were made in the family's blacksmith shop at Walnut Grove plantation near Steele's Tavern in Rockbridge County, Virginia. Robert McCormick, Cyrus's father, and Leander, Cyrus's youngest brother, with aid from their slaves, handcrafted these early reapers from local wood and iron. The McCormicks fabricated all but the steel sickles, or knives, which they purchased from John McCown's nearby hammer works until 1842 and from Selah Holbrook of Port Republic, Virginia, in 1843. Although designed like the machine with which Cyrus had so easily reaped the family's grain in 1839, both 1840-model reapers failed to perform satisfactorily. McCormick's biographer blamed the failure on defective workmanship, a problem Robert and Leander tried to eliminate in subsequent years.[2]

Cyrus firmly believed workmanship was important. In 1846 he wrote his brother that "much you know depends upon having the workmanship right—indeed almost everything."[3] Again in 1841, the father-son team built 2 reapers for Cyrus. The following year they constructed 7 and in 1843, 29. The Walnut Grove smithy turned out 50 machines in 1844 and 1845. By this time, Cyrus had begun selling territorial rights to manufacture and market his machine, and in 1846 the various shops together with the one at Walnut Grove built 190 machines. The reapers made at the McCormick plantation in 1846, the year of Robert's death, were a "pronounced failure." The editor of *Southern Planter,* also an agent for McCormick, noted in his journal that he had received "great complaint of the manner in which it [McCormick's reaper] was gotten up." Despite this failure McCormick continued to sell reapers and territorial rights to manufacture and market them. Cyrus spent virtually all of his time between 1840 and 1846 on the road demonstrating and selling his machine or rights to it. While Robert and Leander built the Virginia reaper, Cyrus's younger brother William kept the inventor's accounts, including those for the widespread advertising upon which McCormick heavily relied.[4]

Although temporarily set back by complaints in 1846, Cyrus McCormick continued to believe something he had realized in 1842: "We find no difficulty in selling as many machines as we can get made." Knowing the limited resources and output of the Walnut Grove blacksmith shop, Cyrus had purposely included manufacturing rights in his contracts with others for marketing the reaper. Some of these contracts specified that McCormick merely receive a royalty for machines made and sold. In other cases McCormick contracted with a company to build a certain number of machines for him at a specified price. The companies or individuals that agreed to build reapers faced the same technical problems in construction that dogged Robert and Leander McCormick at Walnut Grove. Usually Cyrus gave the contractor a model of his reaper (not a model in the sense used in Chapters 1 and 2, but a miniature reproduction of the full-scale machine) and offered technical assistance. In most of his contracts he met with disappointment in workmanship and "incompetent" men. By August 1845 Cyrus realized that the extreme decentralization of production he had established would cause him inordinate problems and perhaps ruin the reputation of his machine. He sought to tighten up matters: "Cin.[cinnati] and W.[estern] N.[ew] York are the most important points perhaps for manufacturing and to consolidate and have all done well is the great matter now."[5]

When he wrote this, McCormick had just contracted with A. C. Brown of Cincinnati and two firms in Brockport, New York (Bachus, Fitch & Co. and Seymour, Chappell & Co.). Evidently he felt optimistic about these firms' ability to produce satisfactory machines. By December 1846, he was even more sanguine about production in the West. He wrote his brother William that he had "come to the conclusion that 100 or 150 Reapers should be made at Cincinnati or some other point in the West" because "every facility can be had, good machinery, and I think means could be had to build at least 100, and probably more."[6] At the same time, he urged Leander to come to Cincinnati to superintend production and offered him either wages or an interest in his business.

Early in January 1847, Cyrus closed an additional contract with Brown for one hundred reapers. The contract included clauses pertaining to Leander's supervision of production in Cincinnati. Cyrus wrote Leander that he was "to have superintendance [sic] of building with the right to discharge hands, reject bad materials, correct defects that may occur & C.—Brown to have and keep all requisite materials provided, & C.—Brown bound to complete them by 15th May—you to receive $14 per week." McCormick also extended his contracts with the two manufacturers in Brockport, New York, writing one of them that "I can't manufacture myself to half supply the demand." To meet the demand for the 1847 harvest, he also contracted for one hundred reapers with Charles M. Gray and Seth P. Warner of Chicago.[7] This latter contract would soon lead Cyrus McCormick to settle permanently in Chicago and to centralize production.

The one hundred reapers made by Charles Gray and Seth Warner apparently excelled any that had been made elsewhere. The close of the 1847 harvest brought the end of McCormick's contract with the Chicago firm except for the $2,500 in royalties which Gray and Warner owed McCormick. Pleased with the quality of the Chicago product, McCormick entered into a partnership with Gray as a means to continue the satisfactory production of his reaper as well as to discharge the debt. The agreement of partnership called for Gray to build and equip a new reaper factory and to purchase necessary materials to produce reapers. In addition, Gray was responsible for the factory's management, including its accounting, for which he received $1,000 annual salary. McCormick handled the marketing end of the business and dealt with patent litigation. He also received a royalty of $30 per machine and $2 per day salary. The partners evenly divided the profits of McCormick & Gray. The new factory (40 × 100 feet, two stories high, with a ten-horsepower steam engine) opened in time to produce five hundred reapers for the 1848 harvest. By the end of the harvest, Gray and McCormick had begun to feud, at first privately and then in court, about financial matters.[8]

In October 1848 McCormick, trapped by Gray's dealings and pressed financially, was forced to form a new partnership with William B. Ogden and William E. Jones under the banner McCormick, Ogden & Co. The company set out to build fifteen hundred reapers for the 1849 harvest in the year-old factory that Gray had set up. The profits and fees of $65,000 which McCormick collected from the sales of the fewer than fifteen hundred machines actually produced provided him with the means to buy out Ogden and Jones. McCormick immediately formed another partnership with Orloff M. Dorman, who contributed some $12,000 to the new firm, C. H. McCormick & Co. Although no hostilities arose, McCormick ended this partnership after a year and struck out on his own.[9]

Cyrus was alone only in a legal sense because in 1848 he had wooed Leander to Chicago to supervise the production of his reaper. The reluctant brother William also finally agreed to move to Chicago in 1849. Cyrus directed the company and pocketed its impressive profits while Leander provided technical know-how and William conducted the day-to-day business.

From the beginning of his undertaking until about 1847, Cyrus McCormick had experienced problems in the production of his patented reaper. Because the family blacksmith shop in Virginia had been unable to make more than fifty machines at most (and those of dubious quality), McCormick had contracted with other makers in Virginia, New York, Ohio, Missouri, and Illinois to build his machines. Under this arrangement, there was no chance for uniformity in the shape, materials, or workmanship of the Virginia reaper either between or within the various shops. By 1851, when he had acquired complete control of the company he had created in Chicago and when the outstanding contracts giving others the right to make the reaper had expired, Cyrus finally gained an opportunity to standardize his product. Only manufacture at one factory would ensure a degree of uniformity in shape, materials, finish, and workmanship. (See Figure 4.1.)

The factory Charles Gray had built and equipped for making McCormick's reaper caught the eye of a reporter for the *Chicago Weekly Democrat*. In the three-story main building a steam engine drove "some fourteen or fifteen machines: viz. a planing machine, two circular saws, a tenent saw, a lathe for turning handles of rakes, pitchforks, etc.: also two lathes for turning iron, . . . two morticing [sic] machines and two grindstones." A blacksmith shop attached to the main building contained ten forges, and there were plans to expand this part of the facility "as it is at present too contracted for the wants of the factory." Altogether the reaper works employed thirty-three men, ten of whom were blacksmiths.[10]

By September 1849, when McCormick purchased Ogden's and Jones's shares of the partnership, the factory had been lengthened to 190 feet and contained three planers, six saws, two wood lathes, seven metal lathes, three boring machines, and sixteen blacksmith forges heated by a single blower. A thirty-horsepower engine drove the machinery and about 120 men worked at the factory.[11] Fire in late March 1851 destroyed part of the McCormick works. When rebuilt, the new four-story factory contained machinery "of the latest design."[12] William wrote a friend that "more than 1000 [reaping] machines nearly ready for shipment very narrowly escaped" the fire. Despite damages, he anticipated that the factory would finish between 1,400 and 1,500 reapers. According to the company's records, however, only 1,004 machines were produced that year. Nonetheless, the McCormick reaper works greatly impressed the "noble citizens" (William's words) of Chicago. The *Daily Democrat* claimed near the end of 1851 that this plant was the largest reaper works in the world.[13]

When McCormick began to build reapers at Chicago—and indeed throughout much of the period to 1890—he relied upon specialty contractors to supply him with several parts. From Aldrich & White of Fitchburg, Massachusetts, McCormick purchased sickles at $1.30 each, made "of the best quality of iron and with full complement of best double shear steel on the edge." A firm in Elizabethport, New Jersey, sold McCormick malleable iron finger bars in which the sickle ran. Thomas Sherry of Chicago provided "all of the castings and cast iron for the manufacture" of the machines. Under contract with McCormick, Elihu Granger made the patterns for Sherry's foundry. In all of these contracts, Cyrus specified that he expected a high grade of workmanship and quality of material.[14]

The Chicago reaper factory brought these various contracted pieces together, drilled and machined some of them, riveted some parts together, welded others, filed and fitted others, and combined these finished metal pieces with the wooden frame that was made by the company's carpenters. Initially Cyrus even contracted to have the machines painted "according to the directions of said McCormicks Agents." His contract with two painters closely resembled those of the inside contracting system that has been discussed before. This was the only such contract because McCormick hired departmental foremen at an annual salary and directly controlled all of the factory employees. Because of the predominance of castings in the metal parts and the specifications given contractors for malleable and wrought parts, McCormick's reapers were roughly uniform with others made for the same harvest but not necessarily with machines made in other years. During

M'CORMICK'S PATENT VIRGINIA REAPER.

R.N. WHITE — ENGLISH PRAIRIE ILL.

CHICAGO,   1851.

Though the farmer may put his Reaper together by the above cut, I give the following

DIRECTIONS:

I. Lap the in hound and cross piece which has two holes for the axle, for high or low stubble. Lower hole placed on the axle for higher stubble.

II. Bolt the angular board (S, in the cut) marked thus ———, to its place with the 8 inch bolt in the back end, which rests close to the platform.

III. Put the axle in place.

IV. Place the wheel frame on the other end of the axle *lapping* the finger beam above or below—suiting the *higher* or *lower* cut at Q.

V. Put on the main brace D, marked thus =, one side of driver's seat on same bolt.

VI. Separator W, (marked O) to its corresponding mark, there is a little wooden pin to be removed to give place to a bolt.

VII. Dividing iron marked thus ✕. If the reel be required *very* low, this iron is taken off, for small wheat.

VIII. Side board (with brace nailed to it) marked U, to its place, removing a small block to give place to it.

IX. Two small posts erected and cloth bar on top marked T.

X. Reel bearer V.

XI. Reel shaft Z to its place, arms put in with blocks (on each end) *forward* as the reel turns, and after braces are in and tightened by a cross *pin* in the one with long tenon (there is one brace for each end with long tenon) the boards are then nailed on the blocks. See d in cut. The boards have to be strained into a shape that will fit, as the arms are not parallel. Each of these boards having a block on the end (and the end without the block is next the wheels) put on, so as to pass ¾ of an inch from the angular board.

XII. If fingers do not fit right, they can be knocked out with a punch, and with the wedges trained right with very little trouble.

XIII. The small ground wheel has a third and lower sett by bolting the block on the *upper* instead of the *lower* side, and the side next the horses may be raised or lowered some two inches by changing the position of the tongue which may be done if required by boring another hole in the tongue.

XIV. The square washer may be placed on either end of the axle to gear deeper or shallower and other washers of leather can be added.

XV. Put in the driver (9 or 12 for lower or higher stubble) and bolt on the guard (small cast-iron piece) to keep it in place, see that the guard does not tighten on the driver. The block *in* the driver to be tightened on the crank, when it wears.

Bolts having beviled heads belong to corresponding places, heads of tongue bolts to the left. The 8-inch bolt (in the box) goes through the back end of angle board (S S in the cut), cross piece and in hound C. The 5½ bolt through cross piece and in hound near the same place. The two 7-inch bolts through in hound and cross piece, near the axle. *Open seds of small ground whut outward.*

OPERATION.

1. Run the Reaper on hard ground, and see that it works free, as it may be tight at first. Work the reel as high as the smallest grain will admit of, as it runs lighter to be higher. The band will stretch at first, and must be taken up, and the end kept tied, or it will fasten, and be injured in the cogs.

2. Grind sickle with a *short* bevil, on the smooth side, when required, (but not often). To be a little broken will not injure it, but may be evidence of good temper.

3. The above cut shows *exactly* how the raking should be done. The Raker can have a little intermission between sheaves, and then with *a very quick sweep,* (catching by the heads) pull the sheaves round, leaving the butts next the grain.

Keep the nutts (washers only put on where bolts are changed, because not so liable to loose nutts) and keys tight, and oil well.

When iron and wood work together, very little oil is best, and I will add, that long experience has convinced me that the main axles of a Reaper are better to run in the wood, (having slow motion) than metal—nothing to get loose, and if *ever* they wear to require it, it will only then be necessary to put boxes in, and the axles will be perfect.

By the Shipper, the cogs may be geared deeper or shallower, or put entirely out of gear when not cutting.

The strip on the angle board may be raised or lowered to suit the reel.

As the Reaper is not wholly put together at the factory, in some cases very slight trimming at the laps may be found necessary.

If the band flies, it will be owing to its stretching, or the shaft not level, or the pulley not being fair.

C. H. M'CORMICK, Patentee.

FIGURE 4.1.   Cyrus McCormick's Instructions for Reaper Assembly, 1851. A major proponent of advertising, Cyrus McCormick used similar illustrations for handbills and newspaper advertisements to show farmers what his patented Virginia Reaper looked like. (McCormick Collection, State Historical Society of Wisconsin.)

the late 1840s and throughout the 1850s, the McCormicks changed virtually every part of their reaper from year to year. In his biography of McCormick, William T. Hutchinson detailed these changes. He noted that "scarcely a single element was left untouched. . . . Although no basic changes were made, the alterations made annually are evidence of the never-ending experimental work in progress and of the presence of competition."[15]

Although market-oriented Cyrus McCormick considered annual changes—and sometimes changes within a model year—important, they extracted certain costs. Perhaps the most important was the actual expense of production. In 1849, for instance, McCormick estimated his manufacturing cost at $55 per machine.[16] The company devoted so much time and effort to perfecting its new model machine each year that it usually got into production too late to satisfy all the demand. As a result, the annual output was lower than it might have been, and agents and farmers were often disillusioned with the company's performance. Another problem bred by annual changes was in the supply of repair parts, or "repairs" as they were called in the nineteenth century. Very early in the company's history, McCormick set up a repair department which maintained a duplicate of every year's model and the patterns that formed its basis. When a customer needed a part, he had to tell the company which one he wanted and the year the machine was made. The head of the repair department took the appropriate pattern to the foundry, which made the casting.[17] Until McCormick adopted a complete set of gauges—and there is no evidence that he did so until 1880—the farmer assumed the responsibility of fitting the new casting or wrought-iron part into his machine. Agents usually kept a large stock of parts to avoid delays in obtaining them from the factory. Apparently the company sometimes accidentally sold pieces for repairs that had been used as patterns, thus forcing it to find an old machine to use as a pattern. Leander wrote Cyrus in 1859 about such a search: "I have been engaged in having all the old parts of different years build of Mach[ine]s hunted up and putting as far as there is near enough for a whole mach[ine] finishing & making such complete." But McCormick did not supply any wooden parts for his machine. When these were requested, the company advised the inquirer to find a local carpenter.[18]

McCormick's policy of annual changes, which he believed were dictated by the market, was supported by the factory's production system. Because no special-purpose machine tools were used and general machine tools included mainly drill presses and metal lathes, changes did not necessitate scrapping costly tools. Presses to punch out sickle plates, to make cold iron nuts (or rivets), and to shear bar stock easily accommodated changes in design. A heavy dependence on blacksmiths who riveted and welded pieces together and who fashioned the few wrought-iron pieces not contracted out to specialty shops also made it easy to make major or minor changes in models. Alterations in woodwork were wrought easily, especially since all of McCormick's woodworkers were skilled carpenters.[19]

Management of McCormick's reaper factory throughout the 1850s, therefore, centered in three areas: deciding what to produce; supplying parts contractors with correct patterns and purchasing an adequate supply of materials; and directing factory workers. Leander carried out most of the experimental work and, with his brothers, settled on the final model. He also managed the factory's workers. William negotiated materials purchases. Cyrus, although he solicited advice, always made the final decision on how many machines the company would make. With this approach to production, the factory turned out reapers (or combined reapers and mowers) whose quality varied annually with the nature of design. Agents often complained about the workmanship of the McCormick reaper; one from New York wrote that the New York State products excelled those of Chicago in workmanship and noted that "$5 expended in workmanship extra would save twice the amount in repairs" and double sales.[20]

McCormick's factory made slightly more than one thousand reapers per year from 1851 to 1853. Sales increased in 1854 when more than fifteen hundred machines were produced. Suddenly, in 1856, the factory experienced great pressure to turn out four thousand machines. (See Table 4.1.) The McCormick family's response to this unprecedented jump in demand provides us with an idea of how the three brothers viewed large-scale production and sales. In the midst of the harvest of 1856, when the factory always felt pressed to get out as many machines as possible, Cyrus wrote a customer who wanted some spare finger bars: "My fingers are made at five different establishments in N. York and N. Jersey, and come along slow. I cannot finish over 40 Machs. a day [even] if they were on hand." Evidently Cyrus considered moving his factory so that output could be expanded to more than forty machines per day. His brother William protested, saying, "I *consider* it out of the question. I wouldn't and neither would you want to bother with this business after a few years when every county is filled up with it [reapers]." A few days later William told Cyrus that despite large sales in 1856 he considered "the propriety of [manufacturing] over 2000 machines for [18]57 doubtful." Thinking further about the matter, he sought to convince Cyrus that "2000 is a *big business* & if they can surely be made & sold need not care for more."[21] Another fire after the production season in 1856 allowed Cyrus to rebuild a larger factory although it seemed to increase William's apprehensions about big business.[22] Cyrus finally urged his brother to calm down: "Don't be scared man!"[23] When considering making changes in reapers for 1857, however, the inventor insisted that "*caution* is *essential* in every particular in changing from what is well."[24]

Cyrus was obviously pleased by the 1856 design and advocated few changes except a heavier reliance on malleable castings rather than the more brittle cast iron. Despite the reaper works' relatively large output, McCormick noted that his "machines *generally* went together well this year I hear, except that all the sickle butts were tight in casting, giving much trouble." Everything seemed promising for 1857, and Cyrus sought to make and sell another four thousand reapers. He wrote his brother that "we must drive & *do more . . . to sell*" and asked him to purchase stock for four thousand machines. By November, shortly before the factory would begin production for the harvest of 1857, William was able to "predict a good time for at least another four years."[25] Now, he, too, wished to make four thousand reapers.

Leander's reaction to increased production of the Virginia reaper is more difficult to establish. While William pointed out that producing "4000 machines makes great work," the youngest McCormick never complained or expressed apprehension about his job. Apparently Leander ran the factory effectively during the 1856 and 1857 harvests. In 1857 William proudly reported to Cyrus that "we have never had the work so far forward" during the height of the rush to produce reapers. Yet Cyrus worried about Leander's performance as superintendent. He urged his brother to show up promptly every day at 7:00 A.M. and to maintain regular hours of work. Shortly before the factory was to begin production for the harvest of 1859 Cyrus finally asked William whether Leander "attend[ed] to business in a *spirited* manner." Obviously concerned about Leander's ability to get out more than forty-five hundred reapers, the inventor even proposed hiring Charles Gray, his former partner with whom he had earlier fought about financial matters, to help Leander prepare for the company's growing business.[26]

That McCormick could consider bringing back Gray, against whom he still harbored

TABLE 4.1.   MCCORMICK MACHINES BUILT, 1841–1885

Year	Hand Rakers	Self-Rakers	Mowers	Droppers	Harvesters	Binders	Total
1841	2						2
1842	7						7
1843	29						29
1844	50						50
1845	50						50
1846	190						190
1847	450						450
1848	700						700
1849	1,494						1,494
1850	1,603						1,603
1851	1,004						1,004
1852	1,011						1,011
1853	1,101						1,101
1854	1,558						1,558
1855	2,534						2,534
1856	4,076						4,076
1857	4,065						4,065
1858	4,565						4,565
1859	5,118						5,118
1860	4,083						4,083
1861	5,491						5,491
1862	4,965	203					5,168
1863	2,259	2,053					4,312
1864	2,027	4,063					6,090
1865		2,503	1,283				3,786
1866		2,519	5,004				7,523
1867		3,998	3,800				7,798
1868		3,522	5,377	609			9,508
1869		6,932	2,494				9,426
1870		7,032	2,001				9,033
1871		8,497	1,500				9,997
1872		2,996					2,996
1873		8,747	1,278				10,025
1874		7,229	2,044	841			10,114
1875		3,338	2,501	632	5,005		11,476
1876		5,987	3,623	852	3,497	64	14,023
1877		1,000	1,146	354	3,053	1,040	6,593
1878		1,822	2,513	776	6,391	6,316	17,818
1879		1,129	3,165	862	7,798	5,806	18,760
1880		2,499	6,098	507	7,205	5,246	21,555
1881		2,513	9,474	1,020	8,618	9,168	30,793
1882		2,739	15,040	1,514	13,210	14,180	46,683
1883		4,255	14,347	552	14,045	14,821	48,020
1884		3,703	13,697	681	18,128	18,632	54,841
1885		2,221	14,436	1,152	15,565	15,528	48,902
TOTAL	47,432	91,500	110,821	10,352	102,515	90,801	453,421

Source: ''McCormick Machines Built since 1841,'' McCormick Estate Papers, M/I, Box 18.

grievances, indicates the depth of his concern about Leander's apparent nonchalance. He never hired Gray, however, and Leander continued to handle production without problems as long as output remained around five thousand reapers per year. Yet once Cyrus began to envision producing ten, twenty, or even forty thousand machines annually Leander protested.

Both William and Leander lacked Cyrus's optimism for prospective reaper sales and his flair for business. When in 1857 early sales lagged and it appeared to William that only about three thousand reapers would be sold rather than the four thousand machines under construction, the two younger brothers panicked. Rather than despair, Cyrus prescribed what he considered would be the best remedy: "I think advertising *extensively* . . . in many . . . papers . . . may be important." He arranged with James Campbell, one of his eastern agents who often purchased materials for the factory, to travel around the country "stirring up" (Cyrus's words) agents and writing advertisements for suitable newspapers. Rather than having a thousand machines left over after the harvest, McCormick's marketing techniques provided sales for all but 154 reapers, which he probably sold the next harvest at a significant discount—perhaps only slightly above his cost of $46.41 per machine without overhead expenses.[27]

Cyrus's marketing abilities notwithstanding, his desire to produce more machines generally met with opposition from his brothers. Paradoxically, after McCormick lost his battle for his second patent extension, William wrote him, "I [have] often said your money has been made not out of your patents but by making and selling the machines." William also thought "more attention should be bestowed upon improvements from year to year."[28] Despite William's recognition that Cyrus had already earned a fortune, he would nevertheless continue to consider five thousand machines per year an "enormous" business that should satisfy Cyrus. Cyrus McCormick must have been well aware of his brothers' business philosophy but believed he could show them differently because in 1859 he took both of them into his business as partners.[29]

McCormick's agreement with his brothers did not constitute a true partnership because neither William nor Leander had enough money to purchase a share in the factory. Under the name C. H. McCormick & Bros., Cyrus agreed to share one-fourth of the profits with each of his brothers, to supply the firm capital at 8 percent, and to equip the factory at cost (but to charge the company $10,000 rent for his factory). In addition to profits, William and Leander each received an annual salary of $5,000. The brothers made this agreement for twelve years. By 1864, however, they had found it unsatisfactory. At this time, Leander and William each purchased a one-fourth interest in the capital of the company. After William's death in 1865, Cyrus and Leander reached a new agreement.[30] (See Figure 4.2.)

The nominal partnership of 1859 fortunately occasioned a thorough inventory of the McCormick factory tools and machinery, part of which remained in the hands of Cyrus McCormick and another part was charged to C. H. McCormick & Bros. This inventory shows the extent of technological development at the reaper works.[31] In the blacksmith shop the company owned the following tools: 123 blacksmith tongs, 19 sledges, 30 hand hammers, 16 cold chisels, 33 punches, 70 section chisels, 31 chipping chisels, 55 swedges, 10 anvils, 4,447 pounds of cast-iron forms for old model reapers, 837 pounds of cast-iron forms still used, and miscellaneous (shovels, rakes, tempering cans, bellows, and so on). Cyrus maintained the following equipment used in the blacksmith shop: 8 blacksmith forges, 8 bolt machines with two sets of dies each (each machine and set of dies valued at $12), and 1 bolt-cutting machine with 3 sets of dies ($50). These bolt

FIGURE 4.2.   McCormick Reaper Factory, Chicago, ca. 1860. This drawing was made about the time Cyrus McCormick brought his younger brothers William and Leander into formal partnership in his reaper business. (From William T. Hutchinson, *Cyrus Hall McCormick: Seed-Time, 1809–1856.*)

machines were probably rudimentary devices for rotating either cutting dies or the bolt shank itself. Eugene Ferguson has suggested that the machines were placed in the black-smith shop because the shanks required heads to be put on, a process carried out by the skilled blacksmiths who worked there.

In the sickle room, where the individual sickle knives were riveted onto the sickle bar, Cyrus owned a riveting machine (valued at $5), a shearing machine to trim the sickle knives ($25), and three drilling machines for boring holes either in the knives, the sickle bar, or both ($25 each). The riveting machine was probably an elementary device, judging from its assessed price of $5 (even if it had been amortized for several years). Here the company owned two sickle gauges—probably simple devices to check for the correct placement of holes in the knives—and three blocks for straightening sickle bars after all knives had been riveted onto the bar.

Some of the heavier turning and drilling operations were done in the so-called engine room, where holes were drilled and reamed in the wheels. Cyrus kept the following machines: 1 key sear [seat?] lathe ($100), 1 upright key sear [seat?] lathe ($25), 1 large upright drill for wheels ($250), 4 horizontal drills ($50 each), 1 large horizontal drill ($50), 6 turning lathes ($175), and 1 turning lathe ($50). C. H. McCormick & Bros. maintained the tools that were used in these machines as well as other tools: 29 dogs, 21 turning tools, 8 scrapers, 3 reamers, 5 wrenches, 1 pair of tongs, 13 chisels, 13 punches, 6 hammers, and 1¼ dozen files. In addition, three different kinds of gauges appear in the inventory of the engine room tools. Sixteen gauges for a wheel (valued at $8 for the lot), nineteen crank gauges (total value of $9.50), and eight gauges for the main wheel (altogether valued at $8) were listed. I have been unable to find any substantive informa-tion on these gauges or even to ascertain whether they were identical gauges or sets of gauges to check, for example, nineteen critical points on a crank. The assessment of value of fifty cents and one dollar per gauge, however, raises serious questions about their quality. Traditionally the preparation of gauges used in the armory version of the American system of manufacturers constituted one of the major expenses of production hardware. Nevertheless, whatever their quality or exact use, McCormick's factory was using gauges for three components as early as 1859.

There is no evidence that in subsequent years Leander expanded the use of gauges in reaper production. Evidence exists that he continued to rely upon skilled machinists to fashion heads on bolts whose threads had been cut by relatively simple devices. From the inventory of the McCormick finishing shops we may also conclude that a significant amount of handwork went into completing the Virginia reapers.

C. H. McCormick & Bros. listed the tools in its finishing shop in 1859: 9 large drills, 1 mach. punch, 92 boring drills, 67 hand drills, 98 cold chisels, 12 finishing hammers, 5 finger sets, 22 turning lathe tools, 10 hand lathe tools, 12 reamers, 12 lathe drills, 25 lathe drills, 33 chassers, 6½ dozen large files, 7½ dozen small files, 15 iron vises, 3 hand screw plates and dies, and 81 [either pounds or actual numbers] dies and taps plus a large list of miscellaneous dies, taps, punching and other tools. The inventory also lists 2,465 pounds of used files. Cyrus owned a number of machine tools used in the finishing department: 5 upright drills (McCormick factory make at $100 each), 3 upright drills (eastern make at $100 each), 1 big press with 11 shears and 6 dies ($400), 1 big press with 19 dies and 34 old dies ($275), 1 big press with 1 set dies ($275), 1 big press with 1 set dies ($75), 1 iron planer ($175), 1 large turning lathe ($750), 1 turning lathe ($175), 1 turning lathe ($75), and 1 turning lathe ($25). Prices for machinery and tools in the inventory probably varied with the quality and size of the tools and also with the number of years they had been owned by the company or by Cyrus.

The McCormick factory operated five woodworking shops in which Cyrus owned extensive equipment. This included: 1 wood lathe, 1 cross-cut saw, 1 Woodworth planer, 5 circular saws and frames, 2 circular saw frames, 2 Daniels planers, 3 horizontal boring machines, 1 upright mortising machine, 1 compress saw & frame, 1 gaining machine with 6 heads, 2 upright boring machines, 1 mortising and boring machine, 2 chamfering machines, and, 1 rake hand lathe [i.e., Blanchard-type lathe]. A variety of hand vises, saw blades, and small tools complemented the wood shops' machinery.

By 1859 the McCormick works had opened its own foundry. Its equipment seems typical of any small foundry. Leander's reliance on cast iron and cast malleable iron dictated against the use of drop-forging equipment, one of the hallmarks of New England armory practice. Not until the early 1880s is there mention of drop-forging at the McCormick works.

To sum up, the factory inventory of 1859 delineates a manufactory not unlike a large general machine or jobbing shop. The McCormicks relied almost wholly on skilled men and general machine tools rather than special-purpose machine tools for the fabrication of their products. There is no comparison between McCormick's machinery and that, say, purchased in 1854 by the Anderson committee for the Enfield Armoury.[32] McCormick's machinery is entirely different from the highly specialized machine tools that characterized New England armories and the American system of manufacturing. Virtually absent is a rigorous gauging system whereby every critical dimension is checked. Completely absent is a rational jig and fixture system.

Year in and year out the McCormick reaper works made reapers without ever establishing a model in the New England armory sense of a paragon or an "ideal form" machine from which all gauges, jigs, fixtures, and cutters were designed. With armory practice, a factory's mission was to turn out production machines that emulated or conformed exactly to the model. With McCormick, there was no model or ideal form, and what it called a model was an ephemeral machine, a marketing rather than a manufacturing concept.

The American Civil War, traditionally seen as a great boon to reaper sales, brought hardship to the McCormicks, at least in their own eyes.[33] Sales vacillated throughout the war. (See Table 4.1.) Following the lead of other reaper manufactures, McCormick introduced more complex self-raking reapers in 1862 (when the factory built 203 self-rakers) and within three years completely phased out hand rakers. (See Figure 4.3.) Difficulties at the reaper works centered about designing and producing the self-raking machine and, more important, procuring skilled machinists, carpenters, blacksmiths, and molders. The company also faced frequent strikes because these workers realized that they could make continued wage demands. These labor problems did not, however, induce Leander to make significant changes in the production process at the works either during or after the Civil War.[34]

Manufacture for the harvest of 1862 caused few notable problems. Yet in preparing for the next harvest season, William and Leander had begun to feel the pressures and uncertainties of the war. William recognized the increasing scarcity of hands, both in Chicago and on farms. The former would hamper production possibilities at the factory, and the latter would probably "make a small number wanted." On the other hand, William reasoned, a shortage of farm hands might induce farmers to purchase the new self-raking reaper, which eliminated the need for a laborer to rake the cut grain off the reaper's platform. William noted that despite labor shortages, the factory could *"easily"* turn out

FIGURE 4.3. McCormick Self-Raking Reaper, 1862. Known as the Reliable, this machine was introduced in a minor way in 1862 and continued to be manufactured until about 1871. (International Harvester Corporate Archives.)

four thousand machines and asked Cyrus to determine how many should be hand and how many should be self-raking reapers.[35]

Much to William's chagrin, Cyrus had gone to Europe to market the company's products. William even accused his brother of "flee[ing] away from this land of blood & death—where we are trodden by abolitionism in the *North*—without liberty of speech—& with utter ruin in the south." William's charges merit consideration; Cyrus probably found it easier to live in Europe than to be torn directly between continuing his prosperous business in the North or giving up all for the support of his South. Cyrus had even lured Leander to Europe to help set up machines and work with an English contractor on production of the McCormick reaper. Leander stayed in Europe only a short time, however, and would later write Cyrus that he considered his trip "one of the blunders of my life."[36]

Strikes in the foundry in September 1862 forewarned of walkouts during the rush to produce machines for the 1863 harvest. Upon his return to Chicago, Leander found all work behind schedule and doubted the possibility of producing four thousand machines. As pressure began to build during the 1863 production season, William began to panic. He warned Cyrus that molders, carpenters, and machinists were about to strike for higher wages and that the factory desperately needed finishers but could not get them. Despite production of only four thousand reapers, he sought to convince his brother that the business and its concomitant problems were "immense."[37]

Leander believed that had he remained in Europe an additional month, the factory would have manufactured few self-raking reapers. Late in the production season, he had had to fashion new patterns, first of wood and then of iron, for the new machine. In addition, he told Cyrus that the factory had had to build "new iron machinery" to make some of the parts of the self-rakers. Leander also noted for the first time in his correspondence that the factory was constructing fixtures for machining some of the parts on the self-raker.[38] Cyrus had written from Europe several times requesting changes in the design of the machine. Each time Leander capitulated but finally protested, noting that it was "impossible to count the cost or trouble of such changes." He had employed extra men at higher than normal pay to press the work forward, and still the factory was late in its production. William added that "strikes have prevailed—men go off to escape [the] D[ra]ft. . . . Workmen such as we most needed are independent." With what work force remained, the factory operated during evenings. Leander also rented some extra machinery. Both brothers in Chicago clearly recognized that the self-raking reaper demanded more material and better workmanship than the old hand raker. William claimed it doubled the work. He later wrote Cyrus that farmers were willing to pay cash for the McCormick machines, but the factory simply could not build them.[39]

When the annual postharvest tranquility began to set in at the reaper works, William took time to reflect on the hectic production season and its problems. He saw a clear need for additional machinery, but more important was an early decision on the final design of the machine for the next harvest. "Experimental" work had severely delayed initiation of production. The company had almost always dragged out its experimental work too long but had usually recovered in time to produce the number of machines Cyrus had requested. One of the unique problems of 1863, wrote one of the clerks in William's office, was "green and obstreperous" hands. The factory had depended on skilled workers, and the Civil War had depleted the supply of this class of labor. In addition, the final design of the McCormick self-raker for 1863 had not compared favorably to that of other manufacturers. McCormick had never succeeded in perfecting an effective combined mower and reaper—nor had any other maker. Yet others had given up and introduced an efficient, single-purpose mower. Now the company faced designing such a machine which would not infringe on others' patents. William suggested to Cyrus that the company should change its policy of carrying out all development in-house. "I believe," he wrote, "we should look to combine good points in other machines with ours & to *buying* a good light mower. *We* should rely on our manufacturing facilities more than upon invention."[40] No doubt William realized—at least subconsciously—that the war made it difficult for the company to rely solely upon its manufacturing facilities, particularly because full utilization of the factory depended upon an adequate supply of skilled molders, machinists, and carpenters.

Leander wrote his brother that despite the late, small production for 1863 he saw no reason why five or even six thousand reapers could not be made for the 1864 harvest. He pointed out that he needed to buy machinery to replace what he had rented during the spring rush. The self-raker's complexity and design required more machining of parts and demanded that the factory prepare more carefully to produce them. William also advocated manufacturing six thousand reapers (two-thirds to be self-raking machines), provided that adequate preparation was made.[41] He complained to Cyrus that the inventor wanted to make too many changes and that the European business had placed too many demands on the Chicago factory.[42] "Let us not be hindered about these [little] changes in this country," William exclaimed and then asked rhetorically, "had you not better *for* Europe manufacture *in* Europe?" He concluded his lecture by reiteration: "I repeat that if we go to tinkering with alterations for this country we shall get behind & fail again."[43]

To prepare adequately for the 1864 harvest, Leander and William negotiated additions to the factory, the largest being the foundry to accommodate the increase in castings required by the self-raker. After Leander purchased additional lathes and mortising machines, he thought the only remaining problem was the great scarcity of labor and its consequent high price. He noted in a letter to Cyrus that "good men [are] very scarce [with the] government offering them from 2 to $3 [per day] for all sorts [of] mechanecks [sic]." Soon he would be writing about a molders' strike, about "troubles in getting work done," and about having "to pay extravigant [sic] prices and hire extra machinery & very many things not worthwhile to enumerate."[44]

Both Leander and William continued to complain to their brother that his European business hampered Chicago factory operations, particularly since he had demanded certain changes from the American production machine, thus requiring different patterns and machining. In December 1863 or January 1864, Cyrus, apparently unshaken by his brothers' objections, asked them to prepare early three hundred self-rakers for Europe. Friction arose over this request. Leander enumerated the changes in woodwork, castings, and machining necessary to make the European machine and tried to explain the resulting interruption of the flow of work at the Chicago factory, which he regarded as a waste of time. William echoed this attitude, calling Cyrus's request "a *serious* drawback to the entire business here . . . where *labour* has the power & is on strike half the time. Our men refused to work the other day under a new foreman & we had to withdraw him." He added that "employing the right men . . . is a most troublesome business." In late February, William wrote Cyrus that the factory had completed seventy-five European reapers and would make no more. William sent patterns for these machines and admonished Cyrus to find a maker in Europe. Reluctantly, Cyrus yielded to his brothers' demands. Before the end of the American harvest of 1864, he returned to the war-torn United States. William wrote at the end of the year, when preparations were under way for the next harvest, that after they refused to make any more European reapers, construction of the six thousand machines for 1864 had gone "right through without a halt" and consequently had cost less than he had anticipated.[45]

Production for the 1865 harvest dropped off significantly from the six thousand of 1864 to fewer than four thousand machines. Perhaps one reason for the decline was the introduction of a mower and the cessation of the hand-raking reaper, which had been produced in various forms since 1841. But in 1866 an unprecedented seventy-five hundred machines were built, including five thousand mowers. The end of the Civil War obviously served as a boon to McCormick's ability to produce such a large quantity of reapers and mowers. The war's end was also followed by William S. McCormick's death. For many years William had suffered from an apparent nervous disorder, no doubt exacerbated by the long hours of work and anxiety of running the reaper business as well as managing the entire McCormick holdings in real estate and stock. Through the years, William had occasionally tried to restrain Cyrus's enthusiasm to produce ever more machines. Perhaps William's death in September 1865 eliminated one hindrance to larger annual production and played a role in the jump of output by almost four thousand machines.

The years between the end of the Civil War and the Chicago fire of October 1871 constitute one of the most interesting periods in the history of the McCormick reaper factory under the direction of Leander McCormick. During this time tension grew between Cyrus and his brother over the perennial issues of how many machines to produce each season, what changes to make from year to year, and what price to charge. These issues define the basic elements of a business philosophy or a production psychology—they must be settled before one can begin to talk about mass production. Cyrus's intimate involvement with the affairs of the company for the first time since establishing his business in Chicago brought those deep-seated issues to the surface.

In the fall of 1865, when the sting of William's death dominated Leander's correspondence, he wrote Cyrus about what changes were necessary for the 1866 production period. An increase in output demanded acquisition of new machine tools, and Leander speculated that he would have to go to New York to buy them. He also advocated construction of a new building at the factory because "the work is greatly retarded by the miserable cramped arrangement of work."[46] Although a new building could not be built for several months, Leander bought six new engine lathes and fitted them into an already overcrowded factory.[47] Few serious problems arose for Leander in getting out seventy-five hundred mowers and reapers for 1866. No longer affected by William's fear that reaper sales would suddenly end, one of the principal clerks in the McCormick offices wrote that despite many years in business and the current production of seventy-five hundred machines, "we know of no spot that is supplied fully with machines, though in some co[untie]s in Ill[inoi]s we have sold from 40 to 50 every year since we began." He added that the McCormicks were no longer, as they once had been, "simple enough to suppose we should overstock the country in less than seven years."[48]

Rather than fearing overproduction, the company wondered whether it could manufacture more than seventy-five hundred machines per year. Additions to the factory, a new steam engine, and a new cupola in the foundry allowed an increase in production to almost eight thousand for 1867 and ninety-five hundred for 1868. (See Figure 4.4.) Charles Spring, who took on almost all of William McCormick's responsibilities, wrote Cyrus that these additions would pay for themselves in one year.[49]

After the harvest of 1867 Leander announced the *"complete success"* of the McCormick mowers and reapers made that year. To Cyrus he expressed his belief that "we can

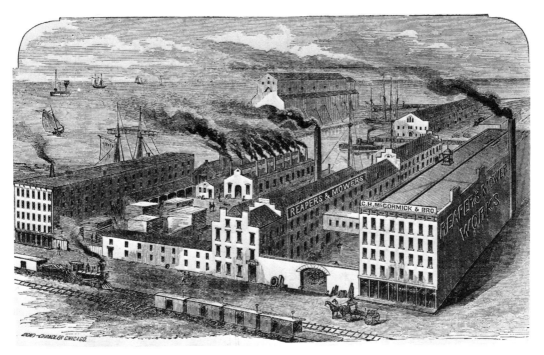

FIGURE 4.4.   C. H. McCormick & Bro. Factory, North Side Chicago, 1868. ("McCormick's Prize Harvester," 1868, McCormick Collection, State Historical Society of Wisconsin.)

sell all that we can make," yet he warned that few would be made if Cyrus kept insisting on "mixing in a new and untried machine." Leander rejected Cyrus's desire for expansion not only because he thought an unproved reaper might jeopardize the standing of the company but also because he had begun to build a production system that could no longer easily accommodate changes. He explained: "We have had much of our work largely under way knowing the difficulty of being late as heretofore and all of our machinery is fixed and made to suit this, & much of it not suited to any other work & would require much alteration, all of our patterns, flasks and *men*, are trained to work at this particular work." Maintaining that the mower need not be changed but merely refined, Leander described for Cyrus how such refinements were being made: "We are taking much paines [sic] to fit up all parts patterns & we have gearing from cut patterns, which will make all work better. . . . With the new cut gear & newly fitted up patterns our machine will run 20 pr. cent lighter than heretofore."[50]

Unfortunately, it is difficult to determine exactly what Leander meant when he stated that all of the factory's machinery had been fixed and was not suitable for any other work. There is no hint during previous years that the factory built or operated to any great extent special-purpose machine tools. All evidence suggests the contrary, that the factory operated a large number of engine lathes and other general machine tools such as planers and presses. Leander must have been alluding to the fixtures and dies used in these machines rather than any highly specialized or single-purpose machine tools. His ambiguous statement that his men had been trained to make one particular product model may suggest that he had instituted a more rigorous division of labor system at the factory. In any case it indicates that the factory depended on workers who, though not so highly skilled that they could easily accommodate design changes, had acquired limited skill through a repetitious learning process. Leander had not totally done away with skilled workers but had boxed himself in by not eliminating them from the production process.[51]

Cyrus McCormick must have yielded to Leander's plea not to change the design of the self-rakers and mowers. For two years the factory turned out the same product (although it attempted to add a new machine to the line in 1868—the "dropper"). But sales declined dramatically in the harvest of 1869; as Leander cried, "We will have *hundreds of them left over.*" Looking back on what went wrong, the vacationing Leander emphasized to Cyrus that "something *new is a popular word.*"[52] The old Reliable, as the self-raker had been dubbed, had outlived its time. Manufacture of a new, improved machine with a lighter draft—the Advance—now seemed necessary to Leander. (See Figures 4.5 and 4.6.) Cyrus felt otherwise and requested that one thousand Reliables be built for the harvest of 1870. Leander, still "satisfied that our old machine (Reliable) has had its day and is *out of date,*" expressed his opinion to Cyrus that "it was wrong to build [a single] one of the old Style but [because the inventor has insisted] we have 1,000 of them on the way." Uncharacteristically, the younger McCormick boasted that the company could sell ten thousand Advances alone and perhaps even more.[53]

Leander's confidence about how many machines might be sold carried over into his attitude toward production. He contemplated taking a trip to Pittsburgh to examine bolt-heading machinery and even thought of traveling to Providence, Rhode Island, to see similar machinery. Considering replacing blacksmiths by these bolt headers, Leander pointed out to Cyrus that these machines had been employed successfully in many large factories. They would cost somewhere between three and five thousand dollars. To complement this machinery, Leander also wanted to buy a press for punching nuts and washers.[54] Although he eventually purchased bolt-headed machinery, there is no evidence that Leander went to Providence or to any other machine tool center in New England. Had he gone, he might have radically changed the manufacturing operations at the McCormick factory. Instead of Providence—or even Pittsburgh—Leander toured establishments in western New York.

After the production season for the 1870 harvest but before the factory geared up for 1871 Leander finally took a trip in search of machine tools. The effect of his visits to several establishments in the area of Utica and Buffalo was profound. Writing to his brother he exclaimed, "I have been in Utica Steam Cotton Mills, and really it is worth a trip from Even Chicago to see the rattle & buz of Machinery, such perfection of work, *perfect* order, Every part performing *its* part of the work. *You ought by all means to stop . . .* [to] *see it.*" Although textile manufacture had nothing in common with reaper production, the order and perfect coordination of the mill offered Leander a sight that he had never experienced and an insight he would never fully grasp. Leander saw other factories, which he did not identify. Writing about one of these he noted, "I know of nothing that is equal to it." In Buffalo, he saw bolt-heading machinery and made arrangements to purchase two machines (at $2,000 each) and one for round heads (at $500).[55]

Returning to Chicago, Leander began to direct the factory's preparations for the 1871 production season. This harvest would see the introduction of a new grain shipper arrangement on the McCormick self-raking machine, a change most pleasing to Leander. The shipper's production presented him with a challenge. In mid-September he advised Cyrus that for production of the shipper the factory needed two lathes, "which we are obliged to build as there are none such to be bought." With them he anticipated that one man would be able to perform the work formerly done by two on regular lathes. Even with

FIGURE 4.5.   McCormick Advance Self-Raking Reaper, 1869. A lighter-weight machine than the Reliable, the Advance was introduced in 1869. McCormick attempted to sell it as a combination reaper and mower, but for mowing, McCormick's Prize mower (See Figure 4.6) was far cheaper and more effective. (International Harvester Corporate Archives.)

FIGURE 4.6.   McCormick Prize Mower, 1869. (International Harvester Corporate Archives.)

more productive tools, Leander anticipated that the cost of the shipper would exceed $100 and could reach upward to $125. According to him, the high setup costs drove up the price of the shipper.[56]

With this new machinery, both purchased and made in the works, the reaper factory turned out an unprecedented 9,997 machines for the 1871 harvest, just three short of the absolute limit upon which Leander had insisted in August 1870. Even before the harvest of 1871 had begun, the production and sales picture looked so promising to Cyrus that he again began to toy with the idea of moving the entire factory to the southern outskirts of Chicago, a move he had considered for two years. Pleased with the factory's current performance, Leander voted against such a move.[57] For one brother good times meant expansion, for the other, satisfaction. Leander's objections notwithstanding, Cyrus purchased property for a new factory. He told Leander that he bought property for buildings "so large as I think they should be—to admit of expanding business." Reflecting further, he wrote, "I may just add that with the new factory with ample room and first rate in all respects would with our means & influence & advantage might yet be making some other leading machinery."[58] Cyrus thought not only about increasing reaper and mower production but also diversifying his product line, perhaps adding threshers to complement the reaper line. This diversification did not take place until after Cyrus's death, but in expansive moments he always considered it.

Cyrus's plans for a new reaper factory were temporarily halted by the report of very poor sales during the 1871 harvest. In mid-July Leander predicted that seven thousand or even eight thousand new and old model machines would be left over. Calculating necessary production for 1872, he could not see manufacturing more than four or five thousand machines. Moreover, even though the design experts at the factory made some important changes in the newer machine, Leander believed that because of the large leftover stock of 1871 machines, "we cant [sic] expose to the public this last improvement, as it w[oul]d injure so large an amt of old work." A month later, Leander again surveyed the situation and lashed out against his brother. "As you well know the prospects in the business were never perhaps as flattering as at that time [March 1871]," he wrote Cyrus. "You and Mr. Spring were figuring on 20,000 machines, you must have new buildings put under way *at once,* while I did all I could in every way *to avoid* it, as you know. It took the hardest sort of pleading to keep you out of it." Expressing pride about his conservatism, Leander concluded, "I mention all of this to call your attention to the fact that all was success & the more work the more money, was the idea. We did not build [a new factory or twenty thousand machines] and well that it is so."[59]

Leander also lamented the increased production cost of reapers and mowers, yet because of the poor sales the company faced reducing the selling price by about 25 percent. Charles Spring, who had become the general manager of the business after William's death, advocated selling the leftover machines at $140 each, a figure Leander believed came close to the actual factory cost. Furthermore, Spring advised Cyrus to produce another large amount and sell them at slightly above cost in order to crush competition, which he believed to be the source of McCormick's poor sales. Leander would hear of no such thing. He complained to Cyrus that "we have the most expensive machine in the country to build, & yet must undersell others & break them down to build up ourselves. . . . I have no opinion of any breakneck arrangement of 14,000 or 20,000 machines."[60] He agreed with Spring, however, in calling for a price reduction.

Cyrus McCormick's response to Leander's survey of the company's situation at the end of the 1871 harvest touched off a continued discussion about the basic philosophy of the business. The founder stated explicitly and categorically for the first time that although he was sixty-two years old, he did not wish to continue reaper manufacture on the same small scale as in past years. Leander said that in principle he agreed but the fact that seven or eight thousand machines were left over from 1871 and four thousand from 1870 dictated against expanding or even maintaining production. Regarding the company's pricing policy, Leander wrote that "it is thought so here, as I understand it, that the price of our machine was so high that others came in & took the field with a lower priced machine." Whatever the price of the machine, he argued, "I have not favored so large an increase of business. If we have built only 8,000 this year it would have been very much better for us. I have had none of the 14000 or 20000 fever myself. Our machine is very complicated having a *great many parts* and *none* of which you can dispense with and accomplish what we do. It has worked well & *stands well* is the universal declaration with the new working arrangement."[61] It is difficult to determine Leander's thinking about how quantity would affect the cost at the factory. From the context of his letter to Cyrus, two arguments could be suggested: (1) Leander implied that increased production actually raised costs; (2) he implied that even if production were greatly increased, the factory cost would remain the same. In any case, there is no clearly articulated notion of economies of scale in Leander's correspondence with Cyrus McCormick.

These issues of how many machines to produce and how much to charge were eliminated by the Chicago fire of October 1871, which destroyed the McCormick reaper works along with two thousand machines stored there and $200,000 worth of materials that had been gathered for 1872 production.[62] Leander's concern about seven or eight thousand leftover machines suddenly turned into a sense of relief because most of these machines had been shipped to agents throughout the United States, where they could be sold for the 1872 harvest. The fire served as a blessing of sorts for Cyrus because now he could rebuild the works where and how he wished without Leander's protestations that the North Side factory was good—and big—enough. Blessings notwithstanding, the fire did cause a great loss to C. H. McCormick & Bro. Charles Spring estimated it as follows:

on stock	$193,157
on machinery	35,057
on reaper	181,695
on buildings	609,000
on real estate	360,475
	$1,379,384[63]

Insurance covered slightly over a quarter of a million dollars of the loss. Setting up temporary shops at the North Side factory site and repairing the engine and machine tools that were repairable, Leander and his men built almost three thousand Advances for 1872 while at the same time planning the new South Side works at the corner of Western and Blue Island avenues. The construction of the new reaper works offered the McCormicks a fresh start, an opportunity to adopt new production technology. As it turned out, Cyrus and Leander lost that opportunity.

Hardly before the remains from the fire had cooled, machinists began reconditioning the machine tools that had not been completely ruined while patternmakers set about making new patterns for the current model reaper and mower. Working through office clerks, Leander began a search for new lathes, gear-cutting machines, and other machine tools. Leander's procurement of these tools offers an interesting commentary on his awareness (or lack of it) of standard practices in buying machine tools. Two examples will suffice. For cutting teeth on the geared part of the reaper wheels, Leander contracted with Lucius W. Pond of Worcester, Massachusetts. When in March 1872 the factory received the machine, Leander expressed his surprise that the fixture for holding the wheels in the cutter had not been sent with the machine. After firing off an inquiry about the fate of this fixture, he received the following reply from Lucius Pond: "We were considerably surprised at your supposing we were making one of them for you as it is not customary to furnish such fixtures any more than it is to furnish mandrills and cutters, and we have no recollection of ever having furnished . . . one [for anybody]." Before contracting with Pond, Leander had corresponded with William Sellers of Philadelphia, asking if he could supply the McCormick factory with a gear-cutting machine that would cut not only internal gears on reaper wheels 26½ inches in diameter but also bevel and spur gears. Sellers wrote back that he knew of no such universal gear-cutting machine. As a second instance, Leander ordered some presses and their dies for punch work from the Stiles & Parker Press Co. of Middletown, Connecticut, one of the leading press and punch tool-makers in the United States. Upon receipt of the tools, the McCormick factory protested not only the price of the dies but also the charge for the bed plates of the presses. The proprietor of this Yankee firm, N. C. Stiles, replied that he had been in business for fifteen years and had worked at the bench for many years before that. Moreover, he was thoroughly familiar with the way "Eastern shops" conducted their business. Never before had he heard of anyone expecting to receive a bed plate without extra charge. "Suppose your job required 50 bed pieces [i.e., 50 different operations]," Stiles asked. "Should we have furnished them without charge[?]" Stiles added that McCormick's offer to pay $200 for the tools in question was insulting because he had $448.60 tied up in them.[64]

These episodes—only two among many—betray a profound ignorance of the way things were done in the New England machine tool industry. Leander presumed that fixtures were supplied with the machine tools he ordered, although he had never specified that he wanted them when ordering the tools. He had merely told the tool builders what he wanted to use the machine tools for (cutting internal gear teeth on the inside of a 26½-inch wheel, for example). This assumption seems just as naive today as it did to the toolmakers in question a hundred years ago. The new McCormick reaper works would continue the basic practices of manufacture upon which it had relied for many years, practices outside the New England tradition of manufacture. Leander equipped the factory with dozens of new belt-fed engine lathes, overlooking the advantages offered by turret lathes and milling machines, both of which were easy to acquire in the machine tool market by 1872.[65]

The rush to produce reapers at the old factory site while also building and equipping a new factory took its toll on the company. Charles Spring wrote Cyrus McCormick that they were "losing more ground than we can regain in years." Leander worked throughout the summer of 1872 on the new factory as well as gathering an adequate supply of repair parts lost in the fire to serve as patterns. While Cyrus seemed always to remain in New York on business, Leander dealt with such problems as making arrangements for the construction of workers' houses near the new factory because none was available in the area.[66]

During this time the brothers continued to negotiate matters of partnership—interest held by each, rental of the factory, ownership of tools, salaries, and responsibilities. Leander wanted to bring his oldest son, Robert Hall McCormick, into the business both as a shareholder and as assistant superintendent. Cyrus consented to Hall's participation in the company as well as to selling his own machinery to the new company, styled C. H. & L. J. McCormick & Co. Yet the brothers had not finalized a written agreement. Pressures of opening the new factory in 1873 (when more than ten thousand machines were produced) as well as these negotiations taxed Leander's nerves. Finally in mid-May he informed Cyrus that he would not continue in the business unless a written agreement were made; he even offered to sell his interest in the new factory. After additional months of negotiations with his brother, Leander laid down the terms on which he would continue: "In the first place I cant [sic] carry more than one sixth [interest] of the whole business under any circumstances, and secondly, I must have a salary of not less than $10,000 covering the past year for my service. *The number of machines to be limited to 10,000 and no other kind of machinery to be built without the consent of parties.*[67] Once again, Leander sought to restrain Cyrus's growing desire for increasing the scale of his business as well as diversifying its products.

Cyrus responded at length to Leander's conditions, noting, "You ask a good deal of me." Leander had specified verbally that any written agreement should be binding for only one year. Cyrus singled out this point and wrote his brother that for the past two years of negotiations they had been talking about a five-year arrangement. He emphasized again that only through a longer contract and vigorous prosecution of the business could they make reaper manufacture *"a grand success."* Responding to another of Leander's verbal demands—that Cyrus should be more active in the business than hitherto—the inventor pointed out that he had kept "at bay" the lawsuits that continually threatened to interrupt the operations of the company. He would continue to work on legal matters, but he could not give "much of my time to looking after the manufacturing or general interests of the business."[68]

Finally, in November 1873, the brothers began to feud and to abuse each other verbally. After an encounter with Leander, Cyrus wrote his brother that "I can not permit you to come to *my room* to continue and perpetuate your calumnious & bullying abuse of an 'old scoundrel' and 'old scamp,' as you have at different times termed me, while I told you that nothing your mouth could utter or fists gesticulate could induce me (at my age!) to foist myself by a personal encounter with you." He demanded that Leander withdraw certain charges and accusations and also admonished his brother to pay closer attention to business. In particular, Cyrus had discovered that the factory was "paying too much for labor all through."[69] Leander finally yielded to Cyrus's wishes, and the brothers declared a temporary cessation of hostilities although not a peace. During the next five years their relationship would grow stormier until finally Cyrus decided that his business would prosper better without Leander. Events during the period leading up to this decision show how little attention Leander paid to the operations at the factory and, consequently, how its production processes remained virtually static until Cyrus fired him as superintendent.

Although the company's advertising literature claimed that the capacity of McCormick's new reaper works exceeded that of the old by three to four times,[70] output during the first few years of its operation hardly exceeded that before the fire. The last full year of Leander's tenure, when he was only nominally the superintendent, the factory produced fewer than nineteen thousand machines, albeit of five different varieties. Production in 1877 waned to a twelve-year low (excepting the year of the fire), largely because a large number of machines were left over from the previous season, when an unprecedented fourteen thousand machines were made, but also partially because of renewed dissatisfaction on Leander's part.[71] Cyrus had hired a patent expert (who also deemed himself an expert mechanic) to buy patents for improvements on reapers and mowers, to try to use the patent system as a means to gain strategic advantage over competition, and to maintain an adequate perspective of the performance of all the reaper companies in the field. The expert, Charles Colahan, toured the McCormick works near the end of the 1877 production season. Immediately he saw problems, particularly with the manufacture of a self-binding reaper, introduced in small numbers the previous season. (See Figure 4.7.) The binding mechanism of this machine required greater precision of machining than other parts of the McCormick reaper. Writing to Cyrus, he stressed that "you must inaugurate more system, order & discipline [at the factory], in order to secure perfect construction previous to shipping; this would save thousands—tens of thousands of dollars."[72] Additional correspondence from Colahan indicates that Cyrus took his criticism of the factory seriously and undertook measures to impress upon Leander the need for more order and discipline.[73]

Leander must have heeded Cyrus's admonitions. The following year the factory turned out a record eighteen thousand machines at an average cost of $28.88 each compared to the sixty-five hundred of the previous year, each of which cost $55.02.[74] Although this was the largest increase in output and the most impressive decrease in unit cost in the factory's history, Charles Colahan still found cause for complaint. He wrote Cyrus that "while at Chicago I examined the Binder in process of construction . . . with [the] view [of] anticipating expert services in the field. . . . I found some work slighted, or roughly done, and called Supts. office, and Mr. Baker's attention to it." Colahan noted, "I fear my 'meddling' incurred the displeasure of those who should (in my opinion) appreciate my intentions better—and I am at a loss to understand why Mr. L. J. McCormick was so unwilling to hear me say anything."[75] These charges against Leander came at a time when he had begun to launch an attack on his older brother. Leander had written and

FIGURE 4.7.   McCormick Harvester and Wire Binder, 1876. Although introduced in a limited way in 1876, the wire binder was first manufactured in quantities by McCormick in 1877 (1,000) and 1878 (6,000). Eventually, in 1881, McCormick introduced the twine binder (See Figure 4.8). (International Harvester Corporate Archives.)

circulated a document in which he claimed that Cyrus's *father* Robert was the true inventor of the reaper and that Cyrus had merely patented those ideas. In addition, Leander had initiated negotiations with other reaper manufacturers to sell them manufacturing rights to a patent he owned but which he planned to use against his own company.[76]

Correspondence from Leander indicates that he had become bitterly jealous of the hero-worship accorded his brother. Cyrus McCormick's name had become synonymous with reaper manufacture in the public's mind, while Leander, who perhaps rightly believed that Cyrus would have been nothing without the help of his father, William, and himself, had remained anonymous. William Hutchinson, in his biography of Cyrus, attributes the major cause of the many years of dispute between the two brothers to intense jealousy on Leander's part.[77]

By the end of the 1879 production season, when the factory turned out almost nineteen thousand reaping and mowing machines (including almost six thousand binders), Leander was waging open warfare against Cyrus. He bluntly stated that he planned to open his own manufacturing plant and actually contracted with at least one foreman at the reaper works to start work for him in September 1880.[78] Cyrus, now seventy years old and increasingly burdened by arthritis, sought to put the McCormick company on a more permanent basis than the partnership, whose formal agreement of 1874 had expired. Apparently still convinced that Leander contributed positively to the company's manufacturing operations and unwilling to purchase Leander's share in the partnership (although he could have easily done so), Cyrus negotiated with his brother to incorporate the business at a capital of $2.5 million. The inventor subscribed to three-fourths of the stock while Leander and his son Hall took the other quarter. The articles of incorporation called for Cyrus to be president of the company and Leander to be vice-president. In addition, they specified that Leander should be elected to the office of superintendent of the manufacturing department for a five-year term. Guaranteeing a salary of $7,500 per year, the articles also stipulated that Leander "shall give reasonable attention to his said office" and provided that Hall McCormick could serve as assistant superintendent with a $4,000 salary.[79] Cyrus's hope that such an arrangement would bring an end to Leander's hostility soon evaporated.

The McCormick Harvesting Machine Company Board of Directors met for the first time in September 1879 and ratified the articles of association.[80] Cyrus McCormick distributed stock to three of his longtime employees, Charles Spring, William Hanna, and James Whedon, an act that welded their loyalty to him even more firmly than before. As directors rather than tattletale functionaries these men could exert more influence on the company—and on Cyrus—than before. In the months subsequent to the first board meeting (at which Leander was formally elected superintendent of the mechanical department), the period traditionally devoted to intense preparations for the coming production season, these men became increasingly aware of Leander's continued absence from the factory. Charles Spring wrote Cyrus in early October that the superintendent had "been at the factory very little lately" and a few days later wrote again that in three weeks Leander had not spent ten hours at the works. Two weeks later, he advised the aging inventor: "Leander came home day before yesterday. He takes no interest in matters at the factory, will get out of the business as soon as he can. (tells this to Whedon). It will never run right at the factory until they [Leander and Hall] are out of it."[81]

James Whedon, who had been elected secretary of the company, gave Cyrus McCormick, Jr., more specific details of Leander's performance: "You will please to say to your father that I was at the factory on Saturday . . . and that I hear on *all sides* reports of the disinterestedness of both Hall and Leander in the business. . . . Baker [one of the foremen] says that neither of them help him or advise with him. Voice [another foreman] tells the same story. Mr. L. J. is not there but *very* little & *then* only to pour his woes into the ears of any who will listen to him. I am told that he does not pretend to direct and superintend things saying 'any Employee has as *much authority as I.* ''[82]

Cyrus and his son, who had taken over many responsibilities from his father, faced dealing not only with Leander's failure to fulfill the obligations of his office but also with his resistance—both as a director and as superintendent—to increasing the output of the works. For 1880, Leander wanted to produce only fifteen thousand machines, a decrease of more than twenty-five hundred from the previous year. Yet Cyrus's expert, Colahan, had predicted an "unequalled" demand for machines in the coming harvest. He encouraged Cyrus to make sure that adequate preparations were made. Moreover, Colahan advocated the introduction of additional styles of reaping and harvesting equipment. Colahan's advice, coupled with reports prepared by Charles Spring about the number of sales the company had lost the previous harvest through its inability to build enough machines, convinced Cyrus that well over twenty thousand machines should be made for 1880. Leander resisted, not because he thought they could not sell that many but because he did not see how the factory could produce so many. Charles Colahan disagreed. As he wrote Cyrus McCormick, "Manufacturing is comparatively an easy matter when the form of construction is decided in time to perfect preparations [in order] to secure perfect work; and we usually *loose* [sic] *the first year* on account of want of energy *no[t] time.* ''[83]

By February 1880 Cyrus McCormick had decided that he could no longer tolerate Leander's and Hall's abuses of their elected offices as well as their refusal to assign the company several patents which they held. He submitted a statement to the Board of Directors about the poor management of the factory and the patent abuses.[84] Touching off a debate that ran for two months, Cyrus's statement eventually led Robert Hall McCormick to resign as assistant superintendent. The responses by Hall and Leander, coupled with Leander's notification to the board that he would be in Europe for six months from mid-April (a critical period at the factory), finally brought Cyrus to propose to the board that "the position of Superintendent of the manufacturing department, occupied by L. J. McCormick, be declared vacated.''[85]

On April 14, 1880, the board ratified Cyrus's motion. At the following meeting, May 4, 1880, it resolved that Lewis Wilkinson be appointed superintendent of the manufacturing department.

By firing Leander and hiring Lewis Wilkinson, Cyrus McCormick initiated one of the most important changes in his business since he had originally decided to settle in Chicago in 1848. This event brought an end to the blacksmith–machine shop–carpenter shop approach to reaper manufacture and signaled the beginning of production under the American system of manufactures. As noted in Chapter 2 in the discussion of technical choice in the sewing machine industry, the person in charge of manufacturing largely determined the production system. Leander J. McCormick in 1880 was still very much the Virginia blacksmith he had been when he first arrived in Chicago in 1848, and the factory's operations reflected his background. I have been unable to document a single instance when Leander made any substantial contact with the New England system of manufactures delineated in Chapters 1 and 2. Not until Lewis Wilkinson took the superintendent's job did McCormick begin to adopt to any great extent the system used in New England armories and other manufactories. Leander's firing also meant the end of a conservative production policy, which in part had been dictated by the system Leander had set up. By 1884, the year of Cyrus's death, the company was making more than fifty thousand agricultural implements (including binders), and under Cyrus McCormick, Jr.'s, leadership the works was turning out more than one hundred thousand implements by 1889.[86] Cyrus's long-held wish to manufacture reapers on a grand scale came true only after he fired Leander.

Cyrus McCormick, Charles Spring, and Cyrus McCormick, Jr., hired Lewis Wilkinson because of his wide background in manufacturing. According to Cyrus, Jr., Wilkinson had worked at the Colt armory in Hartford, the Connecticut Firearms Company, and the Wilson Sewing Machine Company, as well as other manufacturing firms.[87] (The Wilson Sewing Machine Company of Wallingford, Connecticut, had, at its height in the 1870s, made slightly more than twenty thousand machines per year.) Wilkinson immediately began operating the factory at night so that enough machines could be produced for the harvest that was almost upon them. Cyrus McCormick, Jr., served as assistant superintendent at the factory, and the young Princetonian carefully studied Wilkinson's work.[88] His father noted that Cyrus had "applied himself indefatigably to the knowledge and development of the most important part of our business in the office, *at the Works* & in the field." Looking back at the 1881 production season and viewing the intense preparations being made for a record production of more than thirty thousand machines in 1881, Cyrus noted Wilkinson's seven-month performance: "The present Supt. at the Works has also filled his position with great success, having accomplished many improvements in the special departments there thus facilitating to a large degree the extra production which the increase in the business, has necessarily demanded."[89] (See Figure 4.8.)

FIGURE 4.8.   McCormick Harvester and Twine Binder, ca. 1881. McCormick adopted twine for the binder, which required a different tying or binding mechanism than one for wire. As with all McCormick machines, the early harvester and twine binder was made principally of a wooden framework. (International Harvester Corporate Archives.)

Beginning his tenure late in the production season, Wilkinson could not effect many dramatic changes in the factory operations, but he tried to bring more order and discipline to the work and workmen, which Leander had not done for several years. Not until the fall of 1880, normally the period when design changes were made in machines and the factory geared up to make them, did Wilkinson's rich experience in Yankee armory practice begin to show up. For the first time in the company's records the words "gauges," "pattern machines," and "jigs," consistently appear. Cyrus, Jr., recorded these words in his pocket diary as new concepts rather than old hardware.[90] And Charles Spring's correspondence showed a sudden awareness that the design and preparation of patterns and gauges required several months of work.[91] The pattern machine was no longer a field-tested experimental machine that served as a guide for manufacture; it became the basis of the entire production system. The best machinists at the factory carefully filed and scraped critical surfaces so that the model fitted together as perfectly as possible. From this paragon machine, jigs, fixtures, and gauges were made so that production machines would be made like it.[92]

Although Lewis Wilkinson remained with the McCormick company for only one year, he exerted a decided influence on the factory and its personnel, particularly on Cyrus, Jr., who unofficially succeeded him as superintendent. Under Wilkinson's tutelage, the young McCormick learned the rudiments of American armory practice. In addition to grasping the importance of models, jigs, fixtures, and gauges, Cyrus, Jr., seems to have understood the possibilities of special-purpose, or single-purpose, machine tools for large-scale manufacture. He specifically mentioned such tools in an article describing the McCormick reaper works for *Scientific American* in 1881: "For example, the introduction of the inclosed gear frames for reapers, mowers, and droppers necessitated a machine which could bore all the holes required for shafts, etc., at one operation, and several of these are in use."[93] (See Figure 4.9.) During his many years as superintendent, Leander McCormick had designed a limited amount of special-purpose machinery. Yet during the next two decades, under Cyrus, Jr.'s, direction, the company relied more and more upon special machinery. In 1900, for example, the company exhibited eleven different special-purpose tools of its own design at the International Exhibition in Paris.[94]

Cyrus McCormick, Jr., directed additions to the factory during the summer and early fall of 1881 after Wilkinson left. In September, before preparation got under way for the 1882 season, Cyrus, Jr., announced that he would be the new superintendent and that George Averill, a longtime foreman of the McCormick foundry, would be his assistant. With the additional space and a "new regime" (Cyrus, Jr.'s, words), the works produced more than forty-six thousand machines in 1882, an increase of over 50 percent from 1881.[95] (See Table 4.1 for a breakdown of these implements.) Charles Spring commented on the factory operations soon after Cyrus, Jr., had taken over: "I think the works as now organized is in better working trim than it has ever been before since my connexion with the business and I am confident it will show good results at the end of 1882." As production got under way, Spring delineated for the younger McCormick what quantity production meant on a day-to-day basis: "We must make 200 machs a day every day to get the 40000 out by harvest, and we must ship 12 carloads a day from Jan 1st to get them all shipped."[96]

At a board meeting Leander McCormick and his son Hall, who were still directors despite their severance from the company as employees, criticized such a production as "wild and visionary" and voted against producing so many machines. Cyrus, Sr., on the other hand, believed that with forty thousand machines, "the sky seems bright and

FIGURE 4.9.   Setting up Finger Bars, McCormick Factory, 1881. Despite the examples of special-purpose machinery cited by Cyrus McCormick, Jr., in 1881, hand methods such as these were still a significant aspect of McCormick's production. Under the younger McCormick's "new regime" in the 1880s, the factory became increasingly mechanized. (*Scientific American*, May 14, 1881. Eleutherian Mills Historical Library.)

brightening.'' The old inventor's skies brightened even more when his son reported on August 1, 1882, that the total output had exceeded forty-six thousand machines rather than the planned forty thousand and that all of these machines had been sold. Optimism reigned at the factory for the first time in its stormy history. Charles Spring succinctly summed up the situation: "Since the joint Stock Company was formed we have improved the business in a great many respects, but in no place is it more patent than at the works. I believe we could build 60,000 machines [a year] now."[97]

The grand old man of the reaper industry, suffering greatly from arthritis, finally took an opportunity to see the happy workings of his own factory at the height of production for 1883. Cyrus, Jr., recorded the visit in his diary: "Mr. Averill and I rolled him all over the shops and he spoke to [the foremen] Buly; Mr. Hamilton; Voice; Haskins; Schaffer; Pridmore; Wood; Mr. Bishop; and old Barney in the foundry who has been with us for 33 years. Father examined the working of a center draught mower in the yard. . . . He enjoyed the trip very much and it was of the greatest interest to the men at the works to see him—the head of the business, and the founder of our great industry."[98]

Yet the battle for production at the McCormick works was never completely won. It had to be won yearly—perhaps monthly or daily. Only six weeks after his father's death on May 13, 1884, Cyrus McCormick, Jr., recorded in his small pocket diary that "this is the worst day we have known in the business. Agts. writing that they can't get enough machines. . . . Great flood of orders and no machines." The young president had discovered, in this year when almost fifty-five thousand machines were made, something that an office clerk had noted almost twenty years earlier: "It seems that the more we sell the more we can sell."[99] (See Figures 4.10 to 4.15.)

Cyrus McCormick, Jr., as president of the company, continued to take an interest in the factory and its ability to produce machines on a large scale. By 1902 it had become public knowledge that under his leadership, the company had increased its production fivefold.[100] Now the McCormick reaper works produced machines "on a grand scale," a scale that his father had once made a desideratum for remaining in business. Only as an old man, however, when he had fired Leander as superintendent and would no longer listen to his conservative advice did McCormick's company begin to move toward mass production. Without question, there were certain costs to this expansion and the institution of a new system of manufacture. Labor problems ranked foremost. The net result of the changes initiated by Wilkinson and pursued vigorously by the younger McCormick was that the company gained increasing control of work processes from skilled workmen. These workmen, many of whom had spent their careers at the McCormick factory, did not see the "grand object" of the armory system in the same way that the Ordnance Department officers had done when they relentlessly pursued the system in the antebellum period. Consequently, when the McCormick molders went out on strike in 1885, they were joined by machinists and other metalworkers who had traditionally remained at the bench when the molders struck. This series of changes in approach to production between 1880 and 1884, which offered a sharp break with the past, played a fundamental role in the "prelude to Haymarket."[101] The hardware, processes, and customs of the armory system provided powerful instruments of control over labor. When they were installed suddenly, such as at McCormick, strife was bound to result.

The story of the manufacture of the McCormick reaper offers an interesting parallel to that of the Singer Manufacturing Company. Both were leaders in their respective industries and for a long time manufactured products with what B. F. Spalding called a

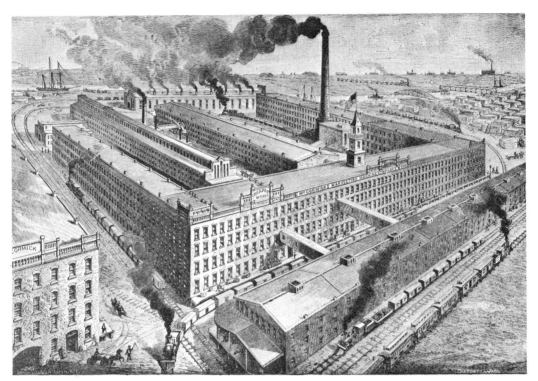

FIGURE 4.10. McCormick Factory, 1885. (*Illustrated Annual Catalogue, McCormick Machines, McCormick Harvesting Machine Company*, 1885, McCormick Collection, State Historical Society of Wisconsin.)

FIGURE 4.11. McCormick Foundry, 1885. The company claimed that this was the "largest molding floor in the world." (*Illustrated Annual Catalogue, McCormick Machines, McCormick Harvesting Machine Company*, 1885, McCormick Collection, State Historical Society of Wisconsin.)

FIGURE 4.12. Lathe and Press Room of the McCormick Factory, 1885. Note the clutter of the aisles and the materials handling methods. (*Illustrated Annual Catalogue, McCormick Machines, McCormick Harvesting Machine Company,* 1885, McCormick Collection, State Historical Society of Wisconsin.)

FIGURE 4.13. Shafting Room, McCormick Factory, 1885. Judging from this and many other illustrations, McCormick had not acquired by 1885 any of the automatic screw machines that other factories, such as that of Singer Manufacturing Company, had become dependent upon. Also, the workmen filing parts at the vises along the windows suggest that as late as 1885 precision manufacture had not become predominant at McCormick. (*Illustrated Annual Catalogue, McCormick Machines, McCormick Harvesting Machine Company,* 1885, McCormick Collection, State Historical Society of Wisconsin.)

FIGURE 4.14. "Setting Up" or Assembling Mowers, McCormick Factory, 1885. (*Illustrated Annual Catalogue, McCormick Machines, McCormick Harvesting Machine Company*, 1885, McCormick Collection, State Historical Society of Wisconsin.)

European approach. The heads of both companies placed more emphasis on sales and marketing than on manufacturing. When they did take interest in factory operations it was usually in response to negative reports about the quality of their machines or an inability to produce enough of them. In both cases company leaders insisted that their products be of the highest quality, and the craft approach to manufacture provided this quality. When the European approach could no longer meet the challenge of quantity production, Singer and McCormick looked to the American system of manufactures.

Yet the two histories also provide sharp contrasts. At Singer, the factory produced the same model sewing machine year in and year out. There was no season of production, no annual model change, no question of how many machines to produce. Rather than how many to produce, Singer was always burdened by the question of how to produce enough machines. The McCormicks—at least William and Leander—worried continuously about their business getting too big, about producing too many machines. Their fears were probably exacerbated by the phenomenon of the "annual model change." This expression must be used cautiously, for unlike the annual model change in the automotive industry, McCormick and the rest of the reaper industry often made changes for reasons of genuine improvement (such as the self-raker, the wire binder, the cord binder, and the harvester) rather than simply for the sake of change. Yet McCormick also made changes for no real reason. Farmers came to expect changes from year to year. Nevertheless, McCormick never developed an explicit, self-conscious marketing strategy based on yearly models.

The historian must often untangle cause and effect. Early in the reaper's history, real improvements took place from year to year, and perhaps these changes set a pattern for

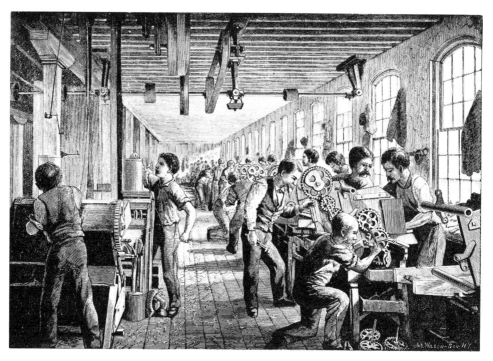

FIGURE 4.15. "Fitting Up Binders," McCormick Factory, 1885. The binder was the most complicated mechanical part of the McCormick harvester. Introduced in 1876, it commanded the greatest
precision in production and quality of materials. "Fitting" was the correct term for the assembly of
binders, as evidenced in this illustration. (*Illustrated Annual Catalogue, McCormick Machines,
McCormick Harvesting Machine Company*, 1885, McCormick Collection, State Historical Society
of Wisconsin.)

both manufacturer and farmer. The industry was drawn into a habitual cycle of change
which, at least in McCormick's case, ruled out any major development in manufacturing
technology. One can speculate that had McCormick adopted the American system of
manufactures early in his business—say, 1851 when this is generally attributed to him—
he might have broken this cycle by producing a large number of cheap(er) machines and,
like Singer, marketed them over a long period of time. In fact, however, his company
continued to follow the vicissitudes of the entire industry. McCormick's history provides
an early example of the "productive dilemma" discussed by William Abernathy, which
forces a manufacturer to choose between change and lower productivity on the one hand
and sameness and higher productivity on the other.[102]

Although it is tempting to draw parallels between annual model changes at McCormick
and those at General Motors, this is probably not advisable. The reason is that between the
histories of these two companies the Ford Motor Company arose. Ford pursued specialization in production to its logical conclusion—gross overproduction—and, in a sense,
created for the first time a genuine necessity for the annual model change as a marketing
strategy. Until Ford, complete saturation of the world market by a single product had
remained only a theoretical concept. The Ford Motor Company achieved unprecedented
productivity partly because of its innovation of the assembly line. The importance of this

innovation is underlined in the following chapter. Although the bicycle industry in the late nineteenth century brought to perfection the American system of manufactures it was unable to solve the problem of finishing and assembly of parts. The Ford assembly line (Chapter 6) overcame this problem, creating in turn problems of its own.

From the American System toward Mass Production: The Bicycle Industry in the Nineteenth Century

Another event having an effect on the designing and manufacturing of machinery entirely unlooked for at the time of its inception was the manufacture of the bicycle. This event brought out the capabilities of the American mechanic as nothing else had ever done. It demonstrated to the world that he and his kind were capable of designing and making special machinery, tools, fixtures, and devices for economic manufacturing in a manner truly marvelous; and has led to the installation of the interchangeable system of manufacture in a thousand and one shops where it was formerly thought to be impractical.
—Joseph Woodworth, *American Tool Making and Interchangeable Manufacturing* (1907)

I never saw a job where the Western man would not beat the Eastern man out every time. You [Eastern men] know too much about tool making, and not enough about making money.
—A Chicago mechanic in *American Mechanist* (1895).

By 1880, when Singer's and McCormick's sales had begun to increase dramatically, forcing them to find new ways to manufacture their products, many sewing machine manufacturers and agricultural implement makers lost their share of the market. Some went out of business, but others turned to a different line of manufacturing—notably to the bicycle—in order to use their manufacturing plant. This transition took place particularly in the 1890s during what contemporaries called the "bicycle craze." Beginning in the late 1880s, this craze reached a feverish pitch by 1895–96 and then collapsed entirely in 1897.

Although the bicycle craze was short-lived, the bicycle and its manufacturing industry exerted a profound influence in America during a period of rapid social and technological change.[1] No better testimony exists about the influence of the bicycle than that of W. J. McGee, one of the era's major figures in American science. In an article titled "Fifty Years of American Science" in the *Atlantic Monthly* in 1898, McGee argued:

A typical American device is the bicycle. Invented in France, it long remained a toy or a vain luxury. Redevised in this country, it inspired inventors and captivated manufacturers, and native genius made it a practical machine for the multitude; now its users number millions, and it is sold in every country. Typical, too, is the bicycle in its effect on national character. It first aroused invention, next stimulated commerce, and then developed individuality, judgment, and prompt decision on the part of its users more rapidly and completely than any other device; for although association with machines of any kind (absolutely straightforward and honest as they are all) develops character, the bicycle is the easy leader of other machines in shaping the mind of its rider, and transforming itself and its rider into a single thing. Better than other results is this: that the bicycle has broken the barrier of pernicious differentiation of the sexes and rent the bonds of fashion, and is daily impressing Spartan strength and grace, and more than Spartan intelligence, on the mothers of coming generations. So, weighed by its effect on body and mind as well as on material progress, this device must be classed as one of the world's great inventions.[2]

In speaking of the bicycle as a progressive technology, McGee overlooked the importance of the bicycle industry in refining the system of production begun in armories in the antebellum period. This refinement of the armory system of manufacture took place almost entirely in New England within firms squarely based in the armsmaking tradition. In this chapter, the history of the leading bicycle manufacturer in this tradition, the Pope Manufacturing Company, will be discussed. Yet bicycle makers outside the armory tradition and located primarily in the Midwest developed an important new method of metalworking. This technique—stamping or pressing—has played a fundamental role in mass production industries of the twentieth century. Although the history of stamping is obscure, the work of the leading bicycle manufacturer that employed this technique, the Western Wheel Works, will also be treated in this chapter.

Taken together, refined armory practice and well-developed stamping techniques provided the technical basis for automobile manufacturing in the early twentieth century. In this sense, the bicycle industry was transitional. Yet the industry proved to be transitional in other ways as well. Despite the refinement of ''old'' manufacturing technology and the introduction of new techniques, the bicycle industry merely exposed and did not solve a fundamental problem in the production of complex consumer durables: assembly. Solution to this problem awaited the Ford Motor Company in the second decade of the twentieth century. The bicycle itself was a transitional technology, for it led many an American—and not a few bicycle mechanics—to a contemplate and to project a faster, more powerful, and less fatiguing form of personal transportation.[3] The automobile, whose early form looked much like a bicycle, fulfilled this objective so dramatically that W. J. McGee could not have anticipated it clearly in 1898.

It might be helpful to review briefly the chronology of the bicycle in late nineteenth-century America.[4] The high-wheel, or ordinary, bicycle came to the United States from England. Albert A. Pope, a Boston merchant who became known as the ''father of the bicycle in America,'' saw a Smith & Starley bicycle at the 1876 Centennial Exhibition in Philadelphia. Captivated by these curious high-wheeled machines, Pope traveled to England to study their manufacture and sales and to take an exhilarating country tour on one of them. Returning to the United States, Pope began to sell imported English cycles. In 1878, however, he contracted with the Weed Sewing Machine Company of Hartford, Connecticut, to manufacture an American version of the English ordinary. The product, the Columbia, initiated the bicycle era in America.[5] (See Figure 5.1.)

FIGURE 5.1.   Columbia Light Roadster High-Wheel Bicycle, 1886. (National Museum of American History. Smithsonian Institution Neg. No. 5843-B.)

FIGURE 5.2.   Columbia Women's Safety Bicycle, 1896. Unlike the high-wheeled era, bicycling with a "safety" bicycle was not for men only. (National Museum of American History. Smithsonian Institution Neg. No. 41230.)

By the time the safety bicycle—the chain-driven bicycle with two wheels of equal size—was introduced in the United States from England in 1887, the Pope Manufacturing Company and a few competitors had produced a total of perhaps 250,000 high-wheelers. But the safety bicycle produced a new and unprecedented wave of enthusiasm. When the safety bicycle boom reached its peak in 1896–97, the industry consisted of more than three hundred companies and produced well over a million bicycles per year. (See Figure 5.2.) The Pope Manufacturing Company continued to be the recognized leader in bicycle production, although the Western Wheel Works outstripped Pope's production in 1896 by ten thousand bicycles.[6] Yet the entire industry was doomed to rapid decline beginning in 1897 because the market for bicycles simply vanished. As Arthur S. Dewing wrote in 1913, "people ceased using bicycles," and this brought "the practical cessation of a manufacturing industry."[7] All efforts to keep the industry afloat, including consolidation and merger, failed.

The Pope Manufacturing Company and Armory Practice

Tradition has it that Pope's first bicycles were made in a corner of the Weed Sewing Machine Company.[8] This tradition leaves the impression that perhaps only a mechanic or two with limited tools set about the task of making a few bicycles. During the summer and fall of 1878, the Weed company produced fifty bicycles, and this small figure may well support a corner-of-the-shop vision. Yet by 1880 the bicycle had become an important item of manufacture for the Weed company, which sought to turn out five hundred cycles a month and had already produced twelve hundred machines. (See Figure 5.3.) By 1881, monthly production had risen to twelve hundred.[9]

The Weed Sewing Machine–Pope Manufacturing Company connection provides an elegant case for Nathan Rosenberg's concept of technological convergence and for the term *armory practice*.[10] It is unclear exactly why Pope chose the Weed company to make his bicycles, but it seems likely that Pope's initial dictum that, unlike the British cycles, Columbias would be made of machine-produced interchangeable parts played a major role in his decision.[11] Organized in 1866, the Weed Sewing Machine Company began selling sewing machines in the same manner that the Pope Manufacturing Company began selling bicycles, that is, without its own manufacturing plant. For ten years the company had contracted with the Sharps Rifle Manufacturing Company, one of the New England small-arms makers that pioneered in the design and construction of special-purpose machine tools, for the production of its sewing machines. The Sharps company originated in Windsor, Vermont, but moved to a new building in Hartford before the Civil War. Significantly, when the Sharps company changed ownership in 1875 and moved west to Bridgeport, the Weed Sewing Machine Company began its own production in Hartford by occupying the Sharps plant, perhaps by purchasing many of its machine tools, and almost certainly by hiring many of the Sharps Hartford employees.[12] Hence the techniques and traditions of manufacturing firearms remained in the same building for the production of sewing machines and bicycles. Later, Pope would make automobiles here.[13] (See Figure 5.4.) Moreover, the superintendent of the Weed company's manufacturing operations, George A. Fairfield, had been an inside contractor at Colt's Hartford armory and an

FIGURE 5.3.   "Bicycle Room" of the Weed Sewing Machine Company Factory, 1880. (*Scientific American*, March 20, 1880. Eleutherian Mills Historical Library.)

associate of Charles E. Billings and Christopher M. Spencer, both of whom had worked for Colt and other armsmakers and had earned reputations as brilliant machinists. In the mid-1860s Billings had even served briefly as the Weed company's superintendent before he was succeeded by Fairfield in 1868.[14]

As in the antebellum arms industry, drop-forging provided the foundation of metal-working at the Weed Sewing Machine Company. This is not surprising given the connections of Billings, Spencer, and Fairfield with Colt. Almost every part of the sewing machine, excluding the cast-iron frame, was first forged in the company's drop-forging room. These parts were then moved to the machining room to be finished by grinding, turning, milling, drilling, or some other metal-removing operation. In addition to its own parts, the Weed company produced forgings for other sewing machine companies as well as for agricultural implement makers and steam engine manufacturers. Employing the same techniques, the Weed company drop-forged and then machined the major parts for the Columbia bicycle, such as hubs, cranks, and steering heads.[15] (See Figures 5.5–5.9.)

Although drop-forging techniques had been extant for more than half a century, manufacturers such as the Weed company continued to refine the process. By 1880 forging was far more complex than a simple drop of the die. The Weed company relied upon a three-step process for making fine forgings which minimized the amount of metal removed during machining. This process included a rough-forming stroke in a drop forge, a pressing operation to trim off the unwanted "flash," and a finishing stroke in another drop forge. For complex parts, however, such as several pieces of the Columbia bicycle, as many as five different operations were necessary. The only new problem in forging bicycle parts, as opposed to those for sewing machines, was posed by their size. Adoption of larger drop forges met this demand. Thus at the most basic level, production of the Pope Columbia bicycle was rooted firmly within the New England tradition of manufacturing firearms and sewing machines.

The machining practices at the Weed company during its early years of bicycle manufacture followed the same approach. Many of the machine tools familiar to New England armories were used at the Weed company for its sewing machine parts and for the Pope bicycle—milling machines, turret lathes, screw machines, grinding machines, drilling and boring machines.[16] Machining of drop-forged parts provided shapes and precision not attained in drop-forging while drilling, boring, and other machine work performed operations not possible in the forging shop (for example, completion of the hub after drop-forging and milling by boring the hole for the axle through the hub and by drilling and tapping holes for the spokes in the hub flanges). Yet the techniques employed here varied only in degree, not in kind, from accepted New England armory practice. It is doubtful whether the Weed company initially built or bought any special-purpose machine tools for machining bicycle parts; set up differently with special fixtures and cutting tools, the machinery used for sewing machine manufacture fulfilled the requirements for production of the high-wheel bicycle.

There is one possible exception to this generalization: ball-bearing manufacture. Use of ball bearings was a major innovation of the bicycle which caused reverberations throughout machine-building. Not long after the introduction and acceptance of the bicycle, most newly built machines turned on ball bearings. Weed company mechanics set up a separate ball-bearing production room and built special machinery for their manufacture.[17] Although no specific documentation exists on this operation, it is reasonable to extrapolate from Robert S. Woodbury's account of the grinding machine that the balls and their races were first drop-forged or perhaps turned, then case-hardened, and finally ground on

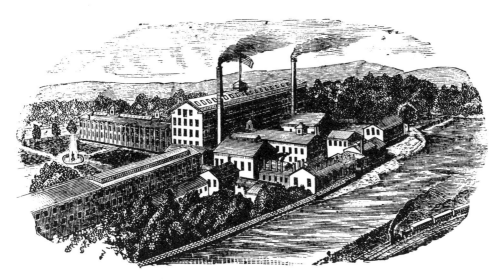

FIGURE 5.4.   Weed Sewing Machine Company Factory, Hartford, Connecticut, 1881. The Weed factory had formerly been owned by the Sharps Rifle Manufacturing Company. Later the Pope Manufacturing Company would buy out the Weed Company and expand its factory operations. (''A Great American Manufacture,'' *Bicycling World*, April 1, 1881. Smithsonian Institution Photograph.)

FIGURE 5.5.   Machining Rear Hubs for Columbia Bicycles, Weed Factory, 1881. (''A Great American Manufacture,'' *Bicycling World*, April 1, 1881. Smithsonian Institution Photograph.)

FIGURE 5.6.   Inspecting and Gauging Columbia Bicycle Parts, Weed Factory, 1881. (''A Great American Manufacture,'' *Bicycling World*, April 1, 1881. Smithsonian Institution Photograph.)

FIGURE 5.7.   Truing Columbia Bicycle Wheels, Weed Factory, 1881. (''A Great American Manufacture,'' *Bicycling World*, April 1, 1881. Smithsonian Institution Photograph.)

FIGURE 5.8.   Assembling Columbia Bicycles, Weed Factory, 1881. ("A Great American Manufacture," *Bicycling World*, April 1, 1881. Smithsonian Institution Photograph.)

FIGURE 5.9.   Warehouse for Columbia Bicycles and Parts, Weed Factory, 1881. ("A Great American Manufacture," *Bicycling World*, April 1, 1881. Smithsonian Institution Photograph.)

special machinery to eliminate warps and to produce exceedingly smooth surfaces. Here again, though, specialized grinding machinery already existed for optical parts and needle manufacture. More important, by 1860 Joseph Brown had developed a special machine for grinding sewing machine needle bars. Brown & Sharpe sold these machines to other manufacturers; possibly the Weed Sewing Machine Company used such a grinder for sewing machine production and adapted it for grinding bicycle ball bearings.[18] Thus despite the difference in products, the means of production—grinding—appears to have been the same for sewing machine and bicycle making within the Weed Sewing Machine Company.

The form of the high-wheel bicycle posed new production problems for the Weed Sewing Machine Company, and these problems would later be exacerbated by the introduction of the safety bicycle. Finishing (painting and nickel-plating) required larger-scale ovens and baths than those used in sewing machine manufacture and exacting preparation and polishing operations. Assembly, however, proved to be the most serious bottleneck. Whereas in 1853 Joseph Whitworth had witnessed the complete assembly of a Springfield musket in only three minutes, the assembly and adjustment of a single big bicycle wheel took as much as twenty times as long.[19] A wheel assembler strung the finished hub and rim together with machine-made spokes. Once all the spokes were put in, the workmen had to true the wheel, a tedious process requiring careful tightening and loosening of individual spokes. Many of the same problems obtained in the final assembly of the complete bicycle. Even though interchangeable parts were used, assembly was not the simple matter it had been at Springfield Armory.

While the Weed company worked out methods in 1880 to produce twelve hundred Columbias annually, Albert A. Pope unfolded his first major promotional campaign. He encouraged his Boston friend and lawyer, Charles Pratt, to write and publish *The American Bicycler,* which gave a short history of the bicycle and cycling, described the exhilaration of riding an ordinary, and told how everyman could join in the fun. Pope also owned or supported *Bicycling World,* a semiweekly journal begun in 1880 that covered bicycling activities throughout the United States, and a shorter-lived periodical, the *Wheelman.*[20] As part of his promotional activities he was instrumental in the formation and expansion of local cycling clubs and the national League of American Wheelmen.[21] (See Figure 5.10.) In addition, Pope sponsored monthly poster contests, which brought to perfection this popular nineteenth-century advertising technique. By 1881, these activities began to pay off with a jump in demand for cycles. To meet this increased demand, the Weed company expanded its plant and hired additional workmen so that by the middle of the year it could turn out twelve hundred machines per month—roughly fifty per day.[22]

Demand continued to grow. Albert Pope and his fellow bicycle manufacturers sought to nurture this demand by initiating an institution that became an important promotional and sales activity during the height of the bicycle era and later for the American automobile industry: the trade show. Begun in 1883 in Springfield, Massachusetts, bicycle trade shows drew large numbers of exhibitors, sales agents, and paying public. By 1894 both Chicago and New York hosted annual shows and counted their money when, as in 1896, Chicago drew more than 225 exhibitors and 100,000 admissions and New York 400 exhibitors and 120,000 admissions.[23] (See Figure 5.11.)

Pope's acumen as an entrepreneur did not stop with promotional activities. From the outset, he sought monopolistic protection by purchasing virtually every patent connected with the bicycle, some dating from the velocipede craze of the 1860s. By 1881 he had secured a patent monopoly that would not begin to deteriorate until 1886. Pope sold

FIGURE 5.10.   Local Chapter of the League of American Wheelmen, 1880s. (National Museum of American History. Smithsonian Institution Neg. No. 57820.)

FIGURE 5.11.   National Bicycle Exhibition, Madison Square Garden, 1895. (*Scientific American,* February 9, 1895. Eleutherian Mills Historical Library.)

bicycles beginning at $120 and going up in price as size and finish were larger and more elegant.[24] Pope's early recognition of the bicycle's promise and his promotion of bicycling would not come free to others desiring to manufacture bicycles. Until he lost his patent position, Pope extracted a fee of $10 per bicycle from other manufacturers.[25] When he first began selling bicycles, two other firms also made high-wheelers, yet his patent monopoly helped drive both companies out of business.[26] The boom demand which Pope created almost single-handedly, however, supported a half-dozen other cycle companies by 1884–85.[27]

None threatened Pope with serious competition until 1885, when A. H. Overman of Hartford set up a two-story bicycle factory at Chicopee, Massachusetts, long a manufacturing center in New England.[28] Overman originally contracted with the Ames Manufacturing Company of Chicopee to manufacture his Victor, so it is entirely reasonable to assume that he departed little from methods employed by the Weed company or any other New England factory.[29] About the same time, Iver Johnson, a firearms maker in Worcester, Massachusetts, began to produce his Springfield Roadster using the same techniques for bicycles as for rifles and handguns.[30] Other firearms makers followed suit.

Thus the first decade of bicycle making in America produced few startling innovations in manufacturing techniques. Other firms most likely followed the Weed Sewing Machine Company in employing Yankee armory practice. Increased annual output resulted from devoting more of existing plant to bicycle production or to expansion of manufacturing plant along familiar lines. Yet a few innovations did take place. Most important from a twentieth-century perspective was the adoption of electric resistance welding, a process of fusing metals by the direct application of electricity rather than through an arc (as is commonly imagined when one thinks of electrical welding).

Developed between 1886 and 1888 by the well-known electrical inventor Elihu Thomson, electric resistance welding automated what has always been one of the most difficult arts of the blacksmith.[31] (See Figure 5.12.) Almost as soon as the Thomson Electric Welding Company, based in Lynn, Massachusetts, began marketing commercial apparatus, the Weed company purchased one setup and soon after bought another to weld the rims of the Columbias, cutting down significantly on production time.[32] Pleased with this process, the Weed company tried to employ it in assembling cycle frames to replace brazing, a technique requiring a great deal of time and skill. A sufficiently strong weld was not produced, however, and hand brazing of frames continued in cyclemaking well into the twentieth century.[33] Pope used electric welding for rims until about 1894, when wooden rims came into style and temporarily replaced steel rims.

That few other innovations in bicycle and tricycle manufacturing took place in the 1880s is reflected by the *American Machinist,* a weekly journal for machinists, engineers, founders, boilermakers, patternmakers, and blacksmiths. The editors appear to have been unaware of the growing importance of the bicycle until late in 1886, when they published a small article, "Bicycle Engineering," which suggested that firms engaged in the manufacture of light machinery and other products consider taking up bicycles. Little else on cycle manufacture appeared in the *American Machinist* until 1895, when the safety bicycle boom was shaking the established New England armory modes of production. *Iron Age,* a hardware journal, followed this same pattern.

Had the safety bicycle we know today not been developed, the bicycle craze probably would have ended with the ordinary or high-wheel bicycle in the late 1880s and a total production of between one hundred thousand and three hundred thousand cycles. Just as the English had developed the high-wheeler, however, they also pioneered the chain-

FIGURE 5.12. Thomson Electric Welder, 1891. Although this particular resistance welder was designed to butt-weld wire, it illustrates how welding was done. The two pieces to be welded were mounted in the quick-loading clamps at the upper left. When the pieces were loaded, the operator simply pushed the lever in the upper right corner to the right momentarily, and the weld was completed. (National Museum of American History. Smithsonian Institution Neg. No. 74-7158.)

driven safety in the mid-1880s. Only the bravest men rode high-wheelers, but the safety bicycle, when promoted sufficiently, gained a certain universality; the whole family rode safeties—father, mother, and children alike.[34] When the safety bicycle boom reached its peak in the mid-1890s, the industry produced 1.2 million machines annually, a figure that adds meaning to the loose term *mass production*.[35]

A. H. Overman introduced Victor safeties in 1887, and when Albert Pope first marketed the Columbia version in 1889 the safety had already begun to catch hold—even many stubborn high-wheel riders had traded in their ordinaries for a safety. The *Report on Manufacturing* of the 1890 census listed twenty-seven firms manufacturing bicycles, presumably safeties, and best estimates suggest that about forty thousand cycles were made in the census year.[36] During the years 1890–97 a number of developments in cycle manufacturing methods occurred both within and outside of New England armory practice.

Demand for the safety undoubtedly played a major role in the development of new manufacturing techniques. Yet equally important was the form of the safety bicycle, which introduced new problems of manufacture. Visualizing the ordinary and the safety, one can readily see major differences: the safety bicycle has a chain and two sprockets for its driving mechanism, whereas in the ordinary the pedals and cranks are attached directly to the axle of the big wheel; the safety's cranking mechanism required an extra axle and set of ball bearings as well as a structure to support them (crank hanger); and the safety also required about twice as much framing material. With the addition of extra framing, an extra axle and bearings, and chain and sprockets, bicycle designers sought to reduce weight (originally between forty and seventy-five pounds) without sacrificing strength. These efforts culminated in the 1980s with first-class cycles weighing between twenty and twenty-seven pounds.

The promise of the safety bicycle lured a number of arms manufacturers and sewing machine companies into cycle manufacture. Iver Johnson Arms Company switched from ordinaries to safeties and soon found John P. Lovell Arms Company, Remington Arms Company, Winchester, and the Colt armory competing with it. The White Sewing Machine Company, the Elgin Sewing Machine Company, and the American Sewing Machine Company, among others, began to compete with the Weed company. Other New England industries developed a cycle line; the E. Howard Watch and Clock Company of Boston produced bicycles as did a New Haven cutlery manufacturer.[37] Mass production of cycles in a modern sense was another matter. Most of these firms that adopted bicycles as a second or third line of manufacture turned out between five hundred and three thousand machines a year, or roughly one and a half to eleven cycles per work day. Most of these firms seemed satisfied to use their extra plant for cyclemaking, adding a minimum number of tools made by the New England machine tool companies such as Pratt & Whitney and Brown & Sharpe. They developed few noteworthy new techniques.

Not until bicycle making became a first line of manufacture did any significant innovations take place, a fact confirmed by the experience of the Weed Sewing Machine Company's production of cycles for the Pope Manufacturing Company. In 1890, Albert Pope gained control of the Weed company, ended its production of sewing machines, and devoted its full attention to cycles and cycle parts. Significant innovations immediately began to appear, and soon Pope Manufacturing Company became one of the largest cycle factories in the United States, hiring more than three thousand employees, producing more than sixty thousand bicycles annually, and selling a large range of parts to other manufacturers.[38]

Unlike other bicycle manufacturers, Pope integrated backward. The company established in 1893 a pioneering cold-drawn steel tubing plant to produce tubing (for frames) formerly obtained from England. Moreover, Pope purchased and enlarged the Hartford Rubber Works to manufacture pneumatic tires, which were another major contribution of the bicycle to modern transportation. Both the tubing mill (which was substantially enlarged in 1896) and the rubber factory produced supplies well in excess of the Pope Manufacturing Company's needs; the surplus was sold to other bicycle manufacturers. Both of these plants represented substantial capital outlays and required an unprecedented level of know-how to operate successfully. Pneumatic tire making was an entirely new manufacturing art in 1890. Similarly, Pope's seamless steel tube mill was the first of its kind in the United States, which probably explains why it worked "only after long and costly effort, with many failures."[39]

The Pope company also moved horizontally. About the same time that Albert Pope dissolved the Weed Sewing Machine Company, he founded the Hartford Cycle Company, although he did not publicly announce his connection with this firm. George Pope, Albert's cousin, assumed the presidency of the new concern.[40] Pope's motive was simple. He prided himself that the Pope Manufacturing Company's Columbia was the finest—and usually the most expensive—bicycle made in the United States. Pope seems always to have been uncompromising on the quality of the Columbia, yet he recognized that if the bicycle was to be more than just a fad for wealthy gentlemen, he must bring out a medium-grade, medium-priced bicycle that was within reach of the middle class. As had Singer with the Domestic sewing machine, Pope felt that a medium-priced Columbia would jeopardize the reputation and harm the sales of his first-class bicycle. Therefore, he established the Hartford company to reach the middle-class market and to compete head-on with the less expensive bicycles of other manufacturers.[41]

Hartford Cycle Company set up its own machinery to produce its cycles with the exception of ball bearings, which it obtained initially from the Pope company. Theoretically, the Hartford company operated independently of the Pope Manufacturing Company, but existing manuscripts show that Albert Pope made its critical decisions and that George Day, successor to George Fairfield as superintendent of both the Weed and Pope companies, often gave technical help and advice.[42] Harry Pope, another of Albert's cousins, filled the position of superintendent at the Hartford company and seems to have had some previous experience in manufacturing. Finally, in 1895, the Hartford Cycle Company became officially affiliated with the Pope company, although it maintained its separate factory. To understand the entire approach to and problems in cycle manufacture by the Pope company, developments in both factories should be considered.

Pope seems to have integrated in every direction except forward into marketing. Unlike Singer with sewing machines and Ford with automobiles, the Pope Manufacturing Company did not establish retail stores to market its products. No information exists to answer the question of whether Pope consciously decided not to integrate forward or failed in an attempt to do so. Whatever the reason, the Columbia and Hartford bicycles were sold through hardware and general merchandise stores, through bicycle shops carrying more than one maker's products, and through individual agents.[43]

Despite this conservative marketing strategy, Pope continued to employ many of the same advertising schemes and to support institutional arrangements that he had favored in the 1880s. Bicycle posters grew more numerous and more elaborate. In fact, Pope's poster contests and exhibitions proved to be a spawning ground for commercial artists who later became famous. Maxfield Parrish, among others, completed artwork for Pope and won a Pope-sponsored contest in 1895. These posters toured the country in exhibits sponsored by Pope and local retailers of his bicycles.[44] Pope also solidified his commitment to the League of American Wheelmen (LAW). With substantial financial backing from Pope, the LAW waged legal battles in New York and other urban areas to get the bicycle classified as a transportation vehicle, a critical problem because cities like New York had prohibited bicycles on city streets. Pope and the LAW also launched the long and effective Good Roads campaign to bring about the construction of roads. This movement was crucial in the creation of both state and federal legislation resulting in a more extensive highway system in America.[45]

In manufacturing parts for the Columbia safety bicycle, the Pope Manufacturing Company initially remained within the tradition of New England armory practice.[46] Yet even within this tradition, the range of choice for some production processes was remarkably wide. One example—the manufacture of a bicycle hub—clearly demonstrates this point and shows how Pope slowly adopted new production techniques.

Traditional armory practice dictated drop-forging the shape of the hub out of steel, then taking it to the machining room to bore the hole for the axle, to recess the ends for the bearing cups, to drill the spoke holes in the flanges, and finally to machine the outside of the hub either on a lathe or on a milling machine. (See Figure 5.13.) Each machining

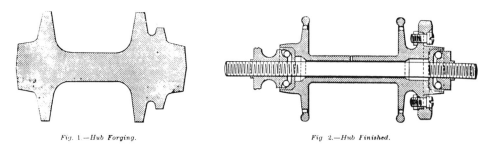

Fig. 1.—Hub Forging.   *Fig 2.—Hub Finished.*

FIGURE 5.13.   Sections of Hub Forging and Finished Hub for Columbia Bicycle, 1892. (*Iron Age,* June 1, 1892.)

operation could be carried out on a variety of machines. Boring the axle hole, for instance, could be performed on a lathe, in a drill press, or in a boring mill. With the development of heavier machine tools and with screw machines, many bicycle manufacturers moved away from drop-forging altogether. These makers used a plain round bar of steel as the raw material for hubs. Rather than forging the bar to shape, it was fed through the hollow spindle of a lathe or screw machine and then machined to shape. This process entailed bringing to bear on the bar a number of different tools: a rough forming tool, a finishing tool, drilling and recessing tools, and a cutoff tool. Despite the choice of screw machines (semiautomatic or automatic), the hub could not be completed in a single machine tool. It required rechucking in a different lathe, usually a turret lathe. Every shop had its own preferences. As with most parts of the safety bicycle during the cycle craze the machine tool industry adopted and marketed its machine tools for hub production, including hand and automatic screw machines and turret and chucking lathes.[47]

Although for several years the Pope Manufacturing Company advertised that its drop-forged hubs provided greater strength and quality than the straight-machined hubs of other manufacturers, it eventually stopped forging its hubs.[48] In 1895 the company developed special machines to manufacture its hubs from bar steel.[49] Despite this departure from drop-forging, Pope and other manufacturers remained wedded to the tradition of metal-working by removing metal from a workplace in a machine tool. This tradition will be contrasted below with that of stamping sheet steel, which was employed successfully by the Western Wheel Works. Although exact dates are unclear, Pope eventually stopped drop-forging other parts of the Columbia, including the crank hanger, and adopted Western stamping techniques. Nevertheless, drop-forging remained important for the production of pedal cranks and sprockets.[50]

Extremely wide variation occurred among bicycle manufacturers in making hubs and all other parts of the bicycle. No one seems to have conducted a comparative cost study of the various methods of production, yet these must have varied significantly. One reason for the lack of accounting is that mechanics argued that the machining processes did not matter; rather, shop organization and management made the critical difference in cost.[51] Such thinking led Frederick W. Taylor at roughly the same time to develop his system of scientific management, based on the assumption that even though an article could be machined in a dozen different ways, the critical factor was management. Taylor and his followers were, in a sense, suboptimizing manufacturing operations by finding the "one best way" to perform a given or already established machining operation rather than finding the least expensive machine process.

At Pope Manufacturing Company, as well as at other firms manufacturing in the armory tradition, the lack of any rigorous system of cost accounting probably resulted from the institution of the inside contracting system. As discussed in earlier chapters, inside contracting fostered the prevalence of traditional approaches in production and prohibited a manufacturer from ever gaining information about the real costs of production processes. Inside contractors headed some twenty departments at the Pope works, including the drop-forging; the brazing; ball, chain, and spoke; toolmaking; press; machining; and various departments for finishing and assembly. The machining department, the largest, was made up of subdepartments usually identified with specific operations such as the "hub job" and the "crank job."[52] It is of particular interest to consider the operations of the assembly, finishing, and testing departments.

Assembly of safety bicycles involved a great deal of labor, just as it had with the high wheel. For instance, the Stearns Manufacturing Company in Syracuse, New York, employed 250 men to assemble and braze its annual production of five thousand cycles, one-twelfth the Pope Company's output in 1896.[53] Pope's managers had concentrated on assembly problems in all aspects of cycle building. Most subassembly work was carried out in departments that produced the finished parts; for example, the ball, chain, and spoke department assembled cycle chains. This department first machined parts for the chain—five parts per link, the block, two side pieces, and two rivets. Pope mechanics devised an automatic chain assembly machine, which wove the five parts together to produce a fifty-three-link chain. Yet chain assembly was not entirely automatic. The ends of the rivets remained to be spun down to complete the riveting process. This task was completed by using a manually operated machine, the operator feeding the chain rivet-by-rivet under a treadle-controlled riveting machine. Almost every assembled chain came out crooked, often between one-thirty-second and one-quarter inch. Workmen straightened the chains by hand with hammers.[54]

Because wheel assembly involved so much labor, the Pope company established a separate department for this operation. The contractor divided the process of assembly into four steps. The first was stringing up of the wheels, which meant running the spokes through holes in the hub flanges and screwing them into the nipples attached to the rim. Other workmen then trued the wheels in a truing stand in which wheels were spun around by hand. One mechanic commented that the Pope stand was "a thoroughly Yankee machine, adjustable all over, and having every essential for the avoidance of needless labor."[55] After truing, the crossings of the spokes were lightly soldered with either a small gas flame or a copper iron. Finally, because some of the spokes protruded through the rim, they had to be ground flush with the inside of the rim to prevent puncture of the pneumatic tire. After finishing this step, the complete wheel was transferred to another part of the factory where the tire was put on.

Assembly of the frame surpassed wheel assembly in complexity and labor requirements. After all the various parts of the frame had been made—the crank hanger, the joints, the end pieces, and the head piece—and after correct lengths of tubing had been cut, these parts were assembled in a Pope-developed frame-pinning jig.[56] This jig combined the several-step processes used by other cyclemakers. First, it correctly aligned the assembled frame in a horizontal plane. Next, a workman drilled holes for connecting pins with the attached drilling jigs and the swing-arm drilling press. The workmen then inserted steel pins and peened their ends with another attachment until the pin pulled the joint up tightly. Now the frame could be brazed, the most critical of all Pope operations.

Pope's brazing room for frames consisted of thirty stations, one for each joint in the frame. Each workman specialized in brazing one joint of the cycle, a thirty- to ninety-second operation. The frame was thus transferred from station to station for complete brazing. A workman located between stations performed this transfer process and at the same time wiped off superfluous spelter with an iron rod while the brazed joints were still hot, a process that eliminated sandblasting and excessive filing. Pope mechanics had carried out numerous experiments on brazing techniques before they mastered a process that minimized the amount of fire used, heated the joint as little as possible (thus minimizing damage to the quality of the steel), and still produced a strong joint in a minimum amount of time. So critical was the brazing process that, to help guarantee that joints were brazed with skill and care, Pope braziers worked on daywork rather than piecework—the only such instance in the Pope company.[57]

After brazing, another group of workmen filed the frames. Workpieces were held in common vises mounted on workbenches. Judging from accounts of cyclemaking at Pope's Hartford Cycle Company, filing was a bottleneck in the assembly and finishing process which was cured only by adding numbers to the filing corps.[58] Pope practice then called for the brazed and filed frames to be trued or realigned in custom-designed fixtures which were, according to *American Machinist*, "magnificently built and . . . beautiful examples of art in tool-making."[59] The fixtures applied force wherever it was required to true the frame. Workmen used specially designed gauges to determine correct alignment.

Once the frame was aligned, it and the other filed parts underwent the process of enameling, carried out in several rooms containing baking ovens and paint and water vats. Initially the parts were dipped in boiling water, removed, dried, and then smoothed by hand with emery cloth. After washing the smoothed parts with benzine, workmen dipped them in enamel for their first coat and placed them in coal-fired ovens to be baked for several hours. Some parts received as many as four coats of paint, each followed by baking.[60] Pope added a final baking process, which was carried out in special gas-fired ovens usually for eight to twelve hours. The temperature constraints of the final bake made this process critical. Temperatures had to be high enough for proper baking but not too high because the brazing material would run back out the joints (which probably happened more than once).[61] Thus the painting of the Columbia, like its assembly, was a long series of steps involving an appreciable amount of labor in working the parts and in moving them between processes. Nickel-plating of various parts such as the cranks and pedals was equally complex, requiring initial polishing, then cleaning, plating, and finally buffing.[62]

After all the subassemblies and finishing were completed the bicycle was put together and crated. No information exists on this assembly process, and therefore it may be fair to assume that it posed few technical problems.

Many of the aspects of manufacturing the Pope Columbia have been explored here. Mundane processes such as painting and assembly have not hitherto been generally considered part of the American system of manufactures, yet the historical record of firms such as Singer and Pope indicates that they posed serious problems in production. Any study of the development of mass production technology must consider the processes of finishing and assembly. The diffusion of armory practice into other areas of manufacture, such as sewing machines and bicycles, brought new problems that had not existed in the production of small arms. The Pope Manufacturing Company worked on these problems for almost two decades, yet it consistently viewed its efforts as being "in vain."[63] Unlike the Ford Motor Company fifteen years later, Pope's approach to assembly did not cause a revolution in manufacturing and work. Nevertheless, Pope's methods in testing and quality control proved to be of major importance for the automobile industry.

Because Pope prided himself on the quality of the Columbia, he demanded a rigid system of quality control. An inspection department was set up, as in many of the New England armories, with a separate corps of inspectors. Before machining, each drop-forging was inspected. About 5 percent were rejected. After machining, inspectors gauged the critical running parts and examined others for appearance before subassembly. Enameled and nickel-plated parts underwent inspection, and finally the complete cycle

FIGURE 5.14.   Pope Testing Department Apparatus, 1896. Item 1 is a detail of Item 3, an Emery hydraulic-testing machine built by William Sellers Company and designed to test tensile, compressive, and transverse strains. Item 2 is a vibratory strain-testing machine for bicycle tubing. Item 4 is a wheel-testing machine on which destructive tests were carried out under various loads and speeds of rotation. (*Scientific American*, July 11, 1896. Eleutherian Mills Historical Library.)

was checked. The Pope company claimed that its cranks were inspected eight times before being sold and some parts as many as a dozen times.

Albert Pope also established a testing department, which carried out destructive testing on its bicycles and bicycle components. (See Figure 5.14.) Pope mechanics devised simple machinery for some of these tests such as the chain tester, a device that measured the force required to break one. Knowing this average figure, the chain department checked every chain it made on the testing machine by applying a force slightly below the average breaking point. Other instruments were more complex; the frame tester, for instance, could apply and measure compression and tension forces at all parts of the frame. Results from this machine enabled Pope designers to change frame designs to minimize usage of tubing while maintaining frame strength and rigidity. The Pope testing department also adopted, through its tests, a nickel alloy steel tubing because it gave greater strength per unit weight.

Not satisfied with simple data on how much weight a wheel could withstand, Pope testers constructed a vibratory machine that measured how long a heavily loaded wheel could withstand intense vibration. These tests were conducted toward the goal of improving the design of Pope's cycles, especially by cutting down weight without sacrificing strength. Swaging of spokes was one result of such tests. A few other cycle manufacturers carried out similar tests, but Pope appears to have been the innovator of "scientific testing" in the industry. By and large, other cycle builders imitated Columbia styles, always assuming that they were superior in design and construction.[64]

The inspection system and testing department provided benefits but not without costs. When cycle competition began to stiffen, a Pope Columbia/Hartford Cycle dealer complained because the companies would not offer discounts as did other manufacturers. George Pope, president of the Hartford company, responded to the secretary of his company, "He [the agent] well knows that we probably put more money into the experimenting departments and into care of inspection, etc., than any other concern in the business. If he does not give us credit for this, he admits that our goods are no better made than those of the concerns that are giving 40 to 50 percent discount."[65] British cycle manufacturers also conducted rigorous inspections and experiments on their bicycles. An economic historian recently suggested that the British cycle industry did not become competitive until it dropped this rigid system of quality control.[66] Despite Pope's insistence on rigorous testing and inspection, his company remained competitive. But it did not maintain its position as the largest producer of bicycles in America. Rising swiftly from a wooden toy factory, the Western Wheel Works of Chicago became the "world's largest bicycle manufacturer," turning out ten thousand more cycles in 1896 than Pope.[67]

The Western Wheel Works and Stamping Techniques

Whereas bicycle manufacture in New England was taken up largely by armsmakers, sewing machine companies, or similar small-item manufacturing concerns, western bicycle builders emerged primarily from the ranks of carriage- and wagonmakers, wooden toy and novelty specialists, agricultural implement makers, or as totally new enterprises.[68] Some of these companies, particularly the larger ones, built many of their own machine tools. For companies that had previously worked in wood, bicycle manufacture presented entirely new demands for precision production. Of course, as in all machine work, such demands varied according to the quality of the product,[69] and western manufacturers turned out cycles ranging from the lowest grade to some of the highest. Companies that made children's wooden carriages or hobby-horses turned to cycle manufacture in the years between 1893 and 1897, when wooden rims almost totally replaced metal rims. Many of the western bicycle manufacturers initially produced only wooden rims but later turned to making entire bicycles.

When these western companies began to make bicycles they adopted radically different techniques of production than those used in most New England factories. Western companies had not previously employed the tools and techniques of armory practice, and they did not adopt this approach for the manufacture of bicycles.[70] These major differences are superbly exemplified by western manufacturers' adoption and development of sheet steel stamping technology—a revolution in metalworking. The Western Wheel Works was an active participant in this revolution.

Successors to the Western Toy Company (which specialized in toy wagons), the Western Wheel Works began making safety bicycles before 1890. Initially it imported stamped or pressed frame joints from Germany, but soon it developed its own techniques of presswork and found American toolmakers to build its equipment.[71] It extended these techniques beyond the fabrication of frame joints and crank hangers to making hubs and sprockets. By 1896, Western Wheel employed press techniques for almost every part of its Crescent, a cycle that was included in the "first class" of bicycles. These parts included the hubs, steering heads, sprockets, frame joints, crank hangers, fork crown, seats, handlebars, and various brackets. (See Figure 5.15.) Western mechanics reduced machining work to a bare minimum. As noted earlier, in Yankee armory practice drop-forging and machining were the principal processes of metal fabrication; almost all of these were eliminated by the Western Wheel Works. As the company's trade literature noted in 1898: "Our use of stampings in the construction of our bicycles has proven highly satisfactory and practical. . . . The result obtained by us in stamping . . . can never be equalled by the working of forgings. In 1890 we were the first to use a sheet-steel stamping in the construction of the bicycle. It is distinctively Crescent."[72] This new approach to metalworking may be best demonstrated by considering the fabrication of three different components of the Western bicycle.[73]

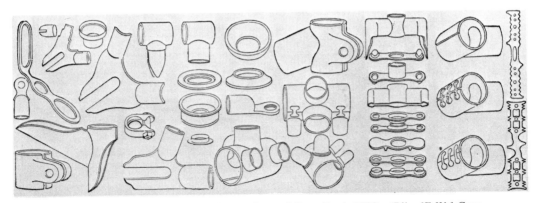

FIGURE 5.15. Examples of Bicycle Parts Stamped out of Sheet Steel, 1890s. (Bliss [E.W.] Company, *Catalogue and Price List*, 1900. Eleutherian Mills Historical Library.)

Crank-hanger production provides a sharp relief between armory practice and Western techniques. The crank hanger is that part of the safety bicycle through which the pedal axle runs and from which the tubing radiates to the steering head, the rear wheel, and the seat. In a sense, the crank hanger is the heart of the bicycle. Manufacturers such as Pope argued that drop-forging provided critical strength. Yet a drop-forged crank hanger required a tremendous amount of machining—mostly boring or drilling—to hollow out the holes of the axle, axle bearing, and tubes. About 80 percent of the metal from the solid forging was removed by cutting operations. The Western technique started with sheet steel. Through a series of punching and pressing operations, carried out in power presses, with periodic annealing (softening of the steel) in between, the crank hanger was formed. The process also usually entailed brazing or electric resistance welding where the ends of the sheets met (See Figure 5.16.) Resistance welding proved to be especially effective for this step because it automated a process that otherwise required highly skilled workers. The technique of stamping crank hangers and other joints became so successful in producing the desired quality, strength, lightness, and cost that it was adopted by almost every bicycle maker, even the venerable Pope Manufacturing Company.

Yet other bicycle parts made in this way were viewed with more skepticism, as is clearly demonstrated in the Western method of making bicycle hubs. (See Figure 5.17.) First, flanges for the spokes were formed from a flat disk of metal through seven pressing operations, each carried out on a different machine with an annealing process performed

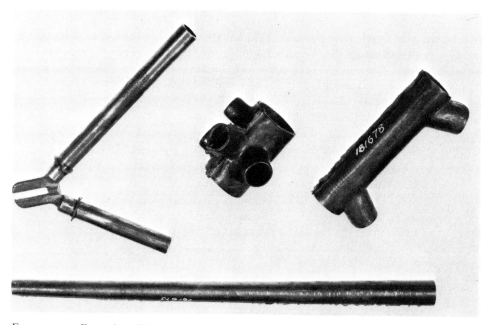

FIGURE 5.16.   Examples of Electrically Welded Bicycle Parts Made from Sheet Steel, 1896. These artifacts were all welded with Thomson Electric Welding Company equipment: tapered tubing (top); steering head (bottom left); crank hanger (center); rear axle holder and frame (right). Except for the rust, which has developed during the past eighty-five years, these artifacts are in the same condition they were when they left the welding machine; the welds have not been filed or ground down. (National Museum of American History. Smithsonian Institution Neg. No. 74-7157.)

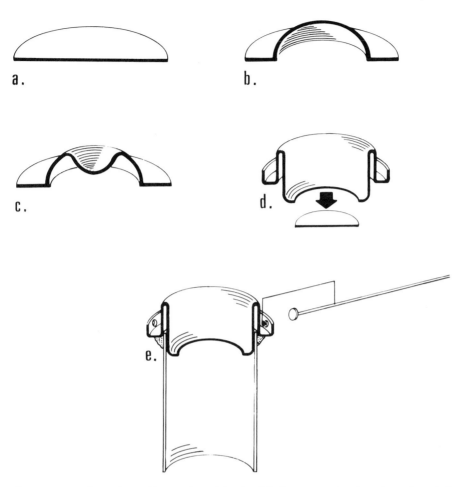

FIGURE 5.17. Steps in Making Hubs. This simplified cutaway drawing shows how hubs were produced from sheet steel. The actual process entailed several more steps than illustrated here. In step *d,* the punch press has formed the inside cup for the ball bearings. Step *e* shows how the steel tubing was brazed onto the flange and how spokes were attached to the flange. The axle and bearings were later inserted similar to Figure 5.13.

between each pressing step. As with crank hanger production, these intermediate annealing steps forced the Western company to develop a smooth system of materials handling in transferring parts between the pressroom and the annealing room. Western mechanics also planned their work so that no pressman ever remained idle. The pressman sat at his machine never having to move because a runner kept him well supplied with parts and removed finished pieces to the annealing room. After the flanges were completed they were brazed onto a body of heavy, drawn tubing and then chucked in a lathe for a fine cutting and truing operation.

Initially, the hubs were considered finished after brazing, but by 1896 Western mechanics had added this small turning step to gain greater precision. When produced in large quantities, press-formed hubs cost significantly less than the machined hubs of Yankee practice. Most Yankee shops, however, never adopted the method, maintaining always that the product was inferior. As George Pope wrote concerning another bicycle part, "Like all Western made stuff, it is awfully cheap looking and of course we do not want it."[74] Undoubtedly, Yankee mechanics were correct that Western hubs were not as strong as the forged or solid-bar machined hubs. There remained, however, the question of whether the Western hub was "good enough" for a bicycle and whether Yankee hubs were "too good." Herein lies a crucial question in American manufacturing technology.

A similar question arose concerning sprocketmaking. With the same ingenuity used for making hubs, Western mechanics developed a novel method of fabricating large sprockets. (See Figure 5.18.) An initial pressing operation lifted the edge of a round, flat steel disk to form a cup. A flat ring of metal the depth of the soon-to-be-cut teeth was then placed inside the cupped disk. Another press bent the rim of the cup over the ring, and another closed it down tightly on the ring. Thus the original disk now had a thick ring (a double thickness of the disk plus a smaller thickness of the inserted ring) about its circumference. Five ribs radiating from the center were then stamped into the disk to stiffen the arms of the sprocket, which were also punched out. Another operation punched out the center hole. Finally, a milling machine (one of the few New England machine tools used by the Western works) cut the teeth of the sprocket. After milling, the ring that was originally inserted in the disk fell out in pieces, leaving a small gap in the sprocket. This is an extremely ingenious method to make a sprocket; its final form, with thin arms stiffened by pressed ribs, cut down on both cost and weight. It represented·a radical departure from armory practice of drop-forging and then machining and served as a preview of automobile manufacturing techniques of the twentieth century.

Many other differences in specific manufacturing processes between the armory tradition and Western practice could be studied, but to do so would be less than fruitful. The examples covered adequately point up the radical departure which the Western company, among others, made from New England armory practice. The Western Wheel Works did use some Yankee machine tools, for example, gear-milling machines. It also used other machine tools, mostly of its own design and make. Almost all of these were fully automatic machines for such processes as cutting chain parts, assembling and riveting chains, and making nipples.[75]

In almost every process using machines, the Western Wheel Works employees sat to operate the machines and did not have to move any material before or after carrying out the operation. Runners handled all such materials flow. Horace L. Arnold, writing under the name Hugh Dolnar in the *American Machinist,* identified Western Wheel Works practice as being along "German locksmithing lines, with regular beer at 9:30 am daily." Yet he added that "the shop is under extremely good management, and no job is allowed to cost more than is needed to produce good, solid, reliable work."[76]

Unfortunately, I have uncovered no information on how the Western company assembled, finished, and controlled the quality of its bicycles, but it is fair to assume that these problems were attacked with the same ingenuity evident in frame-, hub-, and sprocketmaking. Had the bicycle industry continued to flourish rather than collapsing in 1897, the Western Wheel Works' apparent principle of bringing work to the men might have led the company to build mechanical conveying systems to replace the runners who carried work to and from machine operators and assemblers. Yet this kernel of manufacturing productivity had not fallen on good ground, and it remained for the automobile industry, notably the Ford Motor Company, to carry to its ultimate conclusion the idea of bringing work to the men. That the automobile industry picked up where bicycle makers left off is not surprising because the bicycle spawned the automobile.

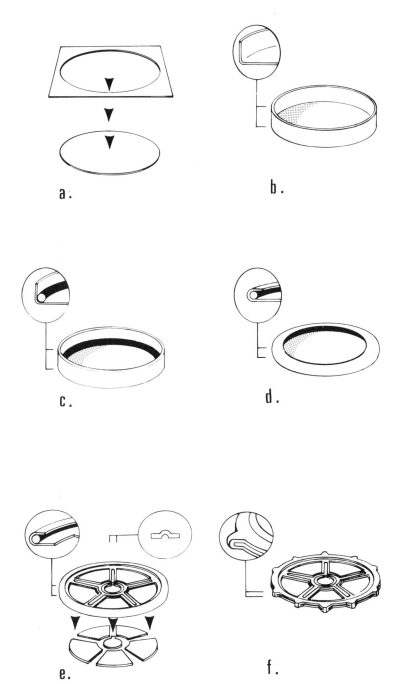

FIGURE 5.18. Steps in Sprocketmaking. This simplified cutaway drawing shows how sprockets were made from sheet steel. From the initial disk (step *a*), a cup was formed (step *b*). Workmen inserted a metal ring (step *c*) and folded the edges of the cup over the ring (step *d*). In step *e*, strengthening ribs have been raised in the radiating arms and around the center hole. Teeth were then cut in the workpiece to complete the sprocket (step *f*). The metal ring fell out in pieces. An example of this type of sprocket can be found in the historic bicycle collection of the Schwinn Bicycle Company.

Writing in an article on the bicycle industry in 1895, Albert A. Pope predicted or anticipated the ''advent of the motor-carriage.'' Good Roads, which Pope and other avid bicyclists had sought, would be essential: ''The day of the horse is already beginning to wane, and as soon as the practical motor-carriage can be had by men of moderate means we must have good roads, not only in and about cities, but throughout the entire country.''[77] The same year these words were published, Pope employed Hiram Percy Maxim to begin building experimental automobiles. In his reminiscences, *Horseless Carriage Days,* Maxim maintained that he had hit upon the automobile idea himself in 1892 while riding his bicycle:

> I saw [transportation] emerging from a crude stage in which mankind was limited to the railroad, to the horse, or to shank's mare. The bicycle was just becoming popular and it represented a very significant advance, I felt. Here I was covering the distance between Salem and Lynn on a bicycle. Here was a revolutionary change in transportation. My bicycle was propelled at a respectable speed by a mechanism operated by my muscles. It carried me over a lonely country road in the middle of the night, covering the distance in considerably less than an hour. A horse and carriage would require nearly two hours. A railroad train would require half an hour, and it would carry me only from station to station. And I must conform to its time-table, which was not always convenient.[78]

At that moment, Maxim related, he tumbled to the notion of a self-propelled vehicle. Maxim tells us that he did not pursue his by-no-means unique idea in 1892; he waited until 1895.

Looking back from the perspective of 1937, Maxim sought to explain his and the entire industry's delay in not constructing automobiles before 1895. ''The reason why we did not build mechanical road vehicles before this,'' Maxim wrote, ''was because the bicycle had not yet come in numbers and had not directed men's minds to the possibilities of independent, long-distance travel over the ordinary highway. We thought the railroad was good enough.'' But the invention and large-scale manufacture of the bicycle ''created a new demand which it was beyond the ability of the railroad to supply. Then it came about that the bicycle could not satisfy the demand which it had created. A mechanically propelled vehicle was wanted instead of a foot-propelled one, and we now know that the automobile was the answer.''[79] The ''demand'' which the bicycle created was, of course, the desire for swift and cheap personal transportation. Whether this transportation was used for ''the ride to work'' or for recreational purposes hardly mattered. With the mass production of the Ford Model T, this demand was not only met but for almost two decades was enlarged and sustained. Both the bicycle and bicycle production technology provided the basis for the age of the automobile in America.

Joseph Woodworth's statement about the importance of the bicycle for production technology, which heads this chapter, emphasizes interchangeability of parts. Based on our understanding of the progress and diffusion of the American system of manufactures in the second half of the nineteenth century, as exemplified at the Singer and McCormick factories, ''the installation of the interchangeable system of manufacture in a thousand and one shops'' was unquestionably an important result of bicycle manufacture.[80] Yet Woodworth had not yet recognized the revolutionary character of the introduction of presswork by western bicycle makers. Perhaps not until the rise of a large-scale automobile industry could this technique's versatility have been evident.

When sufficiently large, the automobile industry also demonstrated the importance of finishing, assembly, materials flow, quality control, and testing, which this chapter has suggested became critical for a few bicycle makers such as the Pope Manufacturing

Company. Henry Ford and the production experts at the Ford Motor Company especially recognized the problems that had arisen as a result of high-volume output, capitalized on some of the solutions offered by the bicycle industry (such as presswork and electric resistance welding), and during two years of intensive work between 1913 and 1915 suggested solutions of their own. When taken together, the Ford methods constituted for the first time the most radical and meaningful system of mass production.

CHAPTER 6

The Ford Motor Company & the Rise of Mass Production in America

Mass production is not merely quantity production, for this may be had with none of the requisites of mass production. Nor is it merely machine production, which also may exist without any resemblance to mass production. Mass production is the focussing upon a manufacturing project of the principles of power, accuracy, economy, system, continuity, and speed.
—Henry Ford, "Mass Production," *Encyclopaedia Britannica* (1926)

Henry Ford had no ideas on mass production. He wanted to build a lot of autos. He was determined but, like everyone else at that time, he didn't know how. In later years he was glorified as the originator of the mass production idea. Far from it; he just grew into it, like the rest of us. The essential tools and the final assembly line with its many integrated feeders resulted from an organization which was continually experimenting and improvising to get better production.
—Charles Sorensen, *My Forty Years with Ford* (1956)

Ford did not have to spend his life, like Oliver Evans, furthering ideas ungrasped by his contemporaries. He may have had the same indomitable energy; but he also had the advantage of coming not at the start but at the end of the mechanistic phase. Success does not depend on genius or energy alone, but on the extent to which one's contemporaries have been prepared by what has gone before.
—Siegfried Giedion, *Mechanization Takes Command* (1948)

Perhaps more than any other historian, Siegfried Giedion has placed the work of Henry Ford, or more correctly the Ford Motor Company, in an appropriate context of technological development. Giedion recognized the important prior developments in production of interchangeable parts, the idea of continuous flow, the rise of an efficiency movement, and the rich suggestion of Chicago slaughterhouse "disassembly" lines. From Giedion's perspective, Ford comes at the end of a long historical process which, in a Hegelian sense, becomes recognizable only at the end, when ever-unfolding historical reason makes itself known.[1] Although this interpretation of Ford deserves careful attention, it underestimates the singular importance of the changes made at the Ford factory in 1913 and 1914 (as well as how they came about) and the way these changes were rapidly diffused throughout the Western world. Both the *act* of mass-producing the Model T Ford and the rapid *diffusion of the techniques* by which it was mass-produced had a profound impact on the twentieth century. Fordism, a word coined to identify the Ford production system and its concomitant labor system, changed the world.[2]

This chapter examines the rise of mass production at the Ford Motor Company between 1908 and 1915. While concentrating on originality at Ford, it also emphasizes the prior developments upon which Ford depended. By contrasting Ford methods with those of Singer, McCormick, and Pope the role of the Ford Motor Company in the rise of mass production technology in America can be properly assessed. Finally, this chapter briefly considers the means by which knowledge of Ford methods diffused rapidly throughout the American technical community.

Mass production at the Ford Motor Company was rooted in the Model T idea and the fruition of that idea. Established in 1903, the Ford Motor Company was Henry Ford's third attempt at automobile manufacture. Not controlled by Ford until 1907, the company sold a number of medium-priced automobiles including the Models A, B, C, F, K, N, R, and S. By 1906, it had become apparent to Henry Ford that, in light of the existing and potential demand for automobiles in the United States, the "greatest need today is a light, low-priced car with an up-to-date engine of ample horsepower, and built of the very best material. . . . It must be powerful enough for American roads and capable of carrying its passengers anywhere that a horse-drawn vehicle will go without the driver being afraid of ruining his car."[3] According to many Ford experts, the Model N possessed some of these characteristics and could rightly be seen as the forerunner of the Model T.[4] Henry Ford, however, found enough fault with the N to decide that a new model was needed, one that would be larger and more powerful but still be called "light" and sell for less than the Model N. Ford battled with other company directors about the desirability of a new model, but when he acquired controlling stock in the Ford Motor Company in 1907, the debate ceased.

Henry Ford ordered that a separate area at the Detroit factory be set aside for the design of what became the Model T, and he started his best mechanics to work on that design. Together, Henry Ford, C. Harold Wills, Joseph Galamb, C. J. Smith, Charles Sorensen, and others arrived at a mechanical synthesis which, if not consciously designed to be, would become a "car for the masses."[5] It fulfilled Henry Ford's vague 1906 prescription for the "most needed" automobile design. A simple block, cast in one piece, provided the foundation for the twenty-horsepower, magneto-fired engine. The engine drove a planetary transmission with two forward speeds and a reverse, which were operated by foot pedals. A liberal use of vanadium-alloyed steel, along with some common-sense structural design, provided the Model T chassis with the desired strength, durability, and lightness. Altogether, the T fulfilled Henry Ford's mandate for simplicity of design and repair. It was destined also to fulfill the *Nation's* prophecy that "as soon as a standard cheap car can be produced, of a simple type that does not require mechanical aptitude in the operator, and that may be run inexpensively, there will be no limit to the automobile market." The world, according to *Harper's Weekly,* stood perched awaiting a car and a manufacturer for the masses: "There is no doubt . . . that the man who can successfully

FIGURE 6.1.   Model T Ford, 1913. Ford Motor Company produced the Model T from 1908 to 1927; some 15 million cars and trucks were made. (National Museum of American History, Smithsonian Institution Neg. No. 44002.)

solve this knotty problem and produce a car that will be entirely sufficient mechanically, and whose price will be within reach of the millions who cannot yet afford automobiles, will not only grow rich but be considered a public benefactor as well."[6] When the Model T left the Ford Motor Company experimental room in 1908, it met all of these mechanical demands. Through an alignment of circumstances that would have been difficult to predict, Henry Ford and the Ford Motor Company put the car within reach of those millions of Americans. (See Figure 6.1.)

Allan Nevins and Frank Hill record the response by Ford agents to the first announcement about the Model T, made March 19, 1908. One agent wrote, "We have rubbed our eyes several times to make sure we were not dreaming"; another exclaimed, "It is without doubt the greatest creation in automobiles ever placed before a people and it means that this circular alone will flood your factory with orders." Even before the factory had turned out a single product, agents had ordered fifteen thousand of the new Model T's.[7] From the beginning of Model T production until the end of World War I the Ford Motor Company, its factory, and its output of automobiles grew dramatically.

An alchemy of circumstances allowed for this growth. The roots of many of these circumstances are found in Henry Ford's business philosophy and its application by the Ford company's financial wizard, James Couzens. The company was financed from within, and after Henry Ford gained control he followed a policy that dictated against taking money out of the company through large dividends (or even large salaries for its top people). The massive profits that began to accrue were consistently plowed back. At the time, as in hindsight, it seemed that the Ford Motor Company did not want to make money as much as it wanted to build cars. With unquestioned financial stability and without any set notions about how automobiles should be made (that is, about the actual manufacturing processes), Henry Ford allowed an extensive amount of experimentation to be carried out in the factory and a surprising rate of scrapping processes and machine tools when they did not suit the immediate fancy of his production engineers. Ford had attracted to his factory a core of perhaps a dozen or a dozen and half young, gifted mechanics, none of whom had developed set ways of doing things. Encouraged by Ford, this group carried out production experiments and worked out fresh ideas in gauging, fixture design, machine tool design and placement, factory layout, quality control, and materials handling. Had the factory been rooted in a definite manufacturing tradition, such as Yankee armory practice or even "western" practice as exemplified by the Western Wheel Works, the Ford company might never have furnished cars for the masses. In a sense, the Ford production engineers took what was best from each approach to manufacture and overcame limitations to these methods by adding their own brand of production techniques. When they were finished, they had created—in Allan Nevins's words—a lever to move the world.

Until about two years before the introduction of the Model T the factory of the Ford Motor Company resembled more closely a poorly equipped job shop than a well-planned manufacturing establishment. Originally working in rented shops, the Ford Motor Company built its own factory in 1904 on Piquette Avenue in Detroit. The three-story plant, 402 × 52 feet, hardly matched the nearby Packard factory or that of Ransom Olds in Lansing. Because the company purchased most of its parts, the Piquette Avenue plant was designed for automobile assembly rather than for accommodating machine tools in large quantities. What tools the company possessed were general machines, operated by hard-to-find skilled machinists. Production during the first year at Piquette—1,745 automobiles—exceeded slightly that at the old factory in 1903–4. On the third floor, pre-assembled engines, frames, and bodies were put together into complete automobiles by teams of workmen. Perhaps fifteen such teams worked at different assembly stations, each demarcated by various piles of parts and by wooden stands upon which the cars were assembled. This method of automobile manufacture continued until the end of 1905, when Henry Ford joined with James Couzens to form the Ford Manufacturing Company as a means of obtaining control of the Ford Motor Company and as a mechanism to begin manufacture of parts for the recently introduced Model N, the light, four-cylinder runabout which Ford planned to sell for $500.[8] The organization and staffing of the Ford Manufacturing Company (consolidated with the Ford Motor Company early in 1907) laid the foundation, or more accurately, established the precedents for the rise of mass production at Ford in the early years of the following decade.

Rather than setting up to produce Model N engines and small parts in the Ford Motor Company's Piquette plant, the Ford Manufacturing Company rented a factory on Bellevue Avenue in Detroit and began to equip the shop. In purchasing machine tools Henry Ford came in contact with Walter E. Flanders, a machine tool salesman whom Charles Sorensen regarded as a "roistering genius."[9] Rather than a genius, Flanders was simply a genuine Yankee mechanic, a breed unknown to the young machinists around the Ford shops. A native of Vermont, Flanders had mastered the machinist's trade almost before he reached manhood and had witnessed quantity manufacture as an employee of the Singer Manufacturing Company. Before selling machine tools for Potter & Johnson, Landis Tool Company, and Manning, Maxwell & Moore (at the same time), the Yankee had built tools for the Landis company, one of the important pioneers in precision automotive grinding. Through his salesmanship, Flanders helped shape the approach to engine manufacture at the Ford Bellevue plant. He then suggested to Henry Ford that he hire Max F. Wollering to superintend the plant. Although young, Wollering proved to be the most competent manufacturing mechanic Ford had yet hired. Within a short span of years he had been employed by International Harvester as a toolbuilder and superintendent of gas engine production and by the Hoffman Hinge and Foundry Company of Cleveland, Ohio.[10] Wollering began at the Ford plant in the spring of 1906, and by August of that year Henry Ford had attracted Flanders to fill the post of overall production manager for the two companies.

In planning for the large-scale production of the Model N, Henry Ford caught for the first time that age-old New England contagion for interchangeability. Perhaps Flanders had passed it on to him. In an oral history interview conducted in the early 1950s, Max Wollering said that the common belief that Flanders had played a critical role in bringing the idea of interchangeability of parts into the Ford Motor Company was "all hooey." "There was nothing new [about interchangeability] to me," Wollering contended, "but it might have been new to the Ford Motor Company because they were not in a position to have much experience along that line." Whatever the origin of the idea, he emphasized that Henry Ford firmly grasped the importance—if not the techniques—of achieving interchangeability. "One of Mr. Ford's strong points was interchangeability of parts," Wollering said later. "He realized as well as any other manufacturer realized that in order to create great quantity of production, your interchangeability must be fine and unique in order to accomplish the rapid assembly of units. There can't be much hand work or fitting if you are going to accomplish great things." As Wollering reiterated, Ford "stressed that point very, very much."[11]

"We are making 40,000 cylinders," the Ford company advertised, "10,000 engines, 40,000 wheels, 20,000 axles, 10,000 bodies, 10,000 of every part that goes into the car . . . all *exactly alike*."[12] Although he advertised uniformity before his factory had actually achieved it, Henry Ford essentially gave Wollering and Flanders carte blanche to fulfill that which he had promised. When Wollering began his tenure he set the mechanics under him designing and building fixtures, jigs, and gauges for all the parts made at the Bellevue plant (devices he called "farmer tools" because with them he asserted that he could make a farmboy turn out work as good as that of a first-class mechanic). Wollering supervised the heads of each of seven departments: block, crankcases, and axles; bushings and small parts; engine assembly; second floor machinery; toolmaking; engine testing; and overall inspection.[13]

Flanders's arrival four months later initiated changes in machine tool placement, production departments, and materials purchasing policy. The Yankee mechanic placed machine tools according to sequential operations on various parts rather than by the types of machine (such as milling machines all in one department).[14] If hardening or softening or any such nonmachining operation needed to be carried out during this sequence, Flanders placed a furnace or whatever in the correct sequential location if possible. With regard to machining operations, Flanders impressed upon Ford and all the mechanics at the factory the desirability of interchangeable parts and the notion that absolute in-

FIGURE 6.2.   Static Assembly, Model N, Ford Motor Company Piquette Avenue Factory, 1906. The cramped condition of the Piquette Avenue factory would soon lead Henry Ford to expand the plant in 1907 and build the Highland Park plant, which opened in 1910. (Henry Ford Museum, The Edison Institute. Neg. No. 833-37306.)

terchangeability would become imperative in high-volume production. Flanders, as well as Wollering, also showed the Ford production mechanics the productivity gains possible through the use of special- or single-purpose machine tools. In October 1906, the Vermonter wrote a policy statement for manufacturing operations at Ford which dictated long-term purchasing of materials while at the same time requiring the supplier to carry the inventory. Flanders demanded that the factory keep on hand only a ten-day supply of these materials.[15] Being a bold fellow, he even made suggestions to Ford about sales policy. Charles Sorensen aptly summarized Flanders's contributions to the Ford company, saying that he ''created greater awareness that the motorcar business is a fusion of three arts—the art of buying materials, the art of production, and the art of selling.'' Clearly, as Sorensen recognized, Flanders—particularly in his rearrangement of machine tools— ''headed us toward mass production.''[16] (See Figure 6.2.)

The consolidation of the two Ford companies in 1907 and the enlargement of the Piquette Avenue factory allowed the company to move all of its machinery out of the Bellevue plant into the enlarged works. This move also allowed Flanders and Wollering additional opportunity to refine machine tool placement and the flow of materials throughout the factory. Perhaps at this time simple gravity slides (not unlike rain gutters) were installed in the factory between machine tools to move parts from one machining operation to another, thus expediting the flow of materials.[17]

Walter Flanders remained at Ford less than two years. Accepting a more attractive offer with the Wayne Automobile Company, he also took with him Max Wollering and Ford's advertising manager, LeRoy Pelletier.[18] In hindsight, it appears that Flanders stayed at the Ford Motor Company just long enough to introduce the fundamentals of an admittedly modern version of New England armory practice to the handful of young mechanics Ford had assembled. Had he remained longer he might have indoctrinated them with the belief that this approach was the one best way to manufacture cars. For the next three years the Ford engineers elaborated the basic principles shown them by Flanders, but eventually they moved beyond Flanders, taking only what suited them.

Henry Ford possessed an uncommon gift—or was unusually lucky—in attracting to his company well-educated mechanics who believed that "work was play."[19] C. Harold Wills, Oscar Bornholdt, Carl Emde, Peter E. Martin, Charles Sorensen, and August Degener, among others, formed the backbone of the Ford production team, a backbone given strength by Flanders's and Wollering's brief residence at the Ford factory. Draftsman, toolmaker, and better-than-amateur metallurgist, Harold Wills played a major role in Ford automobile design and factory layout from 1902 until after Highland Park was built. After Flanders left, Ford put Wills nominally in charge of manufacturing operations and machine tool procurement. Wills left these duties almost completely to "Pete" Martin and "Cast-iron Charlie" Sorensen. Henry Ford despised job titles, but Martin functioned as the factory superintendent and Sorensen as his assistant. According to Sorensen, Martin oversaw production while he worked at "production organization and development." Ford had hired both men shortly before Flanders arrived; Martin eventually became general superintendent and a vice-president of the Ford Motor Company while Sorensen became the mastermind behind Ford production plants in Europe, the River Rouge, and the Willow Run bomber factory. Carl Emde, a technically trained German immigrant, assisted Oscar Bornholdt in tool design and construction. Again, Ford had hired both machinists before 1906. Their contact with Flanders and Wollering proved very fertile. When Bornholdt left Ford in early April 1913, Emde took charge of tool design. By this date the Ford shops had arrived at a distinctive approach to machine tool, jig, and fixture design that clearly showed the marks of Bornholdt and Emde. Even before the Ford Motor Company was formed, Henry Ford hired August Degener as a draftsman. By the time the Highland Park factory opened in 1910, "Gus" had become the superintendent for inspection.[20]

Because Flanders left shortly after the company had announced the Model T—and long before it had actually produced one—this team of mechanics suddenly became responsible for "tooling up" for Model T manufacture. They faced more pressing problems than Flanders had encountered because of the rapidly rising demand for the Model T. When Flanders wrote his policy memorandum of October 1906, he called for the production of 11,500 automobiles (in three models) during the year from October 1906 to September 1907. Actual production reached only about 8,250. Nevins argued that quantity production at the Ford Motor Company began in the fall of 1907, but during the year previous to June 16, 1909, the factory turned out only 10,660 automobiles, less than a 30 percent increase from the 1906–7 period.[21] Table 6.1 shows the rapid rise in sales and the decrease in price of the Model T from its beginning in 1908 until 1916.

To P. E. Martin and Charles Sorensen fell the chief responsibility of getting the Model

TABLE 6.1. MANUFACTURING AND MARKETING OF MODEL T FORDS, 1908–1916

Calendar Year	Retail Price (Touring Car)	Total Model T Production	Total Model T Sales
1908	$850	n.a.	5,986
1909	950	13,840	12,292
1910	780	20,727	19,293
1911	690	53,488	40,402
1912	600	82,388	78,611
1913	550	189,088	182,809
1914	490	230,788	260,720
1915	440	394,788	355,276
1916	360	585,388	577,036

Sources: Columns 2 and 4: United States Board of Tax Appeals Reports, vol. 11, p. 1116, as reprinted in Alfred D. Chandler, Jr., *Giant Enterprise*, p. 33; column 3: my compilation based on monthly production reports, Ford Archives.

T into production. Although preparations may have seemed frenzied at times, the two superintendents, with toolbuilders Bornholdt and Emde, approached production methodically. As Henry Leland had done with Willcox & Gibbs sewing machines and Singer had done with its machines, the Ford production men wrote out operations sheets.[22] These detailed the machining operations on various parts, the requisite material inputs, and the necessary tools, fixtures, and gauges (all of which were numbered and referenced to drawings of parts) and suggested how the factory ought to be laid out according to the sequential structure delineated on paper. Preparation of these sheets brought order and clarity to what might have been a chaotic effort to produce the new model. In detailing requirements in machine tools the sheets also suggested possibilities for the design of entirely new machines. Rather than hardening into rigid policy statements, the operations sheets served as guides to production and materials procurement.

With information from operations sheets, Sorensen rearranged machine tools for Model T engine production, following the practice of sequential machining operations that Flanders had suggested.[23] The engine of the Model T differed significantly from that of the N, consisting of a single-cast block and a magneto rather than two castings (each with two cylinders) and a battery-fired ignition system. Sorensen's ability as a pattern-maker was clearly established by his solution to the problem of making a one-piece block. More important, he demonstrated his ability to bring original ideas to overall production when he recommended to Ford that stamping techniques rather than usual casting methods be employed for making crankcases.

Sorensen knew about steel stamping methods because he had grown up in Buffalo, New York, where the John R. Keim stamping company made bicycle crank hangers and other bicycle parts. According to his reminiscences, Sorensen had often prowled around the scrap pile at the Keim plant, picking up pieces that only a boy would find useful. Not long before Sorensen initially advocated pressed steel crankcases, William Smith, part-owner and superintendent of the Keim mills, had called at the Ford factory and suggested that Ford's rear axle housings could be made of pressed steel. Henry Ford encouraged both Sorensen and Smith. Soon Harold Wills and Sorensen went to Buffalo to see the Keim plant. Smith and his team of engineers made a suitable rear axle housing for the

FIGURE 6.3. Punch Press Operations, Highland Park Factory, 1913. Much of Ford's punch press machinery came from the John R. Keim Company of Buffalo, which Ford purchased in 1911 and moved to Detroit. At the far right, stacks of Model T transmission covers and crankcases are visible. (Henry Ford Museum, The Edison Institute. Neg. No. 0-6341.)

Model T and offered them at a cheaper price than cast ones. Before Model T products proceeded too far, Sorensen (with Ford's approval) adopted the use of pressed steel parts wherever possible—crankcase, axle housing, transmission case. Ford purchased the Keim company in 1911 and moved it to Detroit. (See Figures 6.3 and 6.4.)

With the company's equipment also came a group of talented engineers who played a decisive role in the development of mass production at Highland Park. This group included William Smith (who continued engineering work), John R. Lee (who became Ford's welfare department head), William Knudsen (who directed Ford's assembly plant operations in other American cities and eventually became the president of General Motors), Charles Morgana (who worked with Carl Emde as the Ford machine tool purchaser and conveyor of specifications for capital equipment), John Findlater (a diemaker who became Ford's master of presswork), and E. A. Walters (who succeeded Findlater in 1919 as the chief expert in presswork).[24]

While the company's production engineers and machinists worked out details of manufacturing the Model T and Fred Diehl devised a materials purchasing system along the lines suggested by Walter Flanders, Henry Ford and James Couzens concentrated on plans to construct a new factory in which the car for the masses would be built. In 1906, before

FIGURE 6.4.   Punch Press Operations, Highland Park Factory, 1913. (Henry Ford Museum, The Edison Institute. Neg. No. 833-2295.)

the design of the Model T had been completed, Ford had purchased a sixty-acre tract of land at Highland Park on the northern edge of Detroit and had begun work with architects on the proposed factory. Although neither Ford's biographers nor Ford company pioneers mentions Flanders in connection with the Highland Park factory design, the Yankee must have at least told Ford how he would build a factory if he were in Ford's place. Ford, Couzens, Flanders, and others clearly recognized that the Piquette Avenue factory, even when enlarged, was inadequate for the growing production of Model N's and that anticipated for the Model T. Not long after Flanders left Ford he professed that in order for less expensive automobiles ''to equal in quality cars now selling at $700 to $900, it is not only necessary to build them in tremendous quantities, but to build and equip factories for the economical manufacture of every part.''[25] With large profits pouring into Ford's enterprise, it seemed natural to think about erecting a substantial factory along the lines envisioned by Flanders. The directors of the company approved the expenditure of a quarter of a million dollars in mid-1908. The factory opened formally on New Year's Day 1910, although construction at Highland Park continued throughout the next half-dozen years until the sixty acres would hold no more buildings.[26] (See Figure 6.5.)

    The design of the Highland Park factory allowed architect Albert Kahn to elaborate upon work he had started in 1905, when at age thirty-six he designed a new factory for the Packard Motor Car Company, a ''daylight factory'' of extensive windows set in rein-

FIGURE 6.5.   Highland Park Factory, 1923. This aerial photograph was taken at the peak of Highland Park's production. The 8,000-horsepower power plant is in the center of the photograph and the sawtoothed roof of the machining area is visible at the left. This area was connected by a glass-enclosed craneway to a four-story building 865 feet long and 75 feet wide. (Henry Ford Museum, The Edison Institute. Neg. No. 833-34974.)

forced concrete. The principal structure at Highland Park consisted of a four-story building 865 feet in length and 75 feet in width with some fifty thousand square feet of glass (roughly 75 percent of wall area). In a matter of months Kahn placed beside this structure a single-story building with a sawtooth glass roof, 840 × 140 feet, which served as the principal machine shop. Kahn connected these buildings with an impressive, glass-enclosed craneway, 860 × 57 feet. The main building as well as the machine shop opened completely into the craneway on all floors so materials could be moved with ease from one building to the other through the craneway. This craneway would serve as the major distribution point for all raw materials that made up the Model T.[27]

P. E. Martin and Charles Sorensen laid out careful plans for a smooth move into the new factory. Henry Ford simplified their plans in 1909, when he announced that the Ford Motor Company would henceforth make only the Model T and that the runabout, touring car, town car, and delivery car would all consist of identical chassis.[28] Now the plant superintendents no longer had to worry about transferring the Model N production equipment. Besides freeing the usable machine tools, Ford's decision allowed Martin, Sorensen, Emde, and Bornholdt to initiate the design, construction, or procurement of large numbers or special- or single-purpose machine tools. This is what the American system of manufactures was all about. Before moving machine tools to Highland Park, Sorensen and Martin drew up layout boards—scaled figures showing the correct placement of the machinery. They numbered each machine site and then attached brass plates to the machine tools at the Piquette factory. With these plans, the company's millwrights easily designed the electrically driven shafting at the new Highland Park works and correctly placed the machinery. When Highland Park began production, department by department, it was, as Sorensen later wrote, a "progressive" but not a "fully integrated operation."[29] During the next four years Sorensen and his fellow production engineers would effect profound changes at the "Crystal Palace" factory.

In the period between the opening of Highland Park (January 1, 1910) and the installation of the first assembly line (April 1, 1913) the work of the tool department, the move of the Keim pressed steel plant to Detroit, and the six- to tenfold expansion of output (depending on how one counts) distinguish manufacture of the Model T. Historians of the Ford Motor Company, when talking about factory operations, usually make immediate reference to what they and others have called the "classic" work by Horace Lucien Arnold and Fay Leone Faurote, *Ford Methods and the Ford Shops* (1915).[30] Although the work is a classic, few historians have fully understood it, and none has placed it in the context of another series of articles on the Ford factory, written by Fred Colvin in the *American Machinist* in 1913.[31] In many ways the Colvin series surpassed that of Arnold and Faurote in that it gave better details about the machine tools and the fixture and gauging system at the Ford factory. Colvin also compared and contrasted the Ford methods with those of other leading shops, and he grappled with the meaning of large-scale production. Most important, however, the *American Machinist* series described Ford factory operations immediately before the dawn of the assembly line, thus allowing us to see how far Sorensen and others had carried Walter Flanders's Yankee notions and how, once moving assembly was tried, those notions that dealt with assembly were suddenly scrapped.

When Fred Colvin visited the Ford Motor Company plant in the spring of 1913,[32] he was impressed by the way Ford engineers had concentrated on the "principles of power, accuracy, economy, system, continuity, and speed"—Henry Ford's elements of mass production. Noting that Ford manufactured over half the entire United States output of cars, Colvin suggested, "We think of 200,000 automobiles in a single season as being unheard of, if not impossible, as we can hardly imagine such an output. . . . We lose all sense of proportion, and we get to the point where we are quite as ready to accept a million as the proper figure as the paltry(?) 200,000." The well-known technical journalist tried to suggest the meaning of such an output. A million lamps; eight hundred thousand wheels and tires; ninety thousand tons of steel; four hundred thousand cowhides; 6 million pounds of hair for seats; and about 2 million square feet of glass went into the year's production. A complete Model T emerged from the factory every forty seconds of the working day. Five trains of forty cars each left the factory daily, loaded with finished automobiles. In a span of five years the company had gone from producing about six thousand Model T's to roughly two hundred thousand and had lowered costs. "What more could the greatest high priest of efficiency expect?" Colvin asked.[33] Unknown to Colvin, a month before these words appeared in print the priests of efficiency at Ford had made their first experiment with an assembly line.

The power plant at Highland Park, designed by Ford's construction engineer, Edward Gray, and built by the company, consisted of a three thousand-horsepower gas engine, which turned direct current generating equipment. Power was distributed throughout the factory by electric motors, which drove units of line shafting and belting. When Colvin toured the factory, construction was nearing completion on an additional five thousand-horsepower gas engine. The increasing output of Model T's demanded power of this magnitude.[34]

"The Ford testing method is unique and simple," Colvin wrote when he assessed standards of accuracy at Highland Park.[35] Every critical part of the Model T was machined in standard fixtures and checked by standard gauges both during and after the operation sequence. With proper attention by the tool department and the inspection department, the factory maintained essential accuracy. When a unit such as the engine, the transmission, or the rear axle assembly was put together, its bearings were checked with an electric motor. Unlike most automakers, Ford did not run its engine before assembly into the chassis. Not until the car was ready to leave the factory was the engine started. The company did not road-test any Model T. Sorensen, Martin, and others maintained that if parts were made correctly and put together correctly, the end product would be correct.

Principles of economy abounded at the Ford factory. Colvin suggested that the ever-declining price of the Model T served as a testimonial to these principles. He cited numerous instances of economy at Highland Park, all of which were tied to the principles of system, continuity, and speed so evident there. Establishing a theme that would be picked up by other journalists such as Arnold and Faurote, Colvin emphasized the close grouping of machine tools and how this economy of space militated against letting work accumulate in the aisles and made imperative a smooth flow of work throughout the banks of machine tools.[36]

Not long after the Highland Park plant had opened, a newsman from the *Detroit Journal* described the salient feature of the Ford production process as "System, system, system!"[37] In his *American Machinist* series, Fred Colvin reiterated this theme. Only the word "system" could be used to describe the way Fred Diehl purchased materials, their distribution throughout the factory from the main craneway, and the method the company used to handle finished stock. But Colvin was more impressed by the placement of machine tools: "So thoroughly is the sequence of operations followed that we not only find drilling machines sandwiched in between heavy millers and even punch presses, but also carbonizing furnaces and babbitting equipment in the midst of the machines. This reduces the handling of work to the minimum; for, when a piece has reached the carbonizing stage, it has also arrived at the furnace which carbonizes it, and, in case of work to be finished by grinding, the grinders are within easy reach when it comes from the carbonizing treatment."[38] Ford's machine tool expert, Oscar Bornholdt, had likened this sequential operations setup to "the making of tin cans." "At the Ford plant," Bornholdt wrote, "the machines are arranged very much like the tin-can machines"—one right after the other.[39]

Sorensen and Martin had devised a work-scheduling system for the factory. From experience, the average output of each machine tool was recorded and served as the basis for scheduling. If output of a certain class of machine tools in a department was rated at, say, one hundred pieces per machine per day and there were five such machines, total average output would be five hundred pieces. If the production schedule for a single day called for only four hundred pieces, the scheduling system dictated that one machine be shut down while the others turned out their full day's average. Special timekeepers monitored how closely the departments kept to their production schedules.[40] Largely through such systematization, the Ford engineers maintained continuity in the input and output of materials at a calculated rate.

FIGURE 6.6.   Quick-Change Fixture for Crankcase Drilling, 1913. This photograph is one of many examples that Fred Colvin used to illustrate the way Ford machine tool designers had built speed into manufacturing operations at Highland Park. The crankcase was quickly loaded into the fixture and then the entire assembly was rolled under the multiple spindle drill press. (Henry Ford Museum, The Edison Institute. Neg. No. 0-6342.)

The principle of speed was apparent to Colvin everywhere he turned in the Ford plant. He stressed, however, that the most impressive application of the principle was in the design of fixtures and gauges by Oscar Bornholdt, Carl Emde, and others in the tool department. The bulk of Colvin's series concerns the design and use of these devices, whose speed, accuracy, and simplicity epitomized the entire Ford production process. (See Figure 6.6.)

The Ford tool experts designed almost all of the fixtures and gauges so that they could be used by unskilled machine tenders. Simplicity, therefore, was an important concern, yet in certain instances this succumbed to the more important considerations of speed and accuracy. Excited by the rationality of absolute interchangeability of parts and painfully aware of the problems created by noninterchangeability in the troublesome assembly process, Ford's production engineers placed accuracy at the top of the list in fixture and machine tool design requirements. By 1913 Emde and others had achieved simplicity and speed in most of their design work without sacrificing accuracy. This achievement deeply impressed Fred Colvin, who had studied many of the leading factories in the United States. For example, the Ford team engineered milling machine fixtures and tables that

FIGURE 6.7. Machining Engine Blocks, 1915. Ford used multiple head milling machines to machine the blocks and the heads of the Model T engine. Special, easily loaded fixtures held fifteen blocks at a time for accurate machining. (Henry Ford Museum, The Edison Institute. Neg. No. 0-3927.)

held fifteen engine blocks at a time, each easily snapped into place and held rigidly, and similar devices for holding thirty cylinder heads at once. (See Figure 6.7.) Colvin marveled that when brought together the head and block would hold compression with only a plain gasket and without customary—and time-consuming—joint scraping.[41] Readers interested in more details about Ford fixture design should consult the almost countless examples given by Colvin in his series of articles.

Ford tool design depended on a subtle but important interplay with the machine tool industry. Charles Sorensen suggested that Ford men designed all the new machine tools at Highland Park, built a prototype for each, and then relied upon commercial toolbuilders to supply additional machines. He recalled that when Charles Morgana sent out specifications for a Ford-designed machine tool to machine tool manufacturers, the latter often came back to Morgana saying that there must have been an error because the machine could not do what it was supposed to do. Morgana would then show the toolbuilders that no mistake had been made because the Ford-designed and Ford-built prototype could indeed turn out the specified number of units within the specified limits of precision. "So it went on with the thousand pieces of machinery that we bought," concluded Sorensen.[42]

Sorensen no doubt claimed too much in saying that the Ford Motor Company's tool

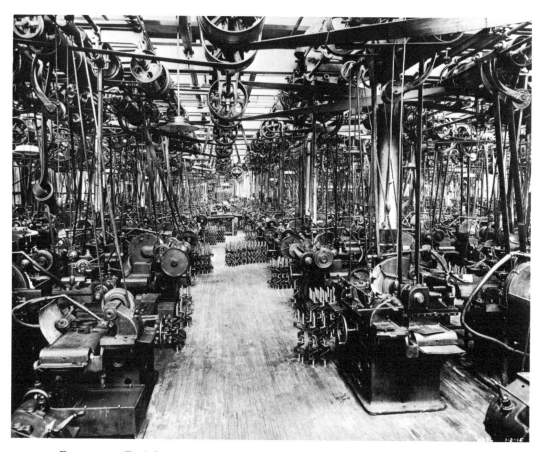

FIGURE 6.8. Ford Crankshaft Grinding Machines, 1915. Developments in grinding technology, such as the machinery for grinding crankshafts, played an important part in the achievement of accurately machined automobile parts. (Henry Ford Museum, The Edison Institute. Neg. No. 833-2296.)

department designed and built at least one of each kind of machine tool in the Ford factory. The Ford team built many of the special machines used for Model T production, but as Ford machine tool purchaser A. M. Wibel maintained, the company relied heavily on midwestern toolbuilders such as Foote-Burt, Ingersoll, and Cincinnati Milling Machine for initial construction, if not design.[43] Unfortunately, we know less about the general development of machine tools between 1900 and 1915 than for the entire nineteenth century, so any assessment of the state of the art must be tentative.[44] One can only speculate that improvements in the accuracy and speed of machine tools during this period, which resulted largely from metallurgical development and greater rigidity, provided a critical component in Ford's—and the entire automobile industry's—rapid expansion of production capability. In view of the assembly problems at the Singer Manufacturing Company caused by inaccurately machined parts one cannot overemphasize Ford's insistence on accuracy. In Chapter 2 the question was raised of whether in the 1870s and 1880s high-volume, economical production of accurate parts was technologically possible. By 1913, when Colvin wrote the series in the *American Machinist* and when Ford

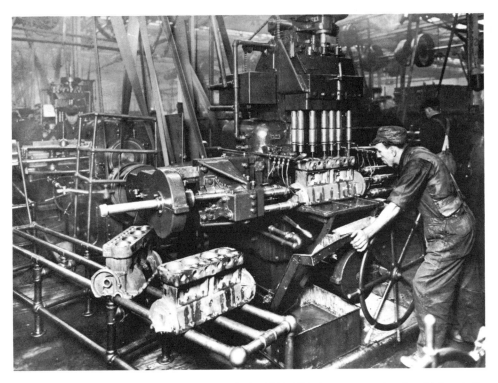

FIGURE 6.9.   Drilling and Reaming Engine Block, 1913. This is one example of multiple spindle drilling and reaming machinery designed to machine the Model T engine block. (Henry Ford Museum, The Edison Institute, Neg. No. 833-219.)

initiated line assembly techniques, the machine tool industry was capable—perhaps for the first time—of manufacturing machines that could turn out large amounts of consistently accurate work.[45] (See Figure 6.8.) From the time the Ford Motor Company moved into the Highland Park factory, its production engineers and its principal owner did not compromise on this issue. As will be seen, this accuracy provided the rock upon which mass production of the Model T was based. Nevins quotes an authority on the automobile industry who argued that the "Ford machinery was the best in the world, everybody knew it."[46]

Henry Ford's determination to produce only the Model T provided his engineers the perfect opportunity to install single-purpose machine tools. The engine department, for example, relied extensively on such machines. Emde's department built special block and head spotting machines, which faced, or machined, the bearing points that were used for locating these parts in subsequent machining operations. Special machines bored out the cylinders and the combustion chambers of the head. Another machine tool drilled at one time forty-five holes in four sides of the block. (See Figure 6.9.) Colvin pointed out that "these spindles are non-adjustable so far as location is concerned." The Ford engine tools provided examples "of the single-purpose machine carried to the limit." Other special-purpose engine tools included a drilling machine for babbitt bearing anchor holes, other types of drilling machines, and broaching machines for valve stem bushings. These examples could be multiplied by the number of other partsmaking departments.[47]

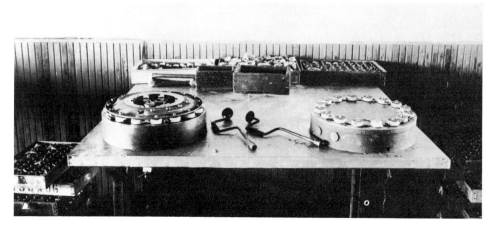

FIGURE 6.10.   Magneto Coil Assembly, Highland Park, 1913. (Henry Ford Museum, The Edison Institute. Neg. No. 0-6337.)

Henry Ford's and his production engineers' constant experimentation with new production methods and their willingness to scrap processes and tools are perhaps best illustrated in pressed steel work. Ford had adopted the use of stamped steel rear axle housings, a process developed in Buffalo by the John R. Keim company. Ford purchased the company and moved its machinery and its leading mechanics to Highland Park in 1911. Not long after setting up at the "Crystal Palace," the old Keim team and the Ford engineers scrapped the rear axle housing stamping process after they developed a superior process that involved flaring out the ends of drawn steel tubing and riveting them to a malleable cast-iron differential housing.[48] The Keim mechanics found other applications for punching, pressing, and stamping, however, such as setting three shafts for transmission gears in the flywheel by a single punching operation.[49] As Colvin pointed out, machine tools and other production processes were constantly undergoing examination and change at Ford. This was clearly the case with processes by which the Ford automobile was put together.

Assembly of Model T components and the entire automobile greatly impressed the *American Machinist* writer. He wrote that after he had watched the entire assembly process he could "see that the production of 800 cars a day is not merely guesswork."[50] Colvin detailed either in the text or through photographs the motor assembly department, rear axle assembly, magneto assembly, radiator assembly, and finally the overall assembly. Ford engineers had set up simple workbenches for putting together the field windings of the magneto. (See Figure 6.10.) At the back of each bench and on its sides, small bins held the various parts that made up the field assembly. A workman stood at the worktable, putting together the parts of this important subassembly.[51] For assembling engines, Ford engineers also used workbenches, but instead of being located against a wall, the engine assembly benches were placed in the open so that men could work on both sides of them. (See Figure 6.11.) Parts bins were placed in the middle of the tables, easily reached from either side.[52] Colvin emphasized that there was no fitting—and therefore no fitters—in any Ford assembling department.[53] The rear axle assembling department relied upon assembly stands almost identical to those used in many New England shops, particularly the Pope Manufacturing Company during the bicycle craze of the 1890s. (See Figure

FIGURE 6.11. Engine Assembly, Highland Park, 1913. Individual workmen assembled entire engines by themselves. Unlike most of the photographs used by Fred Colvin in 1913, the original print of engine assembly no longer survives in the Ford Archives. (*American Machinist*, June 12, 1913. Eleutherian Mills Historical Library.)

FIGURE 6.12. Rear Axle Assembly Stands, Highland Park, 1913. Individual workmen assembled rear axles entirely by themselves at these stands. (Henry Ford Museum, The Edison Institute. Neg. No. 0-6336.)

FIGURE 6.13.   Dashboard Assembly Stands, Highland Park, 1913. (Henry Ford Museum, The Edison Institute. Neg. No. 0-6335.)

6.12.) These stands provided the necessary open work area and also held the parts in bins conveniently reached by workmen. The stands were placed far enough apart so that hand trucks could move material to and from them. As with other subassemblies, individual workmen performed the entire process of rear axle assembly. Colvin said that the design of the stand "show[s] that motion study has been carefully looked into, whether it is called by that name or not."[54] Faced with laborious threading of radiator fins and tubes together, production engineers designed a simple mechanism that pushed ninety-five tubes through the holes in the strips or fins in a single stroke. After mechanized core assembly was completed, however, the remainder of radiator assembly consisted chiefly of laborious hand soldering of the core to the tank and to the frame.[55] (See Figure 6.13 for dashboard assembly.)

"It is impossible to give an adequate description of the general assembly of the Ford automobiles, as this could only be done with a modern moving-picture machine," wrote Colvin about the final assembly process. "As in the machining department the keynote of the whole work is simplicity, even to the *assembling horses* or *stands* shown."[56] (See Figure 6.14.) Laborers distributed the necessary parts at each station and timed their deliveries so that they reached the station shortly before the parts were needed. While the automobile frames remained static upon the horses, dynamic assembly teams or gangs moved from station to station down the row. Each gang had been programmed to perform

FIGURE 6.14.   Static Assembly of Model T Chassis, 1913. Unfortunately, the moving assembly gangs were not included in the photograph. (Henry Ford Museum, The Edison Institute. Neg. No. 0-1267.)

a specific task or series of tasks. Colvin pointed out that this method resembled that used at the Baldwin Locomotive Works and other shops. When carefully orchestrated—as the Ford assembly team was—the method worked well.[57] One might imagine, however, that problems of correct delivery of materials and of assembly gangs not keeping within their time limits (and therefore getting into each others' way) plagued the Ford factory. These problems were soon eliminated.

In "moving the work to the men," the fundamental tenet of the assembly line, the Ford engineers found a method to speed up the slow men and slow down the fast men. The assembly line would bring regularity to the Ford factory, a regularity almost as dependable as the rising of the sun. With the installation of the assembly line and the extension of its dynamism to all phases of factory operations, the Ford production engineers wrought true mass production.

It has generally been assumed that because Nevins and Hill were given complete access to the Ford Motor Company archives and because Nevins was usually a careful scholar, their account of the development of the assembly line is, if not definitive, at least accurate in its broad outlines. The authors maintain that employment of conveyor systems and gravity slides throughout the Ford factory led almost naturally to the assembly line. Pointing out that "no contemporaneous documentary record of the great innovation exists," Nevins and Hill turned to the recollections of the Ford pioneers, made some forty years after the event.[58] Some of these former employees suggested that conveyor systems

FIGURE 6.15.   Ford Foundry Mold Conveyors, 1913. Molds were carried around on the platforms (lower foreground) past bull ladles from which molten iron was poured. (Henry Ford Museum, The Edison Institute. Neg. No. 0-6338.)

and gravity slides had been in use well before April 1, 1913, and that their elaboration eventually led Sorensen et al. to install an assembly line in the magneto department, which in turn led to lines in the engine assembly department, the rear and front axle assembly departments, and finally to chassis assembly.

Although they were aware of the Colvin series of articles in *American Machinist*, Nevins and Hill ignored the contemporaneous evidence—particularly pictorial evidence—in this gold mine of information. Colvin spent ten days at Highland Park perhaps less than two months before the first experiments were conducted on a magneto assembly line.[59] Throughout his articles, Colvin mentions and documents with photographs piles of parts and hand trucks that carried these parts through the factory. Nowhere does he mention conveyors or gravity slides, and none appears in any photograph. Fred Colvin was too keen an observer, too much an advocate of smooth flow of materials, to overlook gravity slides, gravity rollers, and conveyor systems. He does document fully a monorail system, which moved large trays or platforms of work throughout the factory. This materials handling system was typical of many shops, but the conveyor systems and gravity slides that show up in the photographs of the Ford factory in 1914 and 1915 were different.[60] It appears, therefore, that conveyors and gravity slides were adopted either immediately before the assembly line experiments or resulted from the "work in motion" principle

FIGURE 6.16.   Molding Machines, Ford Foundry, 1913. In the center of this photograph are shown molding machines and the hoppers that fed them. Sand-conveying machinery appears in the left foreground. (Henry Ford Museum, The Edison Institute. Neg. No. 0-6339.)

brought to life by the assembly line. The latter view seems more likely. In any case, these two elements of mass production fed on each other; by 1915 both had reached a maturity unknown in 1913.

In his reminiscences, Charles Sorensen claimed that the idea of an assembly line occurred to him in 1908 and that on consecutive Sundays in July of that year, he, Henry Ford, Harold Wills, P. E. Martin, and Charles Lewis, an assembly foreman, laid out a crude chassis assembly line. Wills and Martin rejected the idea out of hand, Sorensen remarked, and it was buried in the rush to open the new Highland Park plant.[61] Although perhaps apocryphal, Sorensen's account suggests that the Ford engineers had been concerned about the problem of assembly. Faced in 1913 with the task of putting together almost two hundred thousand Model T Fords, however, Wills and Martin consented to more extensive experiments.

Sorensen did play a major role in the development of the assembly line.[62] He contributed his expertise in patternmaking and foundry work to the operations of the Highland Park foundry. In February 1913, a conveyor-type mold carrier began operation in the foundry. (See Figures 6.15 and 6.16.) The mold carrier moved past molding machines (at which point the completed molds were set on the carrier) and around past a bull ladle which allowed for continuous pouring of molten iron. (The engine blocks, however,

FIGURE 6.17. Westinghouse Foundry, 1890. A conveyor system (1) carries machine-made molds past pourers. Men break open the molds after sufficient cooling (2). The sand conveying system is illustrated in (3). (*Scientific American,* June 14, 1890. Eleutherian Mills Historical Library.)

continued to be laid out on a pouring floor rather than being put on the carrier.) As early as 1890, the Westinghouse Airbrake Company had devised a similar mold carrier. (See Figure 6.17.) Integral to both the Westinghouse and the Ford mold carriers was a conveying system that moved sand from the spot where the molds were taken off the carrier and shaken open to a sand-mixing operation and thence to hoppers above the molding machines.[63] This system eliminated wheelbarrows and almost all hand shoveling. The productivity of the Ford foundry astonished technical journalists. Rough calculations show that although the Ford foundry had half the area of Singer's Elizabethport foundry (in 1880), it poured daily more than ten times the amount of iron. By 1914, ten continuous-pouring mold carriers had been installed at Highland Park.

In addition to the mechanized foundry of Westinghouse Airbrake, developments outside metalworking practice played on the minds of Ford production men. Three industries in particular seemed to provide models of efficient and smooth materials handling. In his autobiography, written in collaboration with Samuel Crowther, Henry Ford suggested that the ''disassembly'' lines of Chicago meatpackers served as a model for ''flow production'' at the Ford factory.[64] (See Figures 6.18 and 6.19.) Packing houses had come to public attention with the 1906 publication of Upton Sinclair's *The Jungle*, which described their operations in vivid detail. William Klann, head of the engine department at Ford, recalled that he had toured Swift's Chicago slaughterhouse and had then suggested to superintendent P. E. Martin, ''If they can kill pigs and cows that way, we can build cars that way and build motors that way.''[65] Klann also stressed that the Ford flow production also drew upon the mechanical conveying system of both the flour milling and brewing industries.

In 1904, the year before he joined the Ford Motor Company, Klann worked as a machinist repairing grain elevators and other mechanical conveyors in breweries for the Huetteman & Cramer Machine Company of Detroit. Klann claimed that both breweries and foundries used essentially the same hoppers and conveyors to feed those hoppers for, respectively, malting and moldmaking. Huetteman & Cramer made this equipment. Klann recalled that a fellow employee at Huetteman & Cramer, who also later worked for Ford, first interested Henry Ford in mechanical materials handling ''by showing him a catalogue'' of foundry and brewing conveyors and hoppers.[66] Although relatively new in foundry practice, mechanized conveyance had been used in breweries soon after Oliver Evans developed his automatic flour mill in the late eighteenth century.[67] (See Figure 6.20.)

Since the days Evans first operated his automatic flour mill on Red Clay Creek in northern Delaware, the flour milling industry had used and continued to refine his system of mechanical conveyance. Minneapolis had become the flour milling capital of the world by the late nineteenth century, and many informed people were well aware of the sophistication of automatic materials handling equipment in these mills. Indeed, the American system of milling was an object of pride in the United States at this time.[68] Ford production men should have at least heard about these mills. Certainly William Klann had. Klann summed up the importance of all three of these industries: ''We combined our ideas on the Huetteman & Cramer grain [conveying] machine[ry] experience, and the brewing experience and the Chicago stockyard. They all gave us ideas for our own conveyors.''[69] Yet another process technology may have influenced the Ford production men.

Ford's principal machine tool expert, Oscar C. Bornholdt, had in 1913 compared the sequential arrangement of machine tools at the Ford factory to the layout of food canning machinery. An earlier—and striking—illustration of a successful mechanized cannery in Chicago shows that not only were canning machines arranged sequentially but that they were linked by automatic conveyance systems that brought the work to the worker.[70] The illustration is richly suggestive. (See Figure 6.21.) If Bornholdt had established in his own mind an analogy between Ford's machine tools and canning machinery, there is no reason to doubt that he did not make and exploit a similar analogy between materials flow at the Ford factory and the movement of cans in a food-processing facility. Moreover, Klann's tour of a Chicago meatpacking plant could have easily included a cannery. If Bornholdt had seen a mechanized cannery, it is reasonable to assume that other key Ford employees were familiar with similar systems.

The unquestioned success of the mechanization of the Ford foundry, as well as the rich suggestions of meatpacking, milling, brewing, and canning, touched off a burst of experi-

FIGURE 6.18. "Disassembly" Line, Slaughterhouse, 1873. An early example of "flow" production, slaughterhouses such as this one began first in Cincinnati and later became famous in Chicago,
the "hog-butcherer of the world," in the era of Henry Ford. (*Harper's Weekly,* September 6, 1873.
Eleutherian Mills Historical Library.)

FIGURE 6.19. "Disassembly" Line, Slaughterhouse, 1873. Note the ham traveling down the
gravity slide. (*Harper's Weekly,* September 6, 1873. Eleutherian Mills Historical Library.)

FIGURE 6.20. Evans's Automatic Flour Mill, Occoquan, Virginia, 1795. Conveying devices carry wheat from either wagon or boat through all the steps of screening, grinding, and bolting. (*Annals des arts et manufactures*, 1802. Eleutherian Mills Historical Library.)

FIGURE 6.21. Norton's Automatic Canmaking Machinery, 1885. This can line, distinguished by special-purpose machinery and a conveyor system, was the creation of Edwin Norton, who later organized both the American Can Company in 1901 and the Continental Can Company in 1904. (*American Machinist*, July 14, 1885. Eleutherian Mills Historical Library.)

mentation and change at the factory in which every*thing* was put in motion and every *man* brought to a halt. Sorensen claims—and Nevins and Hill corroborate—that he designed a conveyor system for moving radiator work through the machining and assembly processes. He creates confusion, however, by adding that the radiator conveyor took finished radiators all the "way to the assembly line."[71] There was no chassis assembly line until August 1913. If Sorensen did install a radiator conveyor system before the magneto assembly line installation of April 1, 1913, he must have done so in February or March. Fred Colvin visited the factory sometime between January and "the spring of 1913." Had the radiator conveyor been installed or even partially installed, Colvin would have noted it, for he treats in detail the production and assembly of radiators in one of his articles.[72]

This discussion is not solely for the purpose of quibbling with Nevins's and Hill's details of the development of the assembly line. As already noted, these authors suggest that the installation of conveyor systems in radiator assembly and in engine assembly, as well as all the supposed gravity slides and gravity rollers in the machine shop, brought Sorensen et al. logically to the assembly line. All of the contemporary evidence, however, especially that in Colvin's lengthy series on Ford methods, suggests the contrary. No doubt the foundry conveyor system encouraged the Ford engineers to try the magneto assembly line, but it was the rapid rise of the assembly line that brought about the immediate installation of conveyor systems wherever they could be installed. Whatever its origin, its importance is that the Ford production experts were bringing about such rapid-fire changes that none of them could keep straight which came first, the assembly line or mechanized conveyance. Only a devoted diarist could have kept these changes straight, and none of the Ford men was a diarist. They were too busy. (See Figure 6.22.) The adoption and elaboration by Ford of sequential arrangement of machine tools and the dynamism of the foundry's mold carrier may have led logically to full-scale mechanized conveyance and assembly lines, but Nevins's metaphor or rivulets flowing into streams flowing into great rivers is inappropriate.[73] With the speed, magnitude, and impact of change at Ford, this was the Deluge, the Great Flood, which wiped out all former notions of how things ought to be moved and assembled.

As with the question of whether conveyors and work slides or the assembly line came first, there is ambiguity about exactly when and where the assembly line was first implemented at Highland Park. The standard account of Allan Nevins quite rightly relied heavily upon the work of Horace Arnold, which was written about a year after the innovation took place. But Nevins also drew upon the oral history interview of William Klann, who was the foreman of motor assembly at Ford in 1913. Klann's reminiscences are among the most extensive and vividly detailed of any of the Ford Motor Company employees who worked during the Model T era and who were interviewed for the commissioned history of the company. By combining the accounts of Arnold and Klann, Nevins concluded that the first attempt at moving assembly took place in the magneto coil department under the direction of James Purdy. But in discussing the results of this experiment, the historian cited the productivity figures reported by Arnold for the flywheel magneto assembly. Nevins obviously confused the magneto coil assembly, a flat metal disk that supported sixteen coils and was mounted rigidly on the rear of the engine block, and the flywheel magneto, a flywheel with sixteen V-shaped permanent magnets bolted on its front side. Elsewhere in the first volume of his study of Ford, he reproduced a photograph of workmen assembling the magnets onto the flywheel and called this the "first magneto assembly line."[74] (See Figure 6.23.)

Even when this distinction is kept in mind, however, confusion and contradiction

FIGURE 6.22.   Some of the Principal Creators of Mass Production at Ford Motor Company, 1913. This is the superintendent's office at the Highland Park factory. Seated (left to right): Charles Sorensen, P. E. Martin, and C. Harold Wills. Standing directly behind Sorensen is Clarence W. Avery. Note the Model T chassis in the rear of the office. (Henry Ford Museum, The Edison Institute. Neg. No. 833-697.)

prevail. Arnold's account contains inconsistencies which are perplexing to the historian. In the same work he gave both April 1 and May 1 as the date of the first subassembly line. More important, he included a photograph captioned ''West End of Flywheel-Magneto Assembling Line: This is the first of the Ford sliding assembly lines.'' (See Figure 6.24.) Yet this picture shows the transmission mechanism being assembled rather than the assembly of the sixteen permanent magnets onto the flywheel. Fred Colvin later reproduced the same photograph and correctly labeled it ''Assembling the Transmissions.''[75]

In his oral history interview, Klann was asked specifically about the first Ford subassembly line. Obviously having examined Arnold's account of this development, Klann repeatedly and adamantly argued that the magneto coil assembly was the first to be put on a moving line basis and that moving line assembly of the permanent magnets onto the flywheel actually followed placing the engine assembly and the transmission gear and clutch assembly on a moving line.[76] Because he was the foreman of engine assembly in 1913 and because his command of detail was so evident when he was interviewed in the early 1950s, Klann's account is persuasive. Yet it conflicts with nearly contemporaneous evidence. Specifically, a photograph of magneto coil assembly operations which Arnold

FIGURE 6.23.   "The First Magneto Assembly Line," 1913. This is a photograph of what Allan Nevins, among many other historians, called the first magneto assembly line. In his text, Nevins said that the magneto coil assembly was the first subassembly to be put on a line basis, but this illustration shows assembly of the flywheel magneto, the other half of the entire Model T magneto. (Henry Ford Museum, The Edison Institute. Neg. No. 833-167.)

first reproduced in April of 1914 calls Klann's account into question for it shows procedures that clearly precede the assembly line.[77] It is possible that Arnold used an outdated photograph, but it is not probable given the immediacy of the other photographs he used to illustrate his articles.

The historian, therefore, is left with trying to resolve conflicting evidence, some generated a year after the fact and some forty years later. Arnold's and Klann's respective accounts of the first, critical subassembly line at Ford cannot be satisfactorily reconciled. Arnold's account with its substantial data on productivity gains achieved with the moving assembly of the flywheel magneto ultimately stands as the most convincing.[78] The anomaly of the photograph showing transmission assembly on the "first of the Ford sliding assembly lines" can be explained away in a number of ways which need not detain us. What is important is that whichever subassembly came first—magneto coil, permanent magnet, or transmission—the development of the assembly line at Ford was so swift and powerful that it defied accurate, unambiguous, timely documentation by the Ford Motor

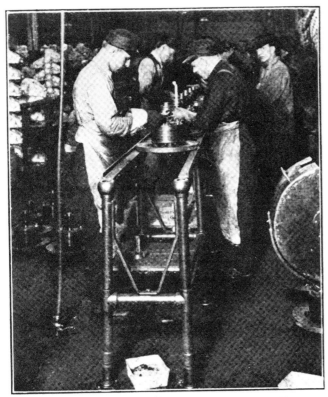

WEST END OF FLYWHEEL-MAGNETO
ASSEMBLING LINE

FIGURE 6.24.   Assembling Transmissions on What Horace Arnold Called "the first of the Ford sliding assembly lines," 1913. Despite the caption, workers are assembling transmissions on the back half of the flywheel. After completion of this operation, the flywheel was flipped over, ready for assembly of sixteen permanent magnets. (See Figure 6.23). Note the pipe nipples at the bottom of the frame used to raise the height of the line because workers complained of backaches suffered from stooping. (*Engineering Magazine*, July 1914. Eleutherian Mills Historical Library.)

Company and its employees.[79] Within a year of the first line, virtually every assembly operation at Ford had been put on a moving line basis, and those early ones had been radically revised. No better example exists than the one that Horace Arnold called the first Ford assembly line.

On April 1, 1913, workers in the Ford flywheel magneto assembling department stood for the first time beside a long, waist-high row of flywheels that rested on smooth, sliding surfaces on a pipe frame. No longer did the men stand at individual workbenches, each putting together an entire flywheel magneto assembly from the many parts (including sixteen permanent magnets, their supports and clamps, sixteen bolts, and other miscellaneous parts). This was no April Fool's joke. The workers had been instructed by the foreman to place one particular part in the assembly or perhaps start a few nuts or even just tighten them and then push the flywheel down the row to the next worker. Having pushed it down eighteen or perhaps thirty-six inches, the workers repeated the same process, over and over, nine hours, over and over. Martin, Sorensen, Emde, and others had designed what may have been the first automobile assembly line, which somehow seemed another step in the years of development at Ford yet somehow suddenly dropped out of the sky. Even before the end of that day, some of the engineers sensed that they had made a fundamental breakthrough. Others remained skeptical. Twenty-nine workers who had each assembled 35 or 40 magnetos per day at the benches (or about one every twenty minutes) put together 1,188 of them on the line (or roughly one every thirteen minutes and ten seconds per person). There were problems, to be sure. The workers complained about aching backs because of stooping over the line; raising the work level six or eight inches would solve that problem. (See Figure 6.24.) Some workers seemed to drag their heels while others appeared to work too fast. Although a piece rate system would probably eliminate the slow ones, the engineers knew that Henry Ford would never tolerate such a system. Soon they found that by moving magnetos at a set rate with a chain, they could set the pace of the workers: speed up the slow ones, restrain the quick. Within the next year, by raising the height of the line, moving the flywheels with a continuous chain, and lowering the number of workers to fourteen, the engineers achieved an output of 1,335 flywheel magnetos in an eight-hour day—five man-minutes compared to the original twenty.[80]

One can only imagine how excited the Ford production engineers were about the problems and possibilities of the assembly line. It became an object of study not only by Martin, Sorensen, and Emde but also by the heads of other assembling departments. Almost immediately after seeing the flywheel magneto assembly line, William Klann, head of the engine assembly, received permission to build an engine assembly line. The rush to implement such a line—beginning with putting the crankshaft in the engine block—led to an accident on the second day of operation which injured a workman seriously enough to bring James Couzens into the factory to inspect this "Goldberg job." Couzens wanted to call a halt to Klann's experiments. But when Klann assured Martin and Sorensen that the line "could be made foolproof," he received their permission to continue. Klann recalled that he started the line again the next day after adding certain safety devices to keep the engines from falling off the conveyors. "In a few weeks we had the job licked," Klann boasted. Arnold wrote that new attempts were not made until November 1913. In any case, productivity gains were enormous.[81] Klann and the Ford production engineers also turned to transmission assembly.

The Model T's transmission consisted of three distinct subassemblies: the transmission mechanism, the flywheel magneto assembly, and the transmission cover. Assembly of the transmission mechanism onto the back side of the flywheel was put on the line soon after the line had been built for the permanent magnets onto the front side of the flywheel. (Or, if one believes Klann, this was done first.) Beginning with the flywheel, workers added the triple gears, the driven gear, three drums, and the numerous parts of the clutch to form a complete subassembly. This line was developed so that when the transmission mechanism was completed, the flywheel was simply flipped over, ready for the magnets to be installed further down the line.[82]

In June 1913, Klann changed transmission cover assembly into a line operation. On this subassembly, the production engineer had to resort to flat-top metal tables instead of rail slides because the shape of the cover did not lend itself to rails. Line operation immediately brought cover assembly time down from eighteen man-minutes to nine minutes and twelve seconds. As Klann pointed out about the adoption of line assembly techniques, "There wasn't any discussion on whether this would work. You couldn't go wrong because the first one worked all right."[83]

By November 1913, Klann, Emde, and others put the entire engine assembly—made up of several subassemblies—on an integrated assembly line. (See Figures 6.25 and 6.26.) This was not one long line but two lines at right angles with several machine tools, babbitting ovens, and other miscellaneous machinery interspersed. Engine line assembly proved to be a matter of constant experiment and refinement. As Klann remarked, "We monkeyed with that thing all kinds of ways before we got it to work on a moving line."[84] By the time Klann and his colleagues had gotten "all of the kinks" worked out, lowering engine assembly from 594 man-minutes to 226 man-minutes,[85] Charles Sorensen and his assistant Clarence W. Avery had tried moving line assembly principles on the chassis. This operation became, in the public's mind, "the" assembly line.

Horace Arnold described the Ford chassis assembly line as "a highly impressive spectacle to beholders of every class, technical or non-technical." Charles Sorensen called it "the most spectacular one." Sorensen may have imagined such a spectacle in 1908 and may have even tried to realize it. But in August 1913 the apparent success of the flywheel magneto line and the unqualified productivity gains of the engine and transmission assembly lines led Sorensen and others to begin experimentation with chassis assembly. Sorensen was not to be denied this time. Appointed directly by Henry Ford, Sorensen's assistant Clarence Avery (who had been Edsel Ford's manual training teacher in high school) proved to be a decisive factor in the success of the assembly line at Ford. As Avery said in 1929, "It was my good fortune to have [been assigned] the problem of developing the first continuous automobile assembly line."[86] Avery was a bright, well-educated young man who had wanted to get out of teaching into the "real world" of manufacturing. When assigned to him by Ford, Sorensen had instructed Avery to master conceptually every manufacturing operation at Highland Park.[87] After eight months of study, Avery was ready to help Sorensen. As Fred Colvin had pointed out in the *American Machinist,* stationary chassis assembly at Ford was not a matter of guesswork. With several assembly gangs moving up and down the rows of chassis and with delivery of parts at each station demanding correct scheduling, the orchestration of the assembly process had required motion and time studies (albeit perhaps elementary ones) to avoid chaos.[88] It was from these studies or from this knowledge about how long certain operations took that Sorensen and Avery laid out the basic plans of the first chassis assembly lines.

The use of time and motion studies for the layout of the final or chassis assembly line at Ford raises an important question: To what extent did Taylorism or scientific management or any other contemporary form of systematic management shape or influence the developments at Ford's Highland Park factory? The Ford Motor Company, after all, arose in the era when Taylorism was approaching the height of its influence. The widely publicized *Eastern Rate* case (1910) and the publication of Frederick W. Taylor's *Principles of Scientific Management* (1911) occurred just before the innovations at Highland Park, and it is natural to assume that there was a connection. Whether that was in fact the case, however, is by no means certain, because the contemporary sources are not adequate to assure a definitive answer. In addressing this issue, the initial problem is arriving at a reasonable definition of Taylorism or systematic management.

If by Taylorism we mean rationalization through the analysis of work (time and motion studies to eliminate wasteful motions) and the "scientific" selection of workmen for prescribed tasks, then we can agree with the recent judgment of Stephen Meyer III that Ford engineers "Taylorized" the Highland Park factory. Indeed, this was the conclusion of Allan Nevins in his standard work on Henry Ford and the Ford Motor Company. Ford's engineers, Nevins suggested, "had doubtlessly caught some of his [Frederick W. Tay-

FIGURE 6.25.   Part of Engine Assembly Line Operations, Highland Park, 1915. (Henry Ford Museum, The Edison Institute. Neg. No. 833-2346.)

lor's] ideas.'' Moreover, Nevins wrote that Clarence Avery, who was clearly critical in the development of the moving chassis assembly line, had ''kept in touch with the ideas of men like Frederick W. Taylor.'' Meyer generalized by arguing that ''Ford managers and engineers may not have followed a specific program [of systematic or scientific management], but they surely followed general principles.''[89]

Unquestionably, Ford engineers standardized work routines at Highland Park after they analyzed jobs and work flow patterns. With the widespread use of special-purpose machine tools at Ford, the engineers hired semiskilled and unskilled workers to operate these machines (scientific selection of workmen, as Taylor called it). As early as 1912 or 1913, the Ford factory had a time study department, although some Ford employees later recalled that it was first known as the work standards department.[90] The very idea of establishing work standards—how much output a manufacturer could expect from a certain machine tool, a work process, or a series of processes if labor did a fair day's work—is the very heart of Taylorism in particular and systematic management in general. Moreover, in the Ford factory, there was a clear division of labor between management and workers along the lines advocated by Taylor in his *Principles of Scientific Management* (for example, machine tenders did not perform any maintenance on their machines but left this to specialists).[91]

Despite these facts, there is much reason to doubt that Taylorism contributed signifi-

FIGURE 6.26. Installing Pistons in Model T Engines, Highland Park, ca. 1914. (Henry Ford Museum, The Edison Institute. Neg. No. 833-832.)

cantly to the new assembly system at Highland Park. Henry Ford himself claimed that the Ford Motor Company had not relied on Taylorism or any other system of management. As Horace Arnold noted in 1914, "In reply to a direct question he [Henry Ford] disclaimed any systematic theory of organization or administration, or any dependence upon scientific management."[92]

Four months before he died, Frederick W. Taylor spoke in Detroit to some six hundred superintendents and foremen of "leading" manufacturers of the city. In reflecting upon his experience in Detroit, Taylor proudly declared that the manufacturers there "were endeavoring to introduce the principles of scientific management into their business and that they were meeting with large success." This especially interested Taylor because it was "almost the first instance, in which a group of manufacturers had undertaken to install the principles of scientific management without the aid of experts." According to Allan Nevins, however, many of those who heard Taylor saw the matter differently. They argued that "several Detroit manufacturers had anticipated his ideas."[93] The Ford Motor Company could have been "Taylorized" without Taylor.

By focusing on those elements of the Highland Park factory that were Taylorized, one runs the danger of misjudging the fundamental differences between the Ford philosophy (Fordism) and that of Taylor (Taylorism). It was Henry Ford himself, or, more accurately his ghostwriter, who pointed up these differences. To explain his system, Taylor often resorted to his tale about Schmidt, the scientifically selected worker who was told how to

FIGURE 6.27.   Engine Drop, Final Assembly Line, Highland Park, 1913. Careful study of this photograph suggests that at least three assembly lines were in operation and that chassis were still being pushed by hand along a single track with the aid of dollys. (Henry Ford Museum, The Edison Institute. Neg. No. 833-198.)

load pig iron scientifically (that is, after time and motion studies had been carried out) and was placed on an incentive wage system. Previously, Schmidt had hand carried each day twelve and a half tons of pig iron up a ramp and dumped it into a railroad car. But after he underwent the magic of scientific management, Schmidt was able to hand carry forty-seven and a half tons of the ninety-two-pound pigs each day. The Taylor approach was to assume that the job of loading pig iron was a given; the task of scientific management was to improve the efficiency of the pig iron carrier. Ford's production experts saw the problem differently. Why, they asked, should pig iron be hand loaded? Could this not be done by some mechanical means? (Ford engineers would later ask why one had to bother with pig iron at all. Why not pour castings directly out of the blast furnace and dispense entirely with handling and reheating pigs?)[94]

The Ford approach was to eliminate labor by machinery, not, as the Taylorites customarily did, to take a given production process and improve the efficiency of the workers through time and motion study and a differential piecerate system of payment (or some such work incentive). Taylor took production hardware as a given and sought revisions in labor processes and the organization of work; Ford engineers mechanized work processes and found workers to feed and tend their machines. Though time and motion studies may have been employed in the setup of the machine or machine process, the machine ulti-

FIGURE 6.28.   End of the Line, Highland Park, 1913. As in Figure 6.27, final assembly operations had not yet been put on the "chain system" when this photograph was taken. Note that Model T car bodies are being put on the chassis on one of the assembly lines. Those not receiving bodies were destined for rail shipment without bodies. (Henry Ford Museum, The Edison Institute. Neg. No. 0-3342.)

mately set the pace of work at Ford, not a piecerate or an established standard for a "fair day's work." This was the essence of the assembly line and all the machinery that fed it. While depending upon certain elements of Taylorism in its fundamentals, the Ford assembly line departed radically from the ideas of Taylor and his followers.[95]

The first attempt at line assembly in August 1913 was crude but phenomenally successful in increasing productivity. At one end of a long open space in the Highland Park

FIGURE 6.29.   Body Drop, Highland Park, 1913. Many historians have argued that this photograph depicts how Model T bodies were first put on the chassis once the assembly line was developed. James O'Connor, who was a foreman for the Highland Park assembly line from 1913 to 1927, argued persuasively in the 1950s that this body chute was used only to drop bodies temporarily onto the chassis to haul them to the loading dock where the chassis, fenders, and bodies were packed separately into boxcars for shipment. Note that the car with the body on it does not have fenders, which lends great credence to O'Connor's statement, as does comparison with Figure 6.28. (Henry Ford Museum, The Edison Institute. Neg. No. 833-917.)

factory, the Ford engineers put a windlass and stretched out a rope 250 feet down the open space. Based on their knowledge about optimal installation times for various chassis components, the engineers placed these components at different intervals along the path. Whereas the man-hour figure had been slightly under twelve and a half with static assembly, the first assembly line attempt (in which six assemblers followed the slowly moving chassis as it made its way past the various components) reduced the figure to five and five-sixths man-hours.

Experiments continued. (See Figures 6.27–6.31.) On October 7, 140 assemblers had been placed along a 150-foot line. Man-hour figures dropped to slightly less than three hours per chassis. By December, Avery and those working with him had extended the line to 300 feet and had increased the assembly force to 177 men. Time: two hours, thirty-eight minutes. After Christmas, 191 men worked along the 300-foot line but pushed the assembly along by hand. Man-hour time increased rather than dropped. Sixteen days later, the engineers had installed a line on which the car was carried along by an endless

FIGURE 6.30.   Radiator and Wheel Chutes, Final Assembly Line, Highland Park, 1914. By this time the final assembly line had been put on the ''chain system'' (see lower left), which controlled the forward progress of the chassis. The frame not only served to carry the chain but also to raise the height of work to a more comfortable level. (Henry Ford Museum, The Edison Institute. Neg. No. 833-895.)

chain. In the next four months, lines were raised, lowered, speeded up, slowed down. Men were added and taken off. As Charles Sorensen wrote, all of ''this called for patient timing and rearrangement until the flow of parts and the speed and intervals along the assembly line meshed into a perfectly synchronized operation.''[96] By the end of April 1914, three lines were fully in operation, and the workmen along them put together 1,212 chassis assemblies in eight hours, which worked out to ninety-three man-minutes. (See Figure 6.32.) Assembly figures became consistently predictable. Horace Arnold noted the effects of these developments: ''Very naturally this unbelievable reduction in chassis-assembling labor costs gave pause to the Ford engineering staff, and led to serious search for other labor-reduction opportunities in the Ford shops, regardless of precedents and traditions of the trade at large.''[97]

Experiment and refinement continued on the existing subassembly lines. These adjustments provided productivity gains comparable to those achieved with chassis assembly

FIGURE 6.31.   Driving Off the Assembly Line, Highland Park, 1914. At the end of the line, workmen filled the radiator and started the engine before driving off the line. (Henry Ford Museum, The Edison Institute. Neg. No. 833-997.)

and led the company to adopt entirely new lines. On June 1, 1914, chain-driven assembly lines began to roll out front axle assemblies. These reduced assembly time from 150 minutes (a January 1, 1913, figure) to 26½ minutes (July 13, 1914). Other subassemblies followed.[98] (See Figures 6.33–6.35.) All of the assembly stands over which Fred Colvin had marveled only months before and which were characteristically Yankee had been taken to the scrap pile.

The Ford engineers next designed and installed conveyor systems to feed these hungry lines. As Arnold wrote in July 1914, ''Besides these almost unbelievable reductions in assembling time [wrought by the assembly line], the Ford shops are now making equally surprising gains by the installation of component-carrying slides, or ways, on which components in process of finishing slide by gravity from the hand of one operation-performing workman to the hand of the next operator.''[99] Reductions in labor costs were thus achieved by assembly lines, conveyor systems, gravity slides, and the like along with the Ford system of machining, which had removed virtually all skill requirements for operation and whose fixtures and gauges allowed foremen to demand speed. But these great achievements had wrought serious labor problems at the Ford factory. Henry Ford's five-dollar day was an attempt to eliminate these problems.

Although the motives behind the five-dollar day are rooted in a sort of industrial beneficence on Henry Ford's part and a consciousness on James Couzens's part that such a wage and profit-sharing system would pay for itself in free advertising, the five-dollar

FIGURE 6.32.   General View of "The Line," Highland Park, 1914. When Horace Arnold toured the Highland Park factory in 1914 and wrote of the assembly line that assembled a car in ninety-three man-minutes, this is the line of which he was speaking. (Henry Ford Museum, The Edison Institute. Neg. No. 833-987.)

day must be seen as the last step or link in the development of mass production. During 1913 the labor turnover rate at the Ford factory had soared to a phenomenal figure. Keith Sward points out that turnover in 1913 reached 380 percent: "So great was labor's distaste for the new machine system that toward the close of 1913 every time the company wanted to add 100 men to its factory personnel, it was necessary to hire 963."[100] Not only did this burden the administrative machinery at Highland Park, but it also affected the operations within the factory. High turnover was also accompanied by growing signs of unionization at the Ford factory. Other Detroit automakers had already experienced strikes. The Ford management sought to relieve these pressures by carrying out labor reforms in 1913.

FIGURE 6.33.   Rear Axle Assembly Line, Highland Park, 1914. (Henry Ford Museum, The Edison Institute. Neg. No. 833-910.)

Jobs were reevaluated and brought into parity with each other. The company gave special raises to efficient employees. And finally, an across-the-board pay increase, averaging 13 percent, was announced on October 1, 1913. The company set $2.34 as the minimum daily wage for every employee.

These reforms, however, did not stem the rising tide of labor problems. The growth in output of the factory, the installation and rigorous improvement in the efficiency of assembly lines in three different departments, and the promise of one being installed in every department added additional force, swelling the tide of labor turnover and dissatisfaction higher and higher in the final months of 1913. Attempting to reward workers who had stayed with the company for three years or more, the Ford directors gave a 10 percent bonus on December 31, 1913. Out of some 15,000 employees only 640 qualified for the bonus, a figure that indicates the extent of worker turnover. The following day, or perhaps a few days later, Henry Ford, James Couzens, P. E. Martin, Charles Sorensen, Harold Wills, John R. Lee (the personnel department head), and Norval Hawkins (the sales manager) met, discussed the labor problems, and considered increasing daily earnings (wages and ''shared'' profits) to $3.00, $3.50, $4.00, $4.50, $4.75, or $5.00. Ford had clearly become concerned about the inequity between the salaries and profits of directors (as well as the salaries and bonuses paid to the production experts) and the wages earned by the majority of workers in the factory. The turnover rate, the signs of unionization, and

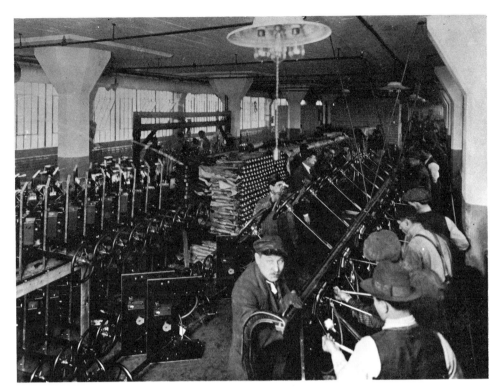

FIGURE 6.34. Dashboard Assembly Line, Highland Park, 1914. Contrast this with Figure 6.13. (Henry Ford Museum, The Edison Institute. Neg. No. 833-326.)

the manifest inequity of income combined in Ford's mind (and Couzens's) to produce a quick solution to all three. Ford, Couzens, and Horace Rackham (a director of the company) met on January 5, 1914, and adopted the five-dollar day. Since Ford owned controlling stock, the meeting was pro forma. Couzens had been convinced of the desirability of the plan—perhaps he had engineered it—so he and Ford encouraged Rackham to make the vote unanimous. Couzens got his free advertising, Ford his hero-worship, "acceptable" workers extraordinarily high earnings.[101] The basic psychology of the plan, however, and its basic effect were that now the company could ask its workers to become for eight hours a day a part of the production machine that the Ford engineers had designed and refined during the past four years.

The five-dollar day assured the company that the essential human appendages to this machine would always be present. This "bonding" effect of extremely high earnings was evident within a month after Ford announced it. As an anonymous housewife of a Ford assembly line worker wrote to Henry Ford on January 23, 1914, "The chain system you have is a *slave driver! My God!*, Mr. Ford. My husband has come home & thrown himself down & won't eat his supper—so done out! Can't it be remedied? . . . That $5 a day is a blessing—a bigger one than you know but *oh* they earn it."[102] As part of the five-dollar day scheme, Henry Ford also scaled up the paternalistic operations of the Ford sociological department, which determined if workers qualified for profit-sharing by investigating their private lives—an extra burden on top of those already imposed by Ford production technology.[103]

FIGURE 6.35.   Upholstery Line, Highland Park, 1916. Even skilled processes such as upholstery were put on a moving line basis at the Ford Motor Company's Highland Park Factory. (Ford Motor Company.)

The story of mass production at the Ford Motor Company was not something that only historians of a later generation would delve into and try to understand. Henry Ford's contemporaries, many of whom were competitors, closely watched the doings at Highland Park, attempting to understand and emulate the revolutionary developments. Henry Ford encouraged their interest. Unlike the Singer Manufacturing Company, the Ford company was completely open about its organizational structure, its sales, and its production methods—at least after Henry Ford was satisfied that his company was on the road to mass production.[104] As Horace Arnold wrote in 1914, "The Ford company is willing to have any part of its commercial, managerial or mechanical practice given full and unrestricted publicity in print."[105] Ford engineers had no skeleton closets in their factory. Proud of their work, they were anxious to have technical journalists tour the shops and write extensive articles about Ford methods. When Horace Arnold was writing the series of articles for *Engineering Magazine* Henry Ford himself devoted attention to the author. Fay Faurote experienced the same cooperation and developed a friendship with Ford over the next fifteen years.

As a consequence of Ford's openness, Ford production technology diffused rapidly throughout American manufacturing. The *American Machinist* series of 1913, *Engineering Magazine*'s series of 1914 and 1915 (which resulted in Arnold's and Faurote's *Ford Methods and the Ford Shops*), a series in *Iron Age* in 1912–13, and occasional but incisive articles in *Machinery* were the primary agents of this diffusion.[106] One can thumb through the pages of these and other technical and trade periodicals in the days after the assembly line appeared in print and find automobile companies that were trying moving line assembly techniques even though they made only one or two thousand cars.[107] Manufacturers of other products also tried the assembly line. Within a decade, many household appliances such as vacuum sweepers and even radios were assembled on a conveyor system.[108] The Ford Motor Company educated the American technical community in the ways of mass production.

Yet exactly one year after the first assembly line experiments at Ford Motor Company, Reginald McIntosh Cleveland wrote an article titled "How Many Automobiles Can America Buy?" Although writers such as Edward A. Rumely and Harry Franklin Porter celebrated Henry Ford, because of his insistence on standardization, as "The Manufacturer of Tomorrow," Cleveland pointed out that already the American automobile industry had succumbed to "the fetish of 'The New Model.'" Manufacturers had resorted to this "creed" in order to sell cars. Ford dogmatically resisted this practice. Through standardization of design and the resulting development of mass production technology, Ford demonstrated a "big lesson" to the entire automobile industry.[109] Yet eventually—long after other manufacturers would have predicted—Ford himself had to resort to model change in order to keep his company from complete collapse. This change came after some fifteen million Model T's had been produced. By this time, Ford production technology had become so highly specialized that the changeover to a new model, the A, brought unimagined problems for the Ford Motor Company. The working out of these problems over a five-year period brought Ford into a new era of mass production technology, that of the annual model change, which demanded "flexible mass production." This was part of what Charles F. Kettering called "the new necessity."[110]

CHAPTER 7

Cul-de-sac: The Limits of Fordism & the Coming of "Flexible Mass Production"

While the bringing out of yearly models results in many disadvantages and, for that reason, we are all against yearly models, I don't see just what can be done about it.
—Alfred P. Sloan, Jr. (July 29, 1925)

General Motors in fact had annual models in the twenties, every year after 1923, and has had them ever since, but . . . we had not in 1925 formulated the concept in the way it is known today. When we did formulate it I cannot say. It was a matter of evolution. Eventually the fact that we made yearly changes, and the recognition of the necessity of change, forced us into regularizing change. When change became regularized, some time in the 1930s, we began to speak of annual models. I do not believe the elder Mr. Ford ever really cared for the idea.
—Alfred P. Sloan, Jr. (1963)

We are going to get rid of all the Model T sons-of-bitches. We are going to get away from the Model T methods of doing things.
—Charles E. Sorensen (1927)

P aradox is part of the stuff of history. The Ford Motor Company's supreme success in mass-producing the allegedly unchanging Model T created in America what Daniel Boorstin has called the "search for novelty."[1] The end result of this search was the institution of the annual model change in the American automobile industry, an epoch rung in by Alfred P. Sloan, Jr., and General Motors, not by Ford.[2] Through unprecedented experimentation, bold moves, and widespread publicity, Ford had given the world the first system, in the fullest sense of the expression, of mass production: single-purpose manufacture combined with the smooth flow of materials; the assembly line; large-volume production; high wages initiated by the five-dollar day; and low prices.[3] Ford effected not simply a technological turning point but, as Peter Drucker pointed out long ago, an economic revolution. Ford showed dramatically that maximum profits could be achieved by maximizing production while minimizing costs and thus turned the theory of monopoly on its head.[4] Yet this revolution was short-lived. As the decade of the 1920s proceeded, Ford saw its 55 percent market share of 1921 dwindle to

263

30 percent in 1926 despite significant body and mechanical changes in the "unchanging" Model T and a deliberate schedule of price reductions.[5] Faced with notably worse sales in the first half of 1927 (less than a quarter of the market), Ford finally announced the end of the Model T era. The market had saturated at the milestone of fifteen million of the mass-produced cars known popularly as Flivvers and Tin Lizzies.

The saturation of the Model T market and the rapid growth of GM's Chevrolet Division were part of a larger movement in the American economy characterized by increased consumer purchasing power to which Ford's earlier work no doubt contributed. Sloan and his managers came to see that growth would occur not by the production of basic needs or by a "car for the masses" but by selling cars whose appearance, if not features, changed annually. As Boorstin described the rationale of this new policy, "Americans would climb the ladder of consumption by abandoning the new for the newer." In this consciously orchestrated economy of change and consumption that stressed style and comfort above utility, mass production as Ford had developed it with the T was no longer suitable. The ground rules had changed. The Ford Model T dictum of maximum production at minimum cost gave way to planning for change. Only when change could be planned satisfactorily was credence paid to the old dictum.[6] This was the era of "flexible mass production."

Thus between 1925, when GM began to discuss the strategy of model changes, and 1932 or 1933, when change had become policy not only at GM but also at Ford, mass production moved through a period of transition. To comprehend mass production in twentieth-century America fully, it is imperative to understand this transition—to understand the process of the "changeover."

The epigraphs that preface this chapter are suggestive of how GM and Ford respectively experienced the transition to producing the annual model. Just as formulation of the overall policy was a "matter of evolution" at GM, so was the development of changeover know-how and procedure. Chevrolet was GM's challenger to Ford's predominance in the low-cost market. During the transition period, Chevrolet production rose from 280,000 passenger automobiles in 1924 to just over a million in 1928. Each year styling changes were made in the model. Unlike Ford's sudden, unprecedented changeover, Chevrolet evolved its changeover know-how while steadily increasing its output. In 1929, however, Chevrolet introduced a major change, increasing the engine from four to six cylinders, and expanded its output to almost 1.5 million units. The changeover took only three weeks, a record that so impressed GM executives that they treated William Knudsen, Chevrolet's president and a former Ford production wizard, to a lavish banquet.[7]

Knudsen certainly deserved the credit for this feat. He had made it possible for Chevrolet not only to change its models frequently without serious delay but also to expand output (and profits) every year since 1924. More important, Knudsen played a critical role in raising the level of General Motors Company's mass production know-how. Although not as central as a Martin or a Sorensen had been to the rise of mass production at Ford between 1910 and 1914, Knudsen had been an important figure in the Ford production organization. It was to the former Keim employee Knudsen that Henry Ford turned in 1918 when the program to mass-produce submarine chasers appeared to be faltering and in 1919 when Ford sought to expand operations abroad. But during the crisis at Ford in 1921, Knudsen resigned from the company because Henry Ford regularly overrode his decisions.[8]

On February 1, 1922, Alfred P. Sloan, Jr., invited him to join General Motors. Well aware of Knudsen's contributions at Ford, Sloan instructed him to evaluate the problems entailed in the production of General Motors' new "copper cooled" (air-cooled) engine. Within weeks, GM president Pierre S. duPont, who had just taken over the command of the Chevrolet Division, named Knudsen vice-president in charge of operations at Chevrolet. DuPont immediately set Knudsen to work on the formulation of a long-range plan for the manufacture of the Chevrolet car, which would be in accordance with recent decisions made by GM's Executive Committee. The premature introduction of the copper-cooled engine set back Chevrolet's program and necessitated its reformulation. In January 1924, Knudsen became president and general manager of the Chevrolet Division and in 1937, president of General Motors.[9]

From his arrival at Chevrolet, Knudsen worked to build Chevrolet's share of the market. Some executives at GM wanted the Chevrolet Division liquidated. They argued that the company could not hope to compete with Ford in the low-priced market, and its loss of almost $9 million in 1921 lent credence to their views. Sloan prevailed, however. The Knudsen-run Chevrolet Division became the foundation for the corporation's strategy in the automobile industry. The wisdom of this strategy was clearly demonstrated in 1932, when Chevrolet's profits helped to keep the entire corporation out of the red.[10]

When Knudsen went to the Chevrolet Division, its production men feared that he would bring in a team of Ford mass-production experts. But Knudsen chose neither to import Ford people nor to create an imitation of the Ford Motor Company. Aware of the havoc that a major change in the Model T would wreak at Ford and recognizing as clearly as any GM executive the advantages of decentralization, he built an organization and production system that could accommodate change and expansion. In 1927, when Chevrolet sold more than a million cars and forced an end to the Ford Model T, Knudsen explained how the company had achieved such progress in the past five years. His analysis also makes it clear how Chevrolet later accomplished so smoothly the changeover from the four- to a six-cylinder engine in 1929.[11]

Improvements in the Chevrolet's design and performance were a major reason for its success. According to Knudsen, the Chevrolet Division "modernized its [product's] appearance so as to remove the inevitable stigma which rests on low priced articles that show it." At the same time, the company overhauled its cost estimates, recognizing that it had to achieve a high volume of output before "the necessary earnings could be produced." Knudsen and his Chevrolet production men achieved the desired high volume by replacing old machine tools with new ones and adopting "sequence lines" or, as Knudsen called it," get[ting] all noses pointed in the same direction."[12] In this latter respect, Knudsen clearly drew upon his experience at Ford Motor Company, but his machine tool purchases reflected a radical departure from the Ford idea of "single-purpose manufacture."

Under Knudsen's direction, as he wrote in 1927, "all old machines were discarded, new heavy type standard machines (not single purpose) were installed, and fixtures strengthened so as to withstand the spring, which is the greater factor [in causing inaccurately machined parts] than wear." This new direction at Chevrolet allowed limits of precision to be lowered, resulting in the reduction of scrapped material. To this end, "gauges and indicators, particularly the latter, were devised for all operations of major importance and the inspection system was given full opportunity to come into its own." Chevrolet also began running careful checks of the raw materials it purchased. At this point, Knudsen reported, "The conveyor came into its own."[13] It is important to emphasize that the entire Chevrolet production system, though led by a former Ford production expert, was based on standard or general-purpose, not single-purpose, machine tools. For this reason, Chevrolet could accommodate change far more easily than could the Ford Motor Company.

Knudsen also pursued decentralization. In 1922, he reported, Chevrolet operated two main manufacturing plants, one at Flint, Michigan, where motors and axles were made, and the other at Toledo, which produced transmissions. At four assembly plants (Tarrytown, New York, Flint, Michigan, St. Louis, Missouri, and Oakland, California), these subassemblies and thousands of parts purchased from vendors were brought together to make the Chevrolet. By 1927, Knudsen had successfully carried out a capital expansion program to enlarge the Flint and Toledo plants and augment manufacturing facilities at Detroit and Bay City, Michigan. Three new assembly plants had been opened following an extensive survey of population centers and the national transportation network. As Knudsen outlined Chevrolet's production facilities in 1927, the Flint works manufactured motors, motor stampings, and car body stampings; Detroit produced forgings, front and rear axles, and axle stampings; Toledo specialized in transmissions; and Bay City made carburetors and parts that were hardened and ground. Each of these plants was run independently by a local manager. To supplement the components made at Chevrolet plants, the division purchased a large variety of parts made by other GM divisions including Remy, Jaxon Wheel, Inland Manufacturing, Muncie Products, New Departure, and Fisher Body Corporation. Knudsen had convinced GM executives that a Fisher Body plant should be attached to each assembly plant so that body production could be coordinated precisely with the daily output of each assembly plant.[14]

With this decentralized production system, the Chevrolet Division could accommodate change, especially when it was carefully planned, such as that from the four-cylinder to the six-cylinder engine in 1929. Planning for this new car actually began in the summer of 1927, when Knudsen informed GM's executives that he proposed to lengthen the wheelbase of his division's car as a prelude to the adoption of the six-cylinder engine. Thus major changes were made in the body one year and in the power plant the following year. When Knudsen and his team of engineers prepared for the changeover from four-to six-cylinder engines, they set up and perfected a pilot line for producing the engines at the division's experimental plant in Saginaw and between September 1 and November 15 made the first two hundred engines before moving the finished machinery, jigs, fixtures, and gauges to the Flint plant. At the same time, additional machinery for six thousand engines per day was ordered for Flint and installation begun. Meanwhile, the Flint plant continued to produce the four-cylinder automobile. Finally, on October 25, Chevrolet closed the Flint motor factory, changed over its machinery to the six-cylinder job, and reopened the plant on November 15. December production reached two thousand engines a day, and by the end of January Knudsen had achieved his production goals for the new automobile. Thus when the new Chevrolet was introduced officially on January 1, 1929, buyers did not have to wait. Within eight months, Chevrolet's factories had turned out over a million sixes, and total production for the year exceeded 1.3 million units.[15]

By contrast, Ford's changeover from Model T to Model A in 1927 occasioned a six-month shutdown and a great upheaval within the company. Charles Sorensen's invective to "get rid of all the Model T sons-of-bitches" and "away from the Model T methods" is a superb description of the vengeance, ruthlessness, and seeming chaos of this unprecedented changeover. For almost two decades Ford production men, with some important exceptions, had sought mainly to produce more Model T's at less cost. Now they were asked to produce an entirely new automobile. The magnitude of the task overwhelmed them. Ford had once produced a record of over two million units annually, yet in 1928 it could not get eight hundred thousand off the new assembly line that had started up in October 1927. The changeover from T to A, however, provided Ford with an important learning experience. In 1932 another major model change, the V-8, was introduced. The company's production men believed that this changeover was less painful and accomplished more quickly than the one of 1927–28 because it was carefully planned. But industry analysts noted that the difference in ease of changeover was of degree, not of kind. Building on the V-8 experience, however, Ford handled subsequent changes more easily, despite, as Allan Nevins and Frank Ernest Hill expressed it, a deteriorating organization.

This chapter will examine in detail mass production in transition at the Ford Motor Company from about 1925 to 1932. General Motors was clearly the major innovator of the annual model change, and it seems to have coped far better with the technical and managerial problems of planned change than did Ford. The primary records of GM's pioneering work on the changeover remain closed to the historian, if they survive at all. By studying Ford, however, one can see in sharpest relief the contrast between the mass production technology pioneered at Highland Park and the new mass production necessitated by the operation of, in Sloan's words, "the 'laws' of Paris dressmakers . . . in the automobile industry." Ford had driven the strategy of mass production to its ultimate form and thereby into a cul-de-sac. It was only with the rapid decline of the Ford Motor Company in the mid-1920s and the changeover to the Model A that Ford learned—painfully, reluctantly, and at great cost—the lessons pioneered at General Motors. "Sloanism" had triumphed over "Fordism"; marketing had triumphed over pure production. Charles Kettering of General Motors had said that it was important "to keep the consumer dissatisfied," and this demanded that a new era of mass production—*flexible* mass production—be initiated in America.[16]

Despite the legends about the never-changing Model T, major changes were made both in the Model T and in the technology of its production between the cluster of developments that brought about "mass production" (1908–15) and the demise of the T (1927). These changes bore heavily on Ford and his company in the year or two preceding the changeover and are, therefore, essential in understanding it.

Hardly had Ford reached an annual output of half a million automobiles at the five-year-old Highland Park plant when he bought a two thousand-acre tract of land east of the River Rouge and southeast of Detroit. Although his short-range goal was to construct blast furnaces and a tractor factory there, he projected the establishment of a deepwater port and an integrated manufacturing operation that would make the Ford Motor Company virtually self-sufficient. Ford told a journalist in mid-1915 that "we may take ten years to bring things to the point where we want them," but he grew increasingly determined to make the Rouge site into what Nevins and Hill have called an "industrial colossus."[17]

Between the end of World War I and the major recession of 1920 Ford moved quickly to develop the Rouge. Huge raw materials bins were constructed alongside the river, work began on the great power house, coke ovens were built, a blast furnace was fired up, and much of the railroad and transportation system was laid out. But most important, the B building, constructed during the war to mass-produce the Eagle boat, was substantially altered to accommodate a bodymaking plant for the Model T. Bodies were produced for the first time in August 1919. Here, also, operations began for the manufacture of the

Fordson tractor. Previously, the Fordson had been produced in Dearborn under Charles Sorensen's direction, and its transfer to the Rouge brought Sorensen to the command of the entire Rouge complex. From this time through the Model A period Sorensen acted as Ford's vicegerent in all matters concerning the Rouge. Sorensen became in fact, it not name, the chief engineer of the Rouge.[18]

Soon after the recession of 1920 turned into the boom of 1921, Rouge operations began to grow "like Topsy" and to gather an enormous momentum, which carried Model T production to the point of rapidly diminishing returns. The thirty thousand-kilowatt power plant went on line, delivering electricity to Rouge facilities and to about a third of Highland Park's operations. Ford also opened the world's largest foundry at the Rouge, an immense structure, 595 feet wide and 1,188 feet long. By 1922, ten thousand men gathered there to make and machine castings for Fordson tractors and initially for Model T engine blocks.[19] Foundry operations achieved, although not without problems, Henry Ford's dream of eliminating waste by pouring iron from cupolas charged directly with molten iron from the blast furnaces.[20] Soon the foundry building would also house all of the machining operations for all Ford castings. Between 1920 and 1923, Ford moved through a series of political and bureaucratic hurdles to finish dredging operations on the River Rouge and a cutoff canal. The successful resolution of this thorny problem, as well as Ford's purchase of the capital-starved Detroit, Toledo & Ironton Railroad, meant that the company could feed its raw-material-hungry Rouge plant with adequate supplies of coal, iron ore, and wood both by water and by rail. Thus, unlike General Motors, Ford had moved far into backward integration. As Nevins and Hill said, at Ford, "the flow of supply fed the flow of production."[21] (See Figure 7.1.)

During these years of development at the Rouge, Ford's "flow of production" proved to be a gusher. Production in 1915 had reached five hundred thousand automobiles. By 1919, the company was manufacturing almost eight hundred thousand cars and, despite a significant drop in 1920, faced a steep climb to the peak of Model T production in 1923, two million cars and trucks.[22] This rapid growth laid bare the inadequacies of the Highland Park plant for the increasingly economical production of the Model T and added momentum to the importance of the Rouge plant in the Ford empire. As the Rouge flourished at the hands of Sorensen and his lieutenants, the birthplace of the assembly line, Highland Park, and its chief production man P. E. (Pete) Martin waned. (See Figure 7.2.) This growing split between the two Ford Detroit plants will become an important part of the changeover story.

In the early 1920s, this split was not apparent to the Ford production men, who were fully occupied with the phenomenal increase in output. To meet the demands of production increases, these men had developed a procedure that became routine in the early 1920s.[23] Ford's engineers had established production schedules based on a sixteen-hour day, or two eight-hour shifts. When increases in output were called for, Sorensen and Martin (and by 1923, Ernest C. Kanzler*) directed the engineering department, headed

---

*Ernest C. Kanzler was Edsel Ford's brother-in-law (they were married to sisters). A Harvard-trained lawyer, he had worked for a legal firm that conducted suits against Ford on behalf of the Dodge brothers and the *Chicago Tribune*. Having been more than an occasional guest at Henry Ford's home, however, he was invited to join the Ford enterprise when Henry Ford initiated his tractor company, Henry Ford & Son. His position was assistant to Charles Sorensen. Having had some technical training in high school and exhibiting a remarkable desire to learn everything about manufacture (by all accounts, he daily donned coveralls and mastered every manufacturing and assembly job in the Ford Company), Kanzler quickly became an outstanding production man for Ford. At one time it appears he was the chief production man. His greatest contributions were in production planning, but his success posed too great a threat to Charles Sorensen, who eventually persuaded Henry Ford to fire Kanzler.

FIGURE 7.1. Ford Motor Company's River Rouge Factory, 1930. (Henry Ford Museum. The Edison Institute. Neg. No. 833-55282-A.)

FIGURE 7.2. Henry Ford and His Chief Production Experts, 1933. During the 1920s, Charles Sorensen (center) and the Rouge became predominant in Ford production while P. E. Martin (right) and Highland Park lost power. The unidentified gentleman with his back turned to the camera may be William Cameron, Henry Ford's spokesman. (Henry Ford Museum, The Edison Institute. Neg. No. 189-10646.)

since 1919 by A. M. Wibel† to procure the necessary machinery and to make arrangements for the manufacture of an additional given number of units per day (a figure that often reached a thousand by 1923). The engineering department commanded information on all aspects of production. Operations sheets, a device whose origins were discussed in Chapter 2, encapsulated much of the critical information. All of the machine tool operations for each part and the type and number of machines required for each operation, given

---

†A. M. Wibel had worked for Ford since May 2, 1912. Following graduation from the University of Indiana, he was hired as an assistant to Oscar Bornholdt, whose role in the development of mass production was assessed in Chapter 6. When Bornholdt retired, Wibel worked for his able successor, Charles Morgana, and when Morgana was lured away from the company by C. Harold Wills in 1919, Wibel became head of the engineering department. Wibel thus was in a position to witness much of the growth of the company and the development of mass production.

a specific production quota, were recorded sequentially. Operations sheets were also maintained for every subassembly and major assembly, listing step-by-step procedures, tools required, and special problems encountered.[24] All of these operations sheets were continually updated as changes in parts were made or as improved manufacturing operations suggested themselves.

When production quotas were raised, the engineering department's machine tool procurers referred to the operations sheets and ordered machine tools accordingly. These men always checked the department's master inventory of all Ford's machine tools to make sure that idle machinery could not be used for the projected operations. Machine tool procurers also pressed makers to increase the productivity of their machine tools, and of course they pushed them to meet delivery dates.[25] Expanded operations could be held up by the absence of a single machine tool. Operations sheets also provided the engineering department with the information necessary to order special Ford machine tools and special cutting tools, jigs, fixtures, and gauges from the Ford tool department. These orders went out at the same time as those to machine tool manufacturers.

Procurement of tools was only the first step. The engineering department also coordinated its work with that of the layout department, which, according to Wibel, ''planned and integrated the overall picture of production and material handling.'' The addition of a single machine tool raised both space and power considerations, which the engineering department and tool department together calculated on a square-foot and horsepower basis. Ultimately both figures could be reduced to dollars and cents to add to the cost of the tools. These space and horsepower requirements were then passed on to the layout department along with specifications of the machine tools on order. The latter department then made the necessary alterations to a given part's manufacturing department, including rearrangement of existing machine tools, furnaces, and conveyors; providing for new power requirements; and construction of any specially needed bases or footings for machinery. Thus, when a new machine tool arrived at the factory, the engineering department received it by attaching a brass tag, bearing its inventory number, to the machine and transporting it immediately to the prepared site. The department's inventory control card would then be filled out, indicating that machine number so-and-so was located in department such-and-such (machinery departments were numbered according to the Model T part number), and was performing operation number so-and-so (which correlated with the operations sheets).[26]

The engineering department, therefore, was critical to Ford's expansion. Its proper functioning depended upon the control of information (about machine tool manufacturers, Ford production processes, and manufacturing operations in general) and the careful coordination of procurement with Ford's layout men and millwrights. Wibel's department, maintaining and improving the system established by his predecessors and mentors Oscar Bornholdt and Charles Morgana, achieved this control and coordination through what can only be described as bureaucratic recordkeeping.[27] Unfortunately, the folk wisdom that surrounds Henry Ford's life and work has precluded any serious scholarship on the day-to-day operations of the Ford Motor Company. But the fact is that the company kept impressive records and controlled and manipulated information far more creatively than any other company dealt with in this study.

Ford's recordkeeping and use of information played an important part in the story of its problems with the changeover from the Model T to the Model A. As demonstrated above, the engineering department's operations were conducted on the basis of ''historical'' information. That is, routine expansion of output was achieved by using information about prior operations and performance in each machining department. The richness and accuracy of this information, properly maintained and used, allowed the engineering department to help the company increase its production in the period between 1915 and 1923 with few serious problems.[28] The changeover to the Model A would essentially cut off the engineering department from the past and make its job of expanding production highly problematical.

This reliance upon past information certainly could have mediated against innovation in production technology, that is, in increasing productivity. Yet by all appearances, the engineering department stayed on top of developments in the machine tool industry. In conjunction with Ford's tool department, headed by the extremely able William Pioch and pushed by Sorensen, Martin, and their lieutenants, the engineering department witnessed achievements in economies of scale.[29] Moreover, its role in the purchase of parts manufactured outside the Ford plants made its members cost-conscious, especially when comparing its cost data with that of in-house operations.

Throughout the accepted wisdom about Ford runs an explicitly stated theme that neither Henry Ford nor anyone else in his company knew the cost of producing one of his automobiles. Nothing could be further from the truth, despite what Ford may have told his numerous interviewers. During the period under immediate consideration, the company assembled cost data and made monthly cost estimates for all minor and major assemblies and body types. These estimates were compared with those of previous years or, in the case of final assembly, with estimates from branch assembly plants.[30] Trends in the cost of materials, labor, and overhead, the three major elements of Ford cost estimates, thus were readily visible, especially when reduced to graphic form. Graphs in the Ford Archives, for example, show the unit cost of materials for particular Model T body types plotted on the same page with curves for labor and overhead unit costs; total factory costs were also plotted with retail sales prices.[31] Ford production men—and, one must believe, Henry Ford himself—knew the costs at the Ford Motor Company, on a monthly and yearly basis, whether it was the cost of a completed automobile, of tool steel used at Highland Park and River Rouge, or of repairs to machine tools.[32]

The engineering department kept much of its own cost data, particularly on automobile components the company purchased from outside manufacturers, which technically—if such was possible at Ford—was the responsibility of the purchasing department. The department often drew graphs consisting of two curves, one for Ford Motor Company's manufacturing costs for a particular part or assembly (for example, the body), the other of prices paid to outside suppliers.[33] This exercise provided production men with valuable information. The department constantly sought prices on components from outside suppliers as a means of checking the company's own production costs. In the process, despite the increasing integration of the Rouge, Wibel and his lieutenants came to realize that "going outside" could be profitable. This experience paid off when the changeover from T to A took place.

Probably the most impressive aspect of the "bureaucratic" way in which the company dealt with day-to-day aspects of production was its use of mechanical drawings. These were paramount in Ford's system of mass production. The company produced and maintained drawings of every part of the Model T, every special tool, jig, fixture, and gauge used in its production, and every master gauge used to check these special devices. Drawings served as the ultimate authority in Ford production, for they specified dimensions, tolerances, gauging points, materials (including shear strength and other metallurgical specifications), and finishes. Used by the design, tool, engineering, and inspection departments, superintendents (Martin, Sorensen, and their lieutenants), and foremen in each of the parts production departments, these drawings served as the medium for exchanging information and for maintaining common understanding. No changes could be made without a change in the drawing.[34]

Yet a change in a part, its method of manufacture, or its placement in an assembly set in motion a series of changes that went well beyond a simple change in the drawing. Change acted like a pebble hitting the middle of a still pond; ripples moved out to the various departments mentioned above. Obviously, a change in the design or construction of a part necessitated altering or rewriting operations sheets for production and assembly in addition to retooling. Maintenance questions also came into consideration. To convey information about changes, whether large or small, to all concerned parties the company adopted as early as 1908 the use of "factory letters." Issued weekly and sometimes even more frequently, these letters listed parts by number and name and then outlined the nature of the changes. The letters often referred to previous issues (they were numbered consecutively), and they always dated the revision. For the theoretically "unchanging" Model T, the number and bulk of these letters are overwhelming. When changes affected the sales and service operations of Ford dealers, the company issued a general letter documenting the changes in parts or assemblies. These letters also contained an abundance of information not related to production.[35]

Taken altogether, the Ford system of drawings, operations sheets, cost accounting, engineering procurement, factory letters, and general letters provided a mechanism for the smooth operation and monitoring of day-to-day events. Change was accommodated through a bureaucratic routine that worked so well that, for the most part, change was not noticed outside the company.[36] Nevertheless, changes were made in the Model T, as is well known by collectors and restorers of this most famous automobile in American history.

Floyd Clymer's book, *Henry's Wonderful Model T, 1908–1927,* contains a useful essay by Leslie R. Henry, which outlines in detail what he calls the Model T paradox— "change in the changeless Model T." Henry suggests the nature of the paradox and lists most of the major changes in the Model T year by year. Similarly, Philip Van Doren Stern's *Tin Lizzie: The Story of the Fabulous Model T Ford* enumerates major changes in the "outward styling" of the Ford automobile.[37] Both lists work to dispel the legend of the unchanging Model T. Obviously, many of these changes were made for the sake of mechanical improvement or easier maintenance. But an enormous number had manufacturing objectives in mind, be it easier or more reliable acquisition of parts, cheaper materials, fewer machining operations, or simpler assembly. Finally, styling played no small role in bringing about change. Of course, even Clio would find it difficult to identify a single motive for each of the changes made in the Model T over its life from 1908 to 1927.

It is not the intention of this chapter to relist changes made in the Model T or to determine a motive for them. Yet it may be useful to identify a few significant changes and relate them to Ford production.[38] Although generalizations must be highly suspect, it appears that the majority of changes made in the Model T from its inception in 1908 until 1915 grew out of production considerations, the most famous example being the decision in 1914 to make only black cars. In 1915, Ford introduced a major styling change, moving away from the "antique" style of straight, flat fenders to "transitional styling," which featured curved or rounded fenders. Transitional styling was introduced on two new body types, the sedan and the coupélet. These additions brought to five the number of different bodies Ford mounted on the Model T chassis in 1915, the year acetylene headlights were replaced with electric ones. Despite these significant changes and many minor ones, Ford increased its output from three hundred thousand to five hundred thousand cars, from 1914 to 1915. Transitional styling was continued the following year, which also marked the beginning of TT—Ford truck—production. Total annual output jumped to over seven hundred thousand.

Model T collectors say that the 1917 models, introduced in September 1916, marked the "new look,"—"streamlining." A new radiator shell, a new hood, crowned fenders, and nickel-plated hub and radiator caps gave the T a snappy look. At the same time, the compression of the engine was lowered, thus decreasing horsepower. Such other changes as replacement of ball bearings by roller bearings on the front wheels and addition of an electric horn also took place. In 1917, probably because the company was occupied with armament for war, Ford production dropped to just over six hundred thousand automobiles.

Ford's postwar output in 1919 of more than eight hundred thousand cars and one hundred thousand trucks almost doubled 1918 figures. Major changes were also made in the engine, including the introduction of an optional electric starter, which necessitated design changes in the cylinder block, the flywheel, the timing-gear cover, the timing gears, and the transmission cover. And, of course, the starter as well as a generator had to be produced and assembled. Although made initially only for cars carrying the starter option, the new engine eventually went into every Model T. Body styles remained constant. The changes made during this production year, when measured with the significant increase in output and the work at the River Rouge site, suggest that the Ford production organization could take change in stride, could, indeed, quicken the pace. The company also made its largest profit in its sixteen-year history.

Between the peak of 1919, the deep valley of 1920, and Ford's Mount Everest of 1923, the company anticipated another major change in body style, similar in scale to that of 1917. As early as 1920, designers introduced an oval gasoline tank, which led in 1923 to a complete line of lowered and "streamlined" bodies. To this line was added the Tudor and Fordor sedans. The production peak of 1.8 million cars and two hundred thousand trucks in 1923 proves that the changes in body style posed no serious threats to output, though cost data suggest an increase in the production cost of the new Model T. The very high overhead of Rouge operations, especially the shifting of machining operations from Highland Park to the new foundry, may have contributed more to this increase in cost than a change in body styles.

The rapid fall from the heights of production and market share in 1923 to the end of Model T production in mid-1927 was accompanied by changes in the model. The most significant, overlooked by both Clymer and Stern, was the change to the "all-metal" body in 1925, a year when Ford produced 1.6 million cars and two hundred sixty thousand trucks. Since 1911, Ford had manufactured cars with a wooden-framed body covered with sheet steel. Fenders and running boards, of course, were stamped out of sheet steel. The initiation of body production at the Rouge precipitated the move to the all-metal body, which took place over a three-year period. When the pressed steel department was moved from Highland Park to the Rouge in 1925, the company took the final step and eliminated wooden framing. Ernest A. Walters, a Ford pressed steel pioneer from the Keim mills in Buffalo and superintendent of pressed steel operations, considered the change to the all-metal body "gradual." He accepted change as routine: "Whenever a minor change took place we . . . would take care of it and would not look for the reason for change. Prints were always gradually coming in for changes in our body parts. The schedule was made up so as to give us necessary time to make these minor changes without our being aware why they were made."[39]

Yet the change to the all-metal body was not as painless or as inconsequential as Walters implied. Ford's chief design engineer for the Model T and the Model A, Joseph Galamb, remembered that "we found that a[n all] steel body needs a lot of tooling, so we stuck [for most of the Model T's history] to the wood job for framework." Coming very late in the life of the Model T, the all-metal body seems to have had two effects on Ford operations and Henry Ford. First, it drove the cost of the car significantly upward.[40] Second, the cost of the tooling and the innovation of the all-metal body probably made Ford far more reluctant to scrap the Model T than had the company continued to make the body in the old manner. As will be seen, Ford's investment in new tooling for the T in 1925 should have been for the A.

Finally, to try to reverse continued and severe slippage in the market, Ford introduced significant style and equipment changes in 1926. Designers lowered the chassis, nickel-plated the radiator shell, put in lightweight pistons in an attempt to beef up the engine for the heavier car, modified the engine-block casting to accommodate a new transmission cover (necessitated by widening the transmission brake bands), enlarged the steering wheel, and changed the steering ratio. The once-optional balloon tire was offered as standard equipment, and the rear brake band was enlarged for better stopping. A dozen other changes graced the new Model T, but the major styling innovation was the re-introduction of color choices after twelve years of black. These changes kept the demand for Model T's from sagging below the 1.3 million mark. They might have severely hindered Ford's ability to mass-produce the automobile, yet as 1926 drew on, the company accumulated the largest inventory of cars in its history. By mid-1927, the Model T era would be over, ushered out with a T whose only innovation was wire wheels as standard equipment.

The record of change at Ford indicates that the company accommodated day-to-day minor changes and some significant style changes with ease. This level caused no breaking of stride, no delay in production, and only two apparent changes in the downward cost trend of the "car for the masses." The paradox of the changing, changeless Model T leads to another. Although the Ford Motor Company handled routine change with apparent ease, Henry Ford resolutely resisted change. That was the viewpoint of the rest of the automobile industry and is the consensus of modern historians.[41] To understand this paradox as well as the vagaries of the automobile industry in the 1920s is to gain additional insight into the story of the changeover to the Model A.

The Model T was as much an idea as it was an automobile. As discussed in Chapter 6, the idea was an unchanging car for the masses. Ford adhered to changelessness for ideological as well as productive or economic reasons.[42] Much of the public bought his car for the same reasons. By 1920, the Model T was expected to be constant. Therefore, change created a peculiar marketing problem for Ford Motor Company. The company could not use change or "improvement" as a selling point. That Ford feared the sales consequences of changing the Model T is attested by the hush-hush atmosphere that surrounded the changes made in 1923 and 1926. The earlier changes were hidden under the ruse of the "English job," and those leading to the 1926 models came to be known as the "Australian job," both names suggesting changes for other markets, not America.[43] Not until mid-1925 did the company use "pronounced changes" in styling and equipment projected for 1926 as a promotional feature. Throughout the substantial coverage of the

Ford Motor Company in the *New York Times* in the 1920s and as late as 1932, the company's position on change, with the exception of the belated Model A announcement, was to deny it—to fear it—for reasons of the anticipated effects of announcements on sales of present models.[44]

But despite the Ford Motor Company's ability to accommodate a low level of change in its automobile and Henry Ford's resistance to any change, automotive America was swept by high winds of change in the 1920s.[45] They struck Ford with violence, reducing the company's market share of more than 50 percent in the early 1920s to less than 15 percent in 1927. The automobile had changed.

By the standards of the mid-1920s, the Model T was outmoded. The ignition, carburetion, transmission, brake, and suspension systems, as well as the styling and appointments, made the T appear antique. Genuine engineering improvements such as battery-powered ignition, electric starters, and shock absorbers had been made by other manufacturers, yet Ford clung firmly to the basic design of the Model T, which had served the company and the public so well since 1908. Throughout the 1920s, automotive industry analysts continually expressed expectations that Ford would incorporate new mechanical engineering developments in the Model T. Until the Model A emerged, they were disappointed. As Ford's decline proceeded, two penetrating studies appeared during 1926 about the condition of the Ford Motor Company in the rapidly changing automobile industry and market.

In an article published in the May 1926 issue of *Motor*, entitled "What Will Ford Do Next?" James Dalton analyzed the changing automobile market. Dalton demonstrated the penetration of other manufacturers' cars into the price class that Ford once monopolized. Clearly, no one competed heads-up with Ford in price, but during the first half of the 1920s, competitors—Chevrolet in 1926—brought the prices of their cars down within 30 percent of Ford's. Dalton plotted price trends of Ford and his competitors on the same graph with a curve of average per capita earnings in America. Between 1921 and 1926 per capita income increased from $551 to $610. This increase contributed heavily to Ford's loss of the market, especially when coupled with information assembled from another graph showing the potential pool of "first-time" automobile purchases from the inception of the Model T to 1930 (by extrapolation). The graph demonstrated that between 1926 and 1930 the pool of "probably first time buyers" of cars would be only 1,940,000 families. Increased per capita income suggested that a growing percentage of these buyers would spend the additional money to buy an "up-to-date" automobile, even if Ford further reduced the price of the Model T. But more important, Dalton suggested that greater wealth and more attractive options in other automobiles surely meant that substantial numbers of families buying an automobile for the second time or purchasing a second automobile would not buy a Ford. Used automobiles made recently by other manufacturers would also provide competition to the Model T.

James C. Young's analysis, published as a feature article in the *New York Times* seven months later, complemented Dalton's. Young agreed that "prosperity . . . caused the Ford decline." He had interviewed Henry Ford to probe the maker's own assessment of that prosperity. Expounding a theme that was familiar to *Times* readers, Ford blasted the rapid expansion of credit buying: "I sometimes wonder if we have not lost our buying sense and fallen entirely under the spell of salesmanship. The American of a generation ago was a shrewd buyer. He knew values in the terms of utility and dollars. But nowadays the American people seem to listen and be sold; that is, they do not buy. They are sold; things are pushed on them. We have dotted lines for this, that and the other thing—all of them taking up income before it is earned."[46]

Ford correctly analyzed the changing character of business and consumption: "Credit, you know, has become a fourth dimension in American business." But this response merely begged the question of whether Ford would join the party. Ford expressed his belief that sales on credit hurt the consumer and that "we have no desire to sell cars at the expense of public benefit." Although the highly competitive Chevrolet could be purchased through the General Motors Acceptance Corporation, Ford steadfastly refused to consider credit as a legitimate instrument of consumption. Ford also defended the Model T:

> The Ford car is a tried and proved product that requires no tinkering. It has met all the conditions of transportation the world over. . . . The Ford car will continue to be made in the same way. We have no intention of offering a new car at the coming automobile shows. Changes of style from time to time are merely evolution. Our colored bodies seem to have found favor. But we do not intend to make a "six," an "eight" or anything else outside of our regular products. It is true that we have experimented with such cars, as we experiment with many things. They keep our engineers busy—prevent them tinkering too much with the Ford car.[47]

Times had changed. But both Dalton and Young—indeed all automotive America—wondered what Ford would do. Would Ford be satisfied with less than a third of the market? Would he go out of business altogether? Dalton considered updating the Model T with such "conventional" engineering features as a three- or four-speed transmission and a nonmagneto ignition system. The cost would be enormous, he argued. "It might be cheaper, therefore, to bring out an entirely new line." Upon further thought, Dalton reached a sobering conclusion: "There can be no junking of present Ford production. . . . His entire organization has been built around the quantity idea and he will have to continue on that basis. Price, therefore, must be the main consideration, just as it always has been." Young reached the same conclusion: "Just two things can assure the old ascendancy of the Ford car. One of them would be a radical cut in prices and the other a period of depression whereby Ford owners would again buy Ford cars."[48]

Yet because Henry Ford and the Ford Motor Company knew what the Model T cost at the factory and realized that Ford dealers had to make a profit, additional price cuts were not a real possibility.[49] The all-metal enclosed body had resulted in the slimmest profit margin Ford experienced in the 1920s. A price reduction in line with the falling costs effected by learning how to produce the all-metal body would prove to be only a stopgap measure—and an ineffective one at that.

Within the company some advocated major change in the Ford automobile—an entirely new model, not a prettified Model T. Ernest C. Kanzler was the only one to confront Henry Ford directly with the necessity of immediate change. On January 26, 1926, Kanzler wrote a memorandum to Henry Ford spelling out in five pages, a summary, and a page of sales data why it was imperative that the Ford Motor Company immediately introduce an up-to-date model. Kanzler had to summon a great deal of courage to deliver the memo. Perhaps company president Edsel Ford, constantly cowed by his father, was a party in its conception and a supporter in its delivery. Kanzler ended his memorandum, "The writing of this has not been pleasing, Mr. Ford, but I have always tried to tell you what I see and feel. These thoughts have been uppermost in my mind the last year and I cannot keep from expressing them any longer."[50]

This famous memorandum, which led to Kanzler's firing within six months, gave the official historians Allan Nevins and Frank Ernest Hill an opening to their chapter "The

End of Model T.''[51] These historians chose the proper document and correctly interpreted its major points, but they disregarded the subtleties of changeover thinking within the Ford Motor Company. Kanzler's plea to Ford was the first of many overly optimistic statements about the ease of achieving a changeover.

Nevins and Hill clearly recognized that during 1925 and 1926, the company's design engineers were busily at work, following Henry Ford's directions, on an automobile that was intended to be as revolutionary as the Model T had been in 1908. Because its eight engine cylinders were arranged in an X-like pattern around the crankshaft, the project was dubbed the X-car. Kanzler's memorandum, as well as testimony of Ford's engineers, makes it clear that Henry Ford viewed the X-car as the automobile that would push back encroaching competition. But Nevins and Hill failed to see that it was not intended to be a replacement for the Model T. Ford's men and Ford himself regarded it as ''an intermediate car'' (Kanzler's words), an important step above the Model T for which Kanzler believed there would ''always be a field for 4000 to 5000 . . . per day.'' Coyly, with the Lincoln on the top and the Model T on the bottom, Ford was contemplating his own version of General Motors with its strategy of ''a car for every purse and purpose.''[52]

Ford's strategy failed to come to fruition because the X-car never panned out. Had it done so, the company might have suffered a far less severe setback than it did in 1927 and 1928. To be sure, as William Abernathy has ably demonstrated, major model changes were overwhelmingly important in gaining (or, for Ford, losing) market share in the automobile industry of the 1920s and 1930s. Because it was different, the X-car would have been important. But it would not have been revolutionary. The Model T was the only revolutionary automobile of the twentieth century. Its design and mass production made people want an automobile. The only other revolution of the American automobile industry in this century was in marketing: GM's explicit, diversified sales strategy and its evolutionary development of the annual model change.[53]

Kanzler sternly warned Ford against putting ''all our eggs in one basket'' with the X-car. He may have realized subconsciously that the engine could not be developed. Openly, he argued that ''there is little chance for the production'' of the car ''within eighteen months which would not be before the summer of 1927.'' Kanzler's proposed solution has been overlooked. He argued that a more conventional engine—explicitly, a six-cylinder one—''be installed in the [apparently already developed X-car] chassis'' as a means to ''hold this market for us against competition until such time as we would sweep all before us with your revolutionary 'X' power plant substituted when its perfection has been achieved.'' The memorandum also implies that more than a chassis was being developed by Ford's engineers, for he alluded to ''all of the difficulties of the . . . new front axle, steering gear, rear axle, spring suspension, body design, and transmission of the intermediate car.''[54] In hindsight, Kanzler's proposal was perhaps the most workable short-run solution to Ford's slippage in the automobile market. Ford could have introduced an intermediate car to entice consumers with increased purchasing power and then updated his Model T. But even the production planner Kanzler misjudged the time needed to initiate production of a new car, even when that car was not projected to be a replacement for Model T.

When in August 1926 the X-engine seemed hopeless and when inventories began to mount as never before, Ford ordered entirely new design work to begin on ''a car for the market, a four-cylinder one.'' Various factors—the precipitous decline of Ford sales; hounding for a new car from dealers, journalists, and a Ford-supporting public; and, according to Sorensen, a fight with Edsel—led Henry Ford to decide not only to build a new car but to cease making the Model T. Ford made his decision sometime before May 25, 1927, the day the company issued its official announcement. If A. M. Wibel's memory was correct, the "official end" of Model T production was announced in the company's executive offices much earlier. Ford's plan had been to stop assembling the Model T and begin at once to make the A. But Wibel said they "miscalculated," that, in fact, the company had to make and procure the parts for an additional fifty thousand Model T's after suppliers had made what they considered final deliveries on their contracts.[55] Such poor planning only multiplied as the Ford Motor Company tried overnight to design and mass-produce an entirely new automobile.

The official announcement that the Ford Motor Company intended to replace the Model T with a new model, "superior in design and performance performance to any low-priced, light car," underlined Ford's plan to move smoothly from the old to the new. The company anticipated no "total shutdown." The *New York Times* quoted Henry Ford: "I am glad that business even now is so brisk that it will not necessitate a complete shutdown. Only a comparatively few men will be out at a time while their departments are being tooled up for the new product. At one time it looked as if 70,000 men might be laid off temporarily, but we have now scaled that down to less than 25,000 at a time. The lay-off will be brief, because we need the men and we have no time to waste."[56]

The day following the official announcement, Henry and Edsel Ford and the eight oldest employees of the company quietly observed the occasion of the assembly of the fifteen millionth Model T. At the Rouge, John F. Wandersee, August Degener, Frank Kulick, Fred L. Rockelman, C. B. Hartner, P. E. Martin, Charles Meida, and Charles Sorensen each stamped a digit of the serial number into the engine's cylinder block. At Highland Park, before the watchful eyes of the Fords, the engine was put into the Model T assembly. When the touring car was finished, with the words "The Fifteen Millionth Ford" painted boldly on its side panels and rear, Edsel drove it off the line with his father beside him. (See Figure 7.3.) Sorensen and Martin joined the Fords by climbing in the back seat. A procession of company people followed them to the Dearborn Laboratory, where they witnessed Henry Ford drive his 1896 quadracycle and Model T #1.[57] The riders in the fifteen millionth Model T offered a portent for the near future, for when all of the wrenching of the changeover ceased, the six other men who had left their imprint on the fifteen millionth engine block and on the company would be gone, fired along with many others on the grounds that they were "Model T sons-of-bitches." The changeover had begun.

Symbolically the last Model T, number 15,000,000, rolled off the famous Ford assembly line nine months after Henry Ford had apparently instructed his designers to begin work on a four-cylinder automobile "for the market" and five months after the first (surviving) sketch of the body layout, clearly identifiable as the Model A, was drawn. At the outset, Ford must have believed that he and his engineers could design an up-to-date automobile and tool up for its production before the public learned that it was coming. That plan obviously failed, for when the announcement came, the Model A's design was nowhere nearly complete. Only a few general statements about the process of designing the Model A are in order here.[58] More careful planning could have circumvented the delay in completing its design. Planning for change was to become as important to the new mass production as planning for production. But his experience with the Model T led Henry Ford to believe that the Model A would be just as easily designed. Also, no one

FIGURE 7.3.   Henry Ford and Edsel B. Ford in the Fifteen Millionth Model T, 1927. Moments after
this photograph was taken, Charles Sorensen and P. E. Martin joined the Fords for a ride from the
Highland Park factory to Ford's Dearborn Laboratory. (Henry Ford Museum, The Edison Institute.
Neg. No. 833-49148.)

knew better than Ford himself what James Dalton had pointed out a year earlier—that a
new Ford model "would necessitate the most careful engineering."[59] Ford could not
botch the job.

Under pressure to design a first-rate car essentially from scratch,[60] Henry Ford broke
away from a traditional Ford approach to design as it related to production. William
Pioch, the company's chief tool engineer, summarized the objective: "Ford was vitally
interested in putting out a car that would stand up better than anything on the road. To
accomplish that, he wouldn't allow any stampings in the chassis. . . . Everything had to
be a forging."[61]

Since Charles Sorensen had first established contact between Ford and the John R.
Keim stamping mill in Buffalo, New York, the entire trend in Ford production technology
had been toward greater use of sheet-steel stampings. The all-metal, enclosed body had
been merely another step in the long process of more and better stampings. A number of
Ford engineers have said that Henry Ford had "always objected to stampings" and
referred to them as "Hungarian stimpings," a play on designer Joe Galamb's pronuncia-
tion of "stampings." Similarly, Ford often called Galamb "shit-iron Joe" to remind him
that he advocated using stamped *sheet* steel wherever possible.[62]

Although the Model A was to be marketed as a recast and updated Model T, Ford wanted the A to be different mechanically and materially. Thus Ford insisted on more castings and forgings, both of which required subsequent machining. Mechanical designer Laurence Sheldrick argued that Ford arrived at this new approach through what might be called a "thermal" argument. Ford knew the steps in the production of sheet steel: ore to pig iron to steel to rolled steel. Because a number of separate heatings were involved here and also in annealing stampings between certain forming operations, Ford thought the process wasteful. "A casting can be made directly from the Blast Furnace," he argued, "which means one heating." Moreover, he assumed that forgings could be made by rough-casting steel and, while still hot but not molten, forging them to final shape between dies. During the initial stages of the changeover, Ford experts worked in vain to perfect Henry Ford's idea of producing forgings.[63]

Ford's autocratic decision to eliminate stampings caused delays in design but more important ones in the start-up of production. Although forgings had not been totally abandoned in Model T production, their number had been reduced so significantly that the company purchased rather than made forgings itself. Because facilities were lacking at the Rouge, the company was forced to rely upon outside suppliers for Model A forgings. Lateness of design of many of those forgings and Ford's tendency to make last-minute changes exacerbated the problem of obtaining them to begin machining and assembly operations. And, as Ford himself was to learn painfully, insistence upon forgings drove up the final cost of the Model A significantly enough that, once it realized the costs, the company moved quickly and quietly back to stampings.[64] Ford's flourish with forgings was a costly, short-lived exercise.

The nation anxiously awaited the completed design of the Model A Ford. Impressive numbers of rumors, most of them wrong, circulated among automotive Americans. In June, sources "close to Henry Ford" said that official details of the new Ford would be announced on July 1. Denying published details of the car on June 22, Edsel Ford said, "As a matter of actual fact, the specifications for the new models are not yet complete, and it would be impossible for any one, even in the Ford organization, to discuss them with accuracy and with authority."[65]

Late in July, the public still had no details, and automotive experts wrote that it would be surprising if production of the new model began before September 1. Reports also circulated, including some on the front page of the *New York Times*, that Henry Ford and his engineers could not agree, that Ford clung tenaciously to T components while engineers pressed for entirely new ones. Yet on his birthday, July 30, Ford promised to reveal details of his new car "in a few weeks." "We have taken our time to design and build this new Ford car," he said, "so that it will be just what a good automobile should be in this day." Unofficially, production of the still-mysterious car began August 4, when the company started rehiring workers a rate of three thousand per day. Finally on August 10, Edsel Ford announced that the "new Ford automobile is now an accomplished fact." He offered an account of the car's performance with no details of its design but assured the public that the company was almost ready to mass-produce the new model. "We know also," he declared, "what is needed as to personnel and factory equipment in order to produce these new Ford cars in greater numbers than any manufacturer has ever attempted before. The work of retooling our plants throughout the country to prepare for the heaviest production schedule we have ever undertaken is now nearly complete."[66]

Ford deceived either himself or the public. Not until the second week of October did Henry Ford finish the design work on Model A to his satisfaction. A week later, the Fords

FIGURE 7.4.   Model A Engine Number One on Test Block, 1927. Largely as a media event, Ford production men gathered to stamp the serial number in the first "production" Model A engine block. Left to right: Charles Sorensen, Edsel B. Ford, Henry Ford, P. E. Martin, August Degener, Charles Hartner. (Henry Ford Museum, The Edison Institute. Neg. No. 833-50046.)

witnessed the first Model A come off the assembly line, and this was actually only a media event because no more Model A's were assembled until November. (See Figure 7.4.) None was shown publicly until December 1, exactly three months after Henry Ford had promised and about six months after the last Model T had left the assembly line. Henry Ford found fault with Model A #1, and ordered certain changes.[67] As will be seen, these and subsequent changes plagued Model A production.

From Edsel Ford's early August assurance that retooling was virtually completed until the first public exhibition in New York City's Waldorf, the company announced on several occasions that a new, more productive system of manufacture was in place, waiting Henry Ford's go-ahead. The *New York Times* gave anticipated production figures: "Through rearrangement of machinery and introduction of higher speed tools the company will be able to turn out 11,000 cars or more daily, as against a maximum of 8,000 under the old scheme of production." Ford's house organ, *Ford News,* described the system more directly, saying that "production machinery has been designed, built, in-

FIGURE 7.5.   Henry Ford, Edsel B. Ford, and the new Model A Ford, Waldorf Hotel, December 1, 1927. Henry Ford always showed his cars at the Waldorf at the same time other manufacturers exhibited at the annual Madison Square Garden show of the Automobile Manufacturers Association, which Ford refused to join. This was the first public showing of the Model A. (Henry Ford Museum, The Edison Institute. Neg. No. 0-4083.)

stalled; assembly and parts conveyors have been constructed and placed in proper relation to one another for the big test of what is in many respects a new production system."[68]

At the Waldorf showing, Henry Ford promised production of one thousand Model A's a day by January 1, 1928. (See Figure 7.5.) This figure was achieved in late February. Only during this month was the company able to supply all of its dealers with a demonstrator but not all body types. Similarly, on March 26, 1928, Ford proposed five thousand cars a day by July 1, but his factory turned out only about three thousand on the appointed day. Early October saw the mass-production of some fifty-five hundred A's per day. Although in February 1928, Henry Ford admitted finally that "you cannot get a great plant converted from one model to another in a day or a week," one may ask the legitimate questions, what caused the serious delay in Ford Model A production and what was learned from this experience?[69]

Of course, design delays have already been identified as a major factor in the ultimate delay of Model A production. Design decisions, notably the minimization of stampings and their replacement by forgings, also caused delay. Ford's insistence on forgings resulted primarily in procurement and cost problems, not purely production or technical ones. But Ford's decision to build the gasoline tank into the cowl of the Model A (that is, to make the cowl serve as the gas tank) wrought a pure production problem. The technical aspects of this chosen design plagued Ford engineers and contributed substantially to the delay in production. Moreover, the gas tank was only a sympton of the production problems associated with the new body, which entailed deeper drawings (that is, sharper bends and deeper recesses in the sheet steel) than the Model T. Finally, some of the same problems arose with the rear axle assembly as with the gas tank. These problems will be discussed in turn.

When the Model T was redesigned in 1926, Henry Ford chose to move the gasoline tank in all bodies (except the Fordor) from under the driver's seat (the front seat had to be lifted to fill the tank) to the space between the firewall and the dashboard. Apparently, the design grew on Ford. He not only retained it for the Model A but designed the tank to be integral with the cowl. As with forgings, Ford's design engineers winced at his scheme but not for the same reasons.[70] Forgings meant higher costs. A cowl tank meant the continued risk, already evident in the Model T, of gasoline leaking into the automobile's interior. Further, its proximity to the heat of the engine heightened their apprehensions. But Ford had his way. He argued—and one cannot help but think of the more modern Pinto—that a tank in the rear presented greater dangers than one placed under the driver's seat or, second best, in the cowl. These locations offered the greatest protection for the tank. Ford also looked upon his design as a way to save a component and therefore weight; if the body was the tank, no separate tank would be necessary.

Had Ford's design engineers and production men foreseen the problems of manufacturing cowl tanks, they might have raised louder objections and might even have convinced Ford to find a different location. A separate tank under the cowl would have been an acceptable compromise. It was easy enough to design a single automobile with a gasoline tank integral with the body; mass production of them was an altogether different story. The master's design was a nightmare to Ford's production men. As with most parts and subassemblies of the prototype Model A, the gas tank was drawn, complete with materials and inspection specifications, by draftsmen in the tool department. In committee with the engineering department, the superintendent's office, and appropriate parts department heads, tool department chief William Pioch or his representative then wrote an initial draft of an operations sheet, choosing suitable machinery. Obviously, the gasoline tank entailed chiefly sheet steel operations, which seemed to be straightforward. Ford designed the tank in two pieces, and the trick in this job turned out to be welding the top section tightly down to the bottom half. Between Pioch's welding machine expert and Wibel's procurement specialist, the company ordered Gibb seam welders—a resistance machine that welded material together by passing it between two rollers—to carry out the major assembly operation. But as Fay Leone Faurote explained, "When it came to actually making this Ford tank two troubles developed. The seam welder would not make a perfect seam-weld." At some points, the sheet steel (which had been coated with an alloy of lead and tin) was burned, but at others "it was not sufficiently heated to weld. The standard seam welding machine, although expected to do the job, did not do it."[71]

As A. M. Wibel always demanded with outside suppliers, the Gibb people sent representatives to straighten out the problem. But these experts were stumped: Ernest Walters recalled that Ford's chief of welding machinery kept his own men "working steadily" on the seam welder "for months."[72] Between the manufacturer and Ford's tool department, the welder was extensively redesigned, incorporating a number of "Ford special'' components.[73] Ultimately, the problem was solved. But as pressed steel superintendent Walters remembered, virtually every one of the tanks made in the first five or six months of A production required hand soldering to seal leaks that were evident when the tanks were tested underwater with thirty pounds of air pressure. Walters remarked that "all of the superintendents visited us daily, worrying about the trouble in not getting production." "Learning by doing" also worked to solve gas tank production delays. Initially, the seam welding operators could produce only 200 tanks in eight hours, but some six months later, this figure stood at about 450. The welding machines, however, posed only one of the major problems in gas tank production. Walters recalled that Joe Galamb had blamed him initially for delays in body production. When called on the carpet by Edsel Ford and Charles Sorensen, Walters showed them a set of blueprints "for making gasoline tanks that had no dimensions on them." Checking revealed that this blunder had resulted from changes made by Joe Galamb even while tooling up was in process. The rushed changeover atmosphere must have manifested itself in such mistakes more than once.[74]

Body production beyond the gasoline tank cowl posed equally serious problems. A. M. Wibel, who was always present at the roundtable luncheon in Dearborn at which Henry and Edsel Ford, Charles Sorensen, P. E. Martin, and a few others met daily, pinpointed the body as a delaying factor: "I think that changes in body styling in the dies and pressed steel had an awful lot to do with the delay in the Model A production."[75] The A design required deep or heavy drawing, which was generally unproven at Ford. Not only were heavier and more sophisticated dies necessary but the men responsible for starting up body production faced a terrific guessing game of how much "spring-back" would occur when a sheet of steel left the dies. Improper fits of body parts after spring-back made the game expensive, which was no object to Ford, and time-consuming, which was all-important. Months of work took place before the body dies were made so that the body parts were the same as those designed by Ford and his engineers in Dearborn.

No diemaker wanted for a job while the Ford Motor Company tooled up for Model A. Thousands of dies were needed. Even with good die prototypes, diemaking was an important factor in delaying production. The company obtained its first Keller engraving (or profiling) machine in early 1926. Although the Keller machine greatly diminished diemaking time, the dies sunk by these machines still required hand finishing. Moreover, Ford could not obtain enough Keller engravers to satisfy the tremendous demands of the changeover. Sometime after the Model A got into production, the company adopted a new technique for making many of its production dies. In crude outline, this process consisted of stamping (rather than engraving) the desired impression into a hot steel die block (engraving was done cold). Ford claimed that dies so made would produce far greater numbers of forgings or stampings before wearing out.[76]

If the gas tank and the remainder of body production posed problems, so did manufacture of the rear axle. Here again design proved to be critical in determining production approaches. Henry Ford sought to make the rear axle assembly both strong and light. Together, he and Laurence Sheldrick arrived at a design far more complex than the Model T rear axle had been. The new axle required not only punch press work (as on the Model T) but also the development of hot metal spinning machines (to form the bell part of the axle housings, which bolted to the differential housing) and electric welders (to weld the two-piece differential housing together and to weld the axle-shaft housing flange to the axle housing).[77] Hot metal spinning necessitated the design and construction of enormous special machines operated by two men. Although the design and operation of these machines appears to have posed no particular problems other than costs of time and money, production of the differential housing haunted Ford production men.

Initially, welding the two pieces of this part must have seemed perfectly straightforward. But when production got under way, they learned that the hand-loaded, hand-operated resistance welders they had designed failed to produce uniformly satisfactory welds. Moreover, these machines achieved only about half of their expected per hour production, an extremely low rate considering that three men operated each welder. To improve the situation, during the early and trying stages of Model A production Ford welding engineers developed an automatic welder operated by a single man. This machine's output eventually satisfied demands for both uniformity and output.[78] Thus the rear axle job and that of the gas tank retarded Model A production by making it a matter of fits and starts in the early stages.

Procedural weaknesses also served to delay and stifle production. The company's long-practiced method of using drawings as the medium for carrying design changes into production realities broke down largely because of the rush for design completion. We have seen that the pressed steel department received dimensionless—and therefore useless—drawings. Laurence Sheldrick, Ford's chief design engineer, recalled that "a great many [Model A components] were taken right from the drawing board to the tool room. They started making tools without any trial whatsoever. A lot of these items went right to the rool room with the expectation that they would work out." This procedure often failed. Designers changed their minds, which often meant that toolmaking and operation sheet writing had been done in vain. More serious, because they lacked a fully tested part, the engineers who did the drawing, both in the design room and the toolroom, found it difficult to specify without error appropriate materials and necessary or desirable tolerances. Even though the limit system of manufacture had become a fine art in American manufacturing, a "model," used in the sense familiar to antebellum firearms makers, remained important, as Ford designers learned.[79]

With the Model A, Ford was not only changing over to a new model, but it was upgrading the precision of its machining work, which made the changeover even more difficult. Designers and production men narrowed limits on the Model A such that when it first appeared, the car was considered one of the finest automobiles made in the United States, especially in its price class. In fact, Ford's competitors believed it impossible to manufacture the Model A below the retail price (which initially was true).[80] Perhaps more than any other aspect of the changeover, Ford had anticipated refinement in precision manufacturing. Wishing to have the best toolroom gauges possible, Henry Ford purchased the famous gaugemaking operation of the Swede C. E. Johansson in 1923 and soon moved it into the laboratory facility in Dearborn. Between 1923 and 1927, the Johansson division supplied "Jo-blocks" to the Ford toolroom and any manufacturer who could afford them. It also made some of the Ford "go" and "no-go" gauges used in production as well as other precision production devices.[81] (See Figure 7.6.) Thus even before the end of the Model T, the Ford Motor Company had established the basis for an upward shift in the precision of its production. Nevertheless, when the Model A emerged, its precision production meant more frequent and finer gauging as well as the unprecedented use of scales and balancing devices. All of these refinements meant greater demands on production time and more expense.

Even if the design of the Model A had not required greater precision, it would have demanded entirely new parts production departments. Up-to-date design of the car dic-

FIGURE 7.6. Johansson Gauge Blocks. (National Museum of American History. Smithsonian Institution Neg. No. P64389-A.)

tated the establishment of entire new water pump, transmission, and shock absorber departments, exacerbating the problems of what William Pioch called the "complicated setup" of Model A.[82]

Pioch identified another bottleneck in the process of changeover, which arose because of what might be called the production theory of the Model T, discussed in Chapter 6. As Pioch explained, "Mr. Ford's idea of a manufacturing plant was to get the machines as close together as possible to save floor space. It was a good idea, but it didn't work out too good [for changeovers] . . . because the machines were in so tight that sometimes if we had to move a machine, we'd have to move four or five different machines to get that one out." For the changeover as Ford planned it, the close-packing of machine tools posed difficulties not only in initial tooling up and starting up but, perhaps more troublesome, in the expansion phase of production. A. M. Wibel explained that because of Ford's monolithic approach to production, expansion took place unilaterally—initial production was to be one thousand units per day, scaled up incrementally over a period to eight thousand or ten thousand units per day. Departments had to be rearranged with each incremental increase (say, one thousand units per day). Rather than building additional lines, with additional superintendents and foremen, Wibel says, "The poor layout man had to rip out all that [department] figuratively and come back literally and rearrange that stuff so that he had the capacity. He would get that all nicely working and along would come an order for

1500 a day; 2000, 3000, 5000, 8000.'' Consequently, ''We moved that machinery around so much that we had round corners on a lot of it.''[83] Close-packing of machine tools heightened this problem.

The sum of all of these design, planning, and layout difficulties, however, hardly compares to the fundamental problem posed by the single-purpose nature of Ford's production machinery in the Model T era. Even before they had fully occurred, the *New York Times* called the changes in machinery at the Ford factories ''sweeping'' and ''probably the biggest replacement of plant in the history of American industry.'' Fay Leone Faurote said in *Industrial Management* that the changeover had brought ''an unparalled example of scrapping machinery.'' As the insiders at Ford were only too well aware, Ford was ''an organization whose every machine tool and fixture was fitted for the production of a single product whose every part had been standardized to the minutest detail.'' Tool chief Pioch would later say that the changeover ''was just like starting out with a new machine shop.''[84]

The changeover in machine tools was of immense consequence. When Model T production ceased, Ford owned forty-three thousand machine tools of which thirty-two thousand were used for production. Model A production necessitated the refurbishing or rebuilding of more than half of these. Half of the remaining machine tools were scrapped; the others remained unchanged. Ford purchased forty-five hundred new machine tools to begin Model A production; Wibel certainly was correct in saying that ''we had a tremendous procurement problem'' with machine tools for Model A. The total cost of the reconstruction and purchase of machine tools, along with the production of new jigs, fixtures, and gauges, was about $18 million.[85]

When Charles Sorensen said in 1927 that he intended ''to get away from the Model T methods of doing things,''[86] he may have had machine tools in mind. Production of Model A was marked by new methods, but innovations reflected development in the machine tool industry far more than any internal event at Ford. Evidence is available to provide a detailed history of the changes in the machine tools used for producing the Model A, but such a study would be of little value. Some general observations are useful, however. First and foremost, Ford continued to rely heavily upon special-purpose machines, as exemplified by the Ford welding machinery and the hot spinning machines.[87] For more general-purpose machine tools that the company purchased, it designed and built special heads, attachments, jigs, and fixtures, as it had for the T. By far, the most radical change Ford made in machine tools appears to have been its abandonment of flat-bedded milling machines, which in the 1912–15 period the company had regarded as the highest achievement in machine tool technology. It will be recalled from the last chapter that these millers had been used for machining cylinder blocks and heads, among other parts, and had required the construction of elaborate fixtures to hold large numbers of castings for the machining operations. These milling machines therefore worked on a batch basis, requiring the loading and unloading of the fixtures at distinct intervals.

For the Model A, the company almost universally adopted continuous-drum milling machines made by both Ingersoll and Bullard.[88] With these machines, the operator loaded castings into a fixture located along the outside of a drum. Once loaded, the casting rotated around and through milling cutters. The operator unloaded a finished casting from another fixture, which had rotated into place, and reloaded it with an unmachined casting. These continuous-drum milling machines allowed for continuous rather than batch operations. Therefore, they better fitted the Ford notion of flow production. In this sense, although the new millers marked a departure from Model T production methods, they stood as truer symbols of Ford production principles, which required that everything moves.

In his desire to discard Model T methods, Sorensen probably was thinking more of assembly operations than machining of parts. Radical changes occurred in assembly that reflected political dynamics within the company as much as any departure in theory or practice. These dynamics were simple: Highland Park versus the Rouge. Parts production operations had been moved gradually to the Rouge from Highland Park during the years before the end of the Model T. Assembly stood as the last holdout, as evidenced by the procedure used to produce the fifteen millionth Model T. But even before the end of Model T production, Sorensen had begun building an assembly line at the Rouge. Of course, no Model T ever came off that line, and even as late as August of 1927 it remained unclear whether the new model would be assembled at Highland Park or at the Rouge.[89] But it was in this crucial period that the full effect of Sorensen's consolidation of power made itself known. Sorensen decreed that the Rouge, not Highland Park, would assemble the first Model A's. As will be seen, Sorensen's move resulted not only in a change of location for the line but a change in technical and supervisory personnel and the very arrangement of the line.

Ford News reported the details of the technical changes in the assembly line. (See Figure 7.7.) At Highland Park, the final assembly line for Model T construction had been 680 feet long. Sorensen shortened the new Rouge line to half that length; yet the company predicted that the line would match the output of Highland Park. Moreover, *Ford News* reported, ''Radical advances have been made in building a body or in transferring it to the assembly line. From first to last the body will be handled by conveyors, hoists, elevators, and transfer tables.'' (See Figure 7.8.) Rouge assembly operations differed significantly from Model T assembly at Highland Park and at branch assembly plants in that it had been designed to handle all body types and trucks as well.[90] Evidently, Sorensen aimed for greater flexibility in assembly rather than cost advantages through a single-purpose approach.

It is unclear whether the 50 percent reduction in the length of the line was accompanied by an equal reduction in the number of assemblers. What was certain, however, was that the demise of the Highland Park line and its replacement at the Rouge provided the means whereby Charles Sorensen could ''get rid of all the Model T sons-of-bitches'' who had been the principal architects of the moving assembly line, particularly Clarence W. Avery and William Klann.[91] When the construction of the Rouge line was complete, the men who had been responsible for Model T assembly assumed that they would be in charge for the Model A. But events proved the contrary. Klann later explained:

> I took [Ernest] Pederson, Al Hussey, Jim Burns, and Ed Gartha out to the Rouge plant with me to run the line. We were getting ready to assemble the first car.
>
> Sorensen said, ''Who are these guys?''
>
> I said, ''Sorensen, you know Pederson. You brought him in the shop yourself twelve years ago. When he flunked as a doctor at U. of M., you brought him here yourself. He is a Swede the same as you are. You know who he is. He has had charge of the line for the last twelve years. You know who Gartha is.''
>
> He said, ''Fire them.''
>
> So I did. I fired them. He didn't tell me why to fire them. He just said, ''Fire them.'' He said, ''Get them out of here and you go and get me a man from [Harry] Bennett [the increasingly powerful Ford personnel officer].''
>
> I said, ''This is a fine how-do-you-do. You bring a brand new job up here and new car and new chassis and all the work and now go and get a new boss for this job.''[92]

FIGURE 7.7.   Model A Final Assembly Line, River Rouge Factory, 1928. Note the comparative size and bulk of the assembly line hardware as compared to that at Highland Park for the Model T. (Henry Ford Museum, The Edison Institute. Neg. No. 833-51079.)

Shortly after Klann's implicit questioning of the wisdom of Sorensen's actions, he was told to take a vacation. While he was driving to the upper peninsula of Michigan with his family, a sheriff stopped him and asked if he were William Klann. Receiving a positive response, the sheriff said he had a message from Detroit. The message: "You're fired." Only a few days earlier, Clarence Avery had been fired in a roughly similar fashion. Avery was perhaps the person most responsible for the development of the assembly line at Ford in 1913 and the entire work-in-motion principle. Henry Ford had then called upon Avery to accomplish new objectives in production technology. For example, when Ford decided to produce safety glass, Avery designed, supervised construction of, and perfected what many claimed to be a pathbreaking glass plant. Avery also engineered facilities for Ford's production of head lamps.[93]

In the stead of these seasoned production men who had made the assembly line the very symbol of mass production, Sorensen, with Harry Bennett's imposition, placed a man named Harry Mack in charge of assembly. At the time he was hired, Mack was head of the Ford box factory, a job in no way related to automobile assembly operations. Mack's arrival further demoralized assembly foremen who had seen their longtime supervisors summarily dismissed. James O'Connor was one such foreman. He later recalled:

I knew what was going to happen. They were all going to be fired, not only me but all the fellows from Highland Park.

FIGURE 7.8. Body Drop, Model A, Final Assembly Line, River Rouge Factory, 1931. (Henry Ford Museum, The Edison Institute. Neg. No. 833-55974.)

> We didn't make any preparation from one day to the other. I didn't make up any special tools, which I would have done if everything had been peaceable. We knew what was going to happen. . . . Of course, they didn't press us for speed in production. So they came to me one day and said, "Is everything all set now if we want to speed this line up?"
>
> I said, "Boy, Mr. Baker [a new supervisor of foremen], all you've got to do is go to that rheostat and open it up." I knew I couldn't build five cars more a day.
>
> He said, "Okay, Harry Mack wants to see you."[94]

O'Connor said that he knew he was about to be fired, and indeed Mack fired him along with the other Highland Park foremen. Finally, even Harry Mack was fired. Amid this "awful turmoil," a former Highland Park assembly foreman was hired by Sorensen's deputy, Mead Bricker, to straighten out assembly. Immediately, this "Model T son-of-a-bitch" rehired as many of his former peers as he could persuade to return to Ford, including O'Connor. O'Connor relates, "They really went to town." Even then, however, the foremen said among themselves that they would rather work for "Billie Klann" than for Sorensen's man Bricker.[95]

Unfortunately, almost none of these signs of turmoil at the Rouge entered the manuscript records of the Ford Motor Company. The oral history reminiscences of many of the principal actors in this comedy of errors stand in universal agreement about the chaos of the changeover once production supposedly got under way. The final assembly operations must be seen as a principal culprit in production delays and extremely low output, for which the "green men on the line" were primarily responsible. Hints about the extent of production problems at Ford do appear in telegrams sent to Charles Sorensen when he was briefly away from the Rouge plant in early January 1928, a critical period in Model A production. On January 10, for example, Sorensen learned that there had been a "low frame production due to hand work." The following day, he received the story in far grimmer detail. Rouge reported a total frame production of forty-six, which was passed off with the following remark: "Considerable trouble this job. Hand riveting and frames out of square." Other news was equally bad: "Branch brake requirements necessitated holding up final assembly line. Sent men home at noon with Mr. Martin's approval."[96]

This particular delay had been caused by a change of a fixture on the brake bracket forging. For a company that had regularly manufactured more than eight thousand cars and trucks daily, such problems can only be regarded as unprecedented and indicative of the profound challenge posed by the changeover. They explain why production stood at an average of barely one hundred Model A's per day in early January rather than Henry Ford's promised one thousand.

Although the ultimate responsibility for such troubles lay with Henry and Edsel Ford, it is impossible not to place much of the blame for the delay in producing (as opposed to designing) the Model A on Charles Sorensen. All of the evidence concurs that the consciously executed plan to purge the Ford Motor Company of all the Model T "sons-of-bitches" originated with and was carried out by Sorensen, thereby resulting in most of the production problems with the Model A. Sorensen sacrificed Model A's in order to get rid of the "high-priced" production men who had made the Highland Park factory the birthplace of mass production. The great irony, which completely evaded Sorensen, was that he was the biggest of all the Model T sons-of-bitches.[97] He had shaped the Rouge with Model T principles. Indeed, despite Sorensen's utterances, Model A production contained more vestiges of Model T production principles than entirely new approaches. P. E. Martin's diary entry of October 21, 1927, the day of assembly for Model A #1, demonstrates the extent of Sorensen's "Model T-ness." Martin recorded the personal predictions of the old-time Ford personnel about how many Model A's would be built. Henry Ford looked for ten million, Edsel twenty million, and Martin fifteen million, but Sorensen expressed his belief that no less than fifty million Model A's would be manufactured at the Rouge and Ford's branch plants.[98] (In fact, fewer than five million Model A's were ever built.) Finally, the person who in 1925 sent Sorensen a post card portrait of Mussolini, writing "M[ussolini] is to Italy what you are to the Ford M[otor] Co.," could not have anticipated how true these words would ring when the production of Model A got under way.[99]

Despite the personnel changes (which left many more Ford executives and their deputies without work than have been identified here) and enormous production problems, the Model A brought the Ford Motor Company almost up-to-date as an automobile manufacturer. The Model A, with the exception of the brakes, won acclaim (and still does) as a well-designed, well-made, well-priced, and "thoroughly up-to-date" automobile. Events subsequent to the initial production of Model A brought the company up-to-date with other manufacturers, particularly with General Motors. These events relate in differing degrees to the changeover phenomenon. First, Ford spent money as never before on advertising the new Model A. Previously, Ford had advertised the Model T nationally in some years and not in others. Like all the other segments of American business, advertisers could never understand Ford. They recognized, of course, that Ford received much free advertising through jokes and cartoons. But Ford coupled the unveiling of the Model A with a concerted national advertising campaign that in the first week was estimated to cost $2 million. The campaign and the Ford reputation soon resulted in more than eight hundred thousand orders, which as we have seen, the factory could not hope to fill for a year.[100] Nevertheless, Ford had discovered what General Motors already knew: advertising—major advertising—was a fundamental part of the changeover strategy.

Second, Ford Motor Company soon moved into the arena of credit financing for its customers. Achieved largely through the efforts of Edsel Ford and Ernest Kanzler, Universal Credit Corporation was perhaps the most revolutionary change (considering Henry Ford's detestation of credit buying) wrought by the changeover. To meet the needs of the new consumerism in America and to compete with General Motors Acceptance Corporation, the Fords established the new corporation to finance Ford automobile purchases by dealers and retail consumers. Universal Credit Corporation allowed Ford dealers to stock new cars by advancing only 10 percent of the retail price of the automobile. Consumers received low-interest, one-year financing, with a down payment of a third of the selling price. Chartered in Delaware in March 1928, Universal Credit Corporation obtained its capital from Ford Motor Company and Guardian Trust of Detroit and New York. Kanzler's association with Edsel proved critical in the establishment of this new venture, for he had become executive vice-president of Guardian Detroit Bank upon his ''resignation'' from the Ford Motor Company.[101]

Although extensive national advertising and establishment of a credit company stand out as departures in the changeover from Model T to Model A, the lessons that Ford production men learned were less distinct but nonetheless of major importance. Before identifying these lessons, one other point is in order. The chaos of the changeover failed to arouse Henry Ford to the point that he established any consistent and clearly understood system of managerial hierarchy. The Ford Motor Company remained a dictatorship. Henry Ford dictated broad policy of the company and details of the car. Charles Sorensen dictated all aspects of production. While Ford held firmly to his place, Sorensen's grip began to be loosened by the snowballing power of Harry Bennett, who would eventually become a virtual dictator within the Ford Motor Company.[102]

Yet below the level of high policy, Ford production men learned important lessons from the changeover. Those who survived this time of trial remained with the company until the 1940s and 1950s, indicating that these lessons were fundamental to the managerially torn Ford company. The changeover to Model A drove home the point to all Ford production men that any changeover could not be accomplished smoothly without adequate advanced planning both of the design of the automobile and its production. When the in-house order for a changeover was made in 1926, the production men probably believed that they could handle its requirements. Tool department head William Pioch recalled: ''We were building Model T's when we were designing the tools for the [Model A] engine. We had a *good start* before we shut down the Model T. I would say that we were about six months in the process of this. By the time we shut down, we had about twenty-five percent of the retooling done.''[103]

As Pioch, Martin, Sorensen, Wibel, and others learned, 25 percent of retooling was not sufficient to remain long in business in the new era of the American automobile industry. Although their recognition of the need for advanced planning could have resulted in the establishment of bureaucratic procedures, this never occurred until the reorganization of the company under Henry Ford II. Nevertheless, when subsequent changes were made in the design of the Ford automobile, Ford engineers planned them far more carefully than they had the Model T to Model A changeover.

Another important lesson Pioch learned with the Model A was the value of establishing pilot lines for testing new approaches in machine work. A pilot line was created for machining the Model A engine block, and as Pioch explained, "We had a pilot line . . . so we knew exactly where we were going before we tore out the old equipment."[104] Time considerations precluded the establishment of other pilot lines for the Model A. But in subsequent model changes, they were important in identifying and solving production problems.

Deep drawing problems with the Model A body also taught Ford production men that body die work required adequate time and room for error. Pioch noted, "Our [Model A] experience told us where our problems were. We'd take a look at a certain shaped stamping. If it looked like it was going to give us trouble, then those were the dies we pushed through first quickly and got the bugs worked out of it before the rest of them came along."[105] Pioch and his peers also learned that he and other production experts should be called in to give their opinions on new designs, particularly regarding body styles for which designers proposed curves that would have proved to be too deep for satisfactory drawing.

Other problems, especially in machine tool acquisition, that arose with the changeover to Model A had actually been identified during the last series of changes made on the Model T for 1926. A. M. Wibel, who was responsible for Ford machine tool procurement, emphasized the importance of the Model T changes on Ford procurement and how the Model A changeover simply reinforced this trend. According to Wibel, until the last important change in the Model T, the company purchased presses with die spaces big enough to handle only specific tasks. When the Model T for 1926 was lowered and lengthened slightly, the company realized that the presses would not handle the larger work. Therefore, Wibel initiated a policy of purchasing machine tools and presses of larger capacity, which could accommodate moderate change in the size of workpieces. Similarly, Ford shifted from using completely specialized, multiple-spindle drill presses to drill presses that simply required a new head to achieve a new arrangement of drills. As Wibel recalled, "This realization struck us gradually, after we paid through the nose for machines we couldn't use on the new models." Thus, over a period of about five years, the Ford Motor Company moved toward what Henry Ford's ghostwriter called "flexible mass production," toward a machine tool system—if not a managerial system—that could accommodate changes in design of the Ford automobile without totally tearing out machinery used to produce the old model.[106]

Finally, the changeover to the Model A taught Henry Ford (or at least reminded him of the days of 1908 and 1909) that a new model in the initial stages of production was a sure target for hundreds of proposed changes. Design engineer Laurence Sheldrick recalled this phenomenon:

> Immediately after the Model A got into production Mr. Ford perhaps realized that there was going to be a deluge of requests for engineering changes. He safeguarded himself very carefully on that. He caused all engineering changes to be cleared through me, whether they were body, axle, transmission, or whatnot. They all had to be channeled through me. I had to present them to him for his signature for quite a period. . . . He knew that on a new product like that [the Model A] a deluge of changes could just bog the whole thing down and he was absolutely right about it.[107]

As described above in detail, despite Ford's supposed iron-handed control, Model A changes greatly bogged down its quantity production. Better planning both in the design stage and in tooling up, especially use of pilot lines, helped to deal with this problem, which was, to a degree, inherent with the new mass production.[108]

The Model A proved to be an immensely successful automobile. Gradually the Rouge worked out its immediate technical problems with Model A production, and in 1929 it achieved new daily and weekly production figures. On June 26, 1928, for example, the company assembled 9,100 Fords, compared with the 8,710 Model T units of October 1925. In July, Ford produced 180,804 Model A's in the United States, a figure larger than any monthly output for Model T. As already noted, substitution of stamping and malleable iron castings for Henry Ford's forgings helped the plant to drive down the production costs of the Model A. By September 1929, the company had produced more than 1.4 million Model A's since the first of the year compared with fewer than eight hundred thousand in 1928. Yet at this very moment, the entire automobile industry witnessed car inventories rising dramatically. By October, rumors began to circulate wildly that Ford would shut down or perhaps change its model.[109]

Despite Ford's recovery in market share during 1929—from slightly over 30 percent in the period January to May to almost 45 percent in October—the automobile industry sank swiftly into what has been called the lean years of the American economy. Black Tuesday, October 29, 1929, closed the door to the fabulously profitable years of the industry. With bullish strength in the first three quarters, 1929 was to be Ford's last good year for profits. Although the company lost $30.5 million in 1927 and $70.6 million in 1928 because of the changeover, its profits in 1929 reached $91.5 million after taxes. This figure would probably have topped the $100 million mark had sales held up during the last quarter.[110] But they did not.

Ford responded to the sagging sales of the Model A by reducing prices in November and by making what could be called an annual model change for 1930. On December 29, 1929, Edsel Ford announced that the Model A had been "re-designed." The body had been lowered some and lengthened by six inches, and change was evident in radiator and grill work and in fender design. Moreover, the new model would be offered in several new colors. The *New York Times* reported that few "mechanical changes" had been made. In his announcement Edsel stressed that "since the Model A was first introduced it has constantly been made a better car. As soon as improvements have been developed and tested, they have been built into cars in production and immediately passed on to the public. That process goes on steadily in the Ford plants . . . [and] is now given expression in the new bodies."[111] The price, Edsel noted, remained the same as for the 1929 models.

Ford's changes for 1930 held much in common with those for 1926. The company responded to decreased demand and increased criticism that the Ford car was "out-of-date" by making cosmetic changes in its automobile. This strategy would have been far better suited for the automobile industry of the 1950s and early 1960s than of the late 1920s and early 1930s. William Abernathy has demonstrated the importance of major changes, as compared to minor ones (the "annual model change"), in gaining or losing market share. As in 1926, the changes in the Model A's body caused no overwhelming production problems. Indeed, not even the slightest hint of problems appears in the records of the company or in the reminiscences of former Ford employees. Ford pursued its marketing strategy vigorously throughout 1930. In March it added a de Luxe sedan and coupe to its line and in August a de Luxe roadster, with "sport treatment throughout."

Later in the year, the company broke with its age-old tradition of lettered models and introduced a car, with the Model A chassis, called the Victoria. Finally, this marketing strategy was pushed with a substantial national advertising campaign costing $8.7 million for the year.[112]

Yet in the midst of the deepening depression and with competitors selling six- and eight-cylinder automobiles (Chevrolet had introduced the six in 1929), Ford's strategy proved to be less than satisfactory. With a production of 1.5 million units, profits for 1930 dropped sharply to $51 million, 45 percent below the 1929 figure. With no change in strategy or in the Model A chassis, Ford's sales fell to six hundred thousand units for 1931 while production was more than seven hundred thousand units. Market share looked even worse—26 percent—and profits turned to losses in excess of $50 million.[113] Walter Chrysler personally demonstrated an improved Plymouth to Henry and Edsel Ford in June of 1931.[114] Soon the automotive world would acclaim the all-new Plymouth, especially its smooth-running engine, which reduced vibration with two-point engine mounts cushioned with rubber. Seeing changes such as these throughout the depressed automobile industry, hearing criticism of his "out-of-date" Model A, and watching his inventories mount rapidly, Henry Ford took steps to right his foundering industrial ship.

On August 1, 1931, the Ford Motor Company sent its seventy-five thousand Rouge employees home for an "indefinite" vacation. The inner circle at Ford pondered changes in the Model A while a "few thousand" men in Detroit, and eleven of the thirty-six branch assembly plants continued production on "curtailed schedules." In early September, the company rehired between fifteen and twenty thousand of its employees to complete the fifty thousand orders on hand for Model A's and expected to hire another fifty thousand men by mid-September to continue manufacture of the Model A.[115] This massive rehiring appears never to have occurred.

Change seemed imminent to those both within the company and outside in the automotive world. Yet the details of this change are obscure.[116] It appears that Ford initially intended to introduce a redesigned Model A body mounted on a chassis with a longer wheelbase, a larger bore four-cylinder engine, and a gasoline tank in the rear. Everyone predicted that the new Ford would appear for the fall market, but it did not. The reasons are less certain. Ford's large inventory may have convinced the company's leaders, father and son, to delay the new model until the inventory was sold. Such a strategy was reasonable given the worsening situation in the entire automobile industry. Delays in retooling also may have been a determining factor. Certainly, Ford's suppliers as well as the public were well aware that Ford's toolrooms were working around the clock in August and the fall months.[117] But there is no indication in the reminiscences of the Ford production engineers that retooling problems delayed the introduction of the V-8 Ford.

An equally viable and perhaps more probable reason for delay in introducing a redesigned Model A lay with engine developments. With the appearance of Chevy's six-cylinder automobile, Henry Ford apparently entertained the idea that his next new engine would be an eight-cylinder job. Ford's thinking in this direction resulted in more than simple talk. Instructed by Ford, engineers designed some twenty or thirty eight-cylinder engines, the first finished in May 1930. At the same time engineers sought to improve the performance of the existing Model A engine, which according to Henry Ford, they did. Delay, therefore, may have stemmed from indecision about what power plant to put in the chassis of the lengthened and restyled Model A. In a February interview with a *New York Times* reporter, Ford alluded to such indecision, saying, "We developed a corking good '4' and were all ready to let it go, but we found it was not the new effort the public is expecting. That's why we're bring out the '8' now."[118] Apparently, Ford delayed his final decision about the introduction of the V-8 until December 7, 1931.[119] The V-8, with fourteen body types, first appeared in showrooms March 31, 1932.[120] At the same time, however, Ford decided not to scrap completely the "corking good" four-cylinder engine that had been developed. The company made its fourteen body types available with either power plant, the sixty-five horsepower V-8 or the fifty-horsepower four-cylinder engine. Such options clearly indicate that Ford had made strides in achieving "flexible mass production." (See Figure 7.9.)

Design and production of the V-8 engine constituted an engineering feat. Unlike other eight-cylinder engines, the V-8 consisted of a unit or single-cast block, which posed enormous problems in core design and molding to obtain usable castings. Such castings could be obtained in small quantities, albeit with a high scrapping rate, but quantity production was another matter. Tool department chief William Pioch helped solve the problem by designing special-purpose fixtures used to place cores, cement them, and allow them to set permanently. According to Pioch, core setting on the V-8 became just another "production line" operation.

Pioch also characterized the machining methods used for the V-8. Although the V-8 was a different type of engine, Ford's previous experience with the Model T and the Model A allowed the factory to move "into high production machinery." Machines were built to bore all eight of the cylinders in the block at once. "In fact," Pioch pointed out, "our machining time on the block wasn't much more than it was on the four-cylinder engine. We had a lot more machining but it didn't take much more time in labor." He also noted that the V-8 brought the widespread use of tungsten carbide tools along with machine tools of higher speeds and feeds and of greater rigidity.[121]

If Ford's decision to produce the V-8 did not occur until December 7, 1931, the changeover to production of the new engine was remarkably short and smooth. Pioch believed that "we made the change much more quickly than we did before. We had most of our production equipment ordered and we had the engine developed. We had a pilot set up for machining the block. We had quite a few blocks machined ahead of time. We had a lot of castings made in the Foundry ahead of time." He estimated that ten thousand blocks had been made before full-scale production got under way. "This time we had planned the changeover," he concluded. Purchasing chief A. M. Wibel agreed, arguing that although the changeover from the Model A to the V-8 was a bigger job, raising bigger problems in layout and machining, it was handled much better. "We did a lot of planning for the V-8 changeover," Wibel recalled.[122] (See Figure 7.10.)

Despite Pioch's and Wibel's assessments, problems reminiscent of the Model T to Model A changeover did occur. The late timing of the model change may, of course, be identified as one cause of the problems.[123] Last-minute changes in design also played havoc in getting the new V-8 automobiles off the assembly line. Ford's official model change announcement appeared in print on February 12, 1932, and indicated that dealer deliveries would be made in the first week of March. But by mid-March, when no deliveries had occurred, it had become evident that delays in production resulted from major or "important" changes being made as late as February 29.[124] (See Figure 7.11.)

Even after V-8 Fords began rolling off the Rouge assembly line, problems persisted, just as they had during the first year of Model A production. In 1932, Ford managed to produce only 287,285 V-8's and Model B's (the four-cylinder option).[125] The company reported that not long after the V-8 was shown publicly, it received deposits on one hundred thousand orders for the new model.[126] With such an initial demand, the company

FIGURE 7.9.   Conveyor Belt, River Rouge Factory, 1932. This photograph suggests how far the Ford Motor Company had moved away from "single-purpose" manufacture in 1932. Both V-8 engine blocks and other parts are en route to the assembly area along with four-cylinder equipment. (Henry Ford Museum, The Edison Institute. Neg. No. 833-57060-29.)

might have turned a loss into profit had it been able to produce more automobiles. Unquestionably, the Great Depression made large-quantity production of V-8's less urgent than had been the case with the Model A. The day had passed when any automobile company would conceive of trying annually to produce two million cars of a single model. In the depth of the Depression, the entire industry could not sell that many cars.[127] Nevertheless, Ford's slowness in bringing postchangeover production up to established prechangeover levels seriously handicapped its competitive position in the market.

The bottom line—the cost—of the changeover to the V-8 suggests that it had not been as great as that to the Model A. In the deep trough of the Depression, 1931 to 1933, Ford lost about $125 million ($71 million for 1932 alone). The changeover to Model A, including experimental and design work, tooling, and loss of profits, totaled about $250 million.[128] Unfortunately, equivalent cost estimates of the V-8 changeover are unavailable, but even these crude figures suggest that the company had gained some control over such costs.

FIGURE 7.10.   Ford V-8 Engine Assembly, River Rouge Factory, 1930s. (Henry Ford Museum, The Edison Institute. Neg. No. 833-68057-105.)

FIGURE 7.11.   Henry Ford and the V-8, March 10, 1932. Ford stamped the serial number on the first V-8 chassis. (Henry Ford Museum, The Edison Institute. Neg. No. 833–57031–2.)

Ford Motor Company achieved this control partly by consciously launching upon a strategy that placed much of the initial capital burdens on outside suppliers. Despite Charles Sorensen's compulsion to maintain the Rouge as an industrial colossus, in mid-1928 he and A. M. Wibel moved toward a strategy—which departed significantly from ''high Rouge'' strategy—whereby Ford came to depend increasingly upon outside suppliers as much as it had once done.[129] General Motors and Chrysler, of course, had been consciously following such a strategy for a long time with obvious success. As Sorensen informed Edsel Ford: ''This week we are working on stimulating the outside buying sources, so that we can make further jumps in production by getting as much help as we can from the outside and at the same time make the minimum expenditure for tools, etc., in our own plants.''[130] Wibel subsequently pursued this strategy with great vigor, and it became even more prevalent at Ford Motor Company when the Depression struck. Ford went to outside suppliers more and more during the early years of the Depression. This strategy may well have played a major role in reducing the time and the cost of the changeover to the V-8.

Although time and costs had been reduced in this second major changeover at Ford, problems remained, stemming largely from the very system of mass production that Ford had created. The writer of an incisive article on Henry Ford and his company, published in *Fortune* in 1933, stated: ''Say competitors [of Ford]: Mr. Ford does not change his car often enough and cannot make changes without disrupting his production schedule. Says Mr. Ford: changing models every year is the curse of the industry. Many a rival agrees with him here, would like to follow suit but is afraid that skipping a yearly model would cost too many sales.''[131]

In the years immediately previous to and during the wrenching changeover from the T to the A, the American public worried greatly about its folk-hero Henry Ford. It celebrated Ford's ''return to genius'' with the mass production of the Model A.[132] Ford's loss of millions of dollars once again in 1931 roused further public sentiment. The editor of the *New York Times* exemplified this concern. He sought to explain Ford's 1931 losses by pointing out that during the year Ford had introduced the V-8. But this simply begged the question, pointed out the editor: ''Was the genius of Henry Ford true to its highest self when it failed to foresee and make provision for new car models, new factory set-ups and new production tempos? Somehow one feels that in passing from Ford T to Ford A—and from a Four to an Eight—the process should have been as uninterrupted, as precise and as effortless as the flow of the separate parts into the making of any Ford Car.''[133]

Henry Ford and the Ford Motor Company had thus, in terms of public criticism and concern, become victims of their own creation, of mass production. It was during these years between the decline of the Model T and the slow rise of the V-8 that the term *mass production* gained currency in the United States. Although Ford and his company were groping their way into the new era of mass production, the era initiated by General Motors in which change had to be planned and carefully executed on a regular basis, the American public had come to expect more, to believe that change itself could be mass-produced and that with mass production anything was possible. Thus in these years what might be called the ethos of mass production was established.

CHAPTER 8

The Ethos of Mass Production & Its Critics

It is agreed by competent observers in this country and in Europe that America's increasing general prosperity and high standards of living are due chiefly to the rapidly increasing use of scientific mass production and distribution.
—Edward A. Filene, "Mass Production Makes a Better World," *Atlantic* (1929)

To the [Tennessee Valley] Authority the solution of this statement [of high prices and low consumption of electricity] seemed to be to apply to the electrical industry the principles of mass production and mass consumption which had proved successful in a number of great private industries in America. The essential element in mass production is a progressive decrease in unit cost—the more items of any commodity a producer turns out, the less each item is likely to cost him.
—Tennessee Valley Authority, *Annual Report* (1936)

To what extent would the mass-production of . . . houses be a solution of the housing problem, and how far would this form of manufacture meet all the needs that are involved in the dwelling house and its communal setting? Those who talk about the benefits of mass-production have been a little misled, I think, by the spectacular success of this method in creating cheap motor cars; and I believe they have not sufficiently taken into account some of its correlative defects.
—Lewis Mumford, *Architectural Record* (1930)

Few expressions have gained greater currency more quickly than did *mass production* in the late 1920s and early 1930s. And few concepts behind such expressions have been so hotly debated as the principles and results of mass production. Henry Ford and the Ford Motor Company were largely responsible for bringing mass production into the American's vocabulary and consciousness. Ford's ghostwritten article by that title, commissioned for the 1926 edition of the *Encyclopaedia Britannica* and published as a Sunday feature in the *New York Times*, defined and focused attention on the expression.[1] (See Figure 8.1.) The unprecedented success of the Ford Motor Company and its staggeringly large output of Model T's lent immediacy and verisimilitude to the article. Yet at this late date in the Model T's history (1926), one could easily entertain fundamental doubts about the benefits and costs of mass production. Americans seemed to be rapidly

303

FIGURE 8.1.   Henry Ford and William J. Cameron in the Dearborn Laboratory of the Ford Motor Company, 1935. In 1926 Cameron wrote the Ford-attributed article, "Mass Production," which was published in the thirteenth edition of the *Encyclopaedia Britannica*. (Henry Ford Museum, The Edison Institute. Neg. No. 0-3171.)

tiring of the standardization that had made it possible more than a decade and a half earlier to achieve single-purpose manufacture—to develop mass production. Were not the costs of "you can have any color you want so long as it's black" too great to bear?

Though Ford long resisted change and though the process of switching from Model T to Model A proved to be tougher than any Ford employee could have imagined, once introduced, the Model A served to reaffirm and to give an entirely new dimension to mass production. To some, Ford's work proved dramatically that mass production was not the antithesis of individuality and aesthetics. The Model A was singled out for its beauty, and it was held up as the prime example of what could be achieved by combining mass production methods with art.[2] As Jeffrey L. Meikle has pointed out, Ford's changeover helped to infect American manufacturers with a fever for industrial design. The leaders of the industrial design movement in the late 1920s and early 1930s shared "a faith in the social benefit of design for mass production" and "the ideal of renewing America through mass production."[3]

Industrial designers were not alone in their vision of the possibilities of the new mass production. Businessmen and social thinkers also saw unprecedented opportunity in combining the productive efficiency of the assembly line with individuality and the aesthetics of designers. Despite the implications of the annual model change, some observers continued to see mass production as, in the words of a contemporary, "a panacea for the industrial and business ills of all nations on both hemispheres."[4] Mass production methods were tried for the first time in various industries, some more successfully than others. America of the late 1920s and early 1930s was pervaded by an ethos of mass production.

While some proclaimed and celebrated this ethos, others severely criticized it and the production system that lay behind it. Journalists, literary figures, filmmakers, labor leaders, and artists identified mass production as a manifestation if not a cause of social ills present in the United States in this period.

Unquestionably, America's production engineers had made a revolution in manufacturing. Whether that revolution could be extended, and whether society was now really better or worse off, however, remained active questions.

When in 1922 Henry Ford published *My Life and Work,* his first autobiographical book written by Samuel Crowther, the expression *mass production* did not appear in the text. But a caption below one of the book's few illustrations—a photograph of a Ford engine line—stood out prominently: "The Ford prosperity recipe is high wages, low prices, and mass production."[5] The implication was that mass production was merely quantity production. Yet by the time that Ford's ghostwritten article on mass production appeared in 1926, mass production had been equated with the "Ford prosperity recipe." The term described a production system characterized by mechanization, high wages, low prices, and large-volume output. As the *Britannica* article stated succinctly, "Mass production is not merely quantity production."

Between the publication dates of *My Life and Work* and *Britannica*'s "Mass Production," a prominent liberal businessman from Boston, Edward A. Filene, began to prophesy the complete "Fordizing" of American business and industry as the solution to the growing economic and social ills in the United States. In *The Way Out* (1925), Filene projected an era of intense competition dawning in the United States primarily because of trade barriers erected by European nations. "This super-competition," wrote Filene, "will ultimately drive us into mass production and mass distribution. It will compel us to Fordize American business and industry. And this application of the mass principle to American industry will bring about the new industrial revolution I have suggested." Filene looked forward to the establishment of "a regime of mass production, [because] with the reduction in prices [that] it will make possible, the handling of the wage problem will be easier than ever in our history."[6]

But what would a "Fordized" America mean? Filene argued that "mass production will force us into a war on waste and compel us to put industry upon a scientific basis. . . . Mass production will mean the increasing standardization of products, and an increasing mechanization of the process of production. And the mass distribution which will follow mass production as a matter of course will find the biggest total profit in selling an enormous number of articles at the lowest possible profit per article." Filene fully acknowledged that "most of the theoretical students of modern industry think that a Fordized America would be hell on earth." But for a long list of reasons, he believed that "a Fordized America built upon mass production and mass distribution will give us a finer and fairer future than most of us have dared to dream."[7]

In conclusion, Filene warned American businessmen to "*Fordize or fail.*" He added, "If this belief is sound—and I am staking my personal business future and investments upon its soundness—it is only a question of time until we shall be living in a Fordized America." A pioneer of the chain department store and founder of the Twentieth Century

Fund and the Consumer Union, Filene took pains to point out that by Fordizing America he did not mean installing Henry Ford as "the political, intellectual, and social arbiter of American life." Filene said he disagreed with everything Ford had said and done save "his principles of mass production, mass distribution, and his primary emphasis upon service to the consumer."[8]

But Filene's 1925 vision of a "Fordized" America never came to pass because the decline of the Model T and competition from General Motors and Chrysler forced Ford to change his mass production system. Faced with reasonable alternatives, the consumer no longer accepted Ford's—and Filene's—precepts of mass production. As a writer in the *New York Times* reflected on this early period:

> When mass production started, individuality stopped. In order to reduce manufacturing costs and turn out automobiles in sufficient quantity to supply popular demand, producers had to evolve factory methods that permitted economical, high-speed operation. They had to concentrate the forces of men and machinery on the production of standard, stock-stamped motor cars. Instead of making different cars, each manufacturer simply made the same car over and over again. Automobiles that came from the same plant had less individuality among themselves than a nest full of eggs from the same hen.[9]

This was precisely the result of mass production that Henry Ford celebrated and described in detail in the *Encyclopaedia Britannica* article. When it appeared in the *New York Times,* the paper's editors chose to comment at length about one of the major contentions of the article. Under the title, "The Super-Factory System," the editors focused on the difference between the early factory system and mass production. William J. Cameron, the true author of the *Britannica* article, had contended:

> The early factory system was uneconomical in all its aspects. Its beginning brought greater risk and loss of capital than had been known before, lower wages and more precarious outlook for the workers, and a decrease in quality with no compensating increase in the general supply of goods. More hours, more workers, more machines did not improve conditions; every increase did but enlarge the scale of fallacies built into business. Mere massing of men and tools was not enough; the profit motive, which dominated enterprise, was not enough.

"If by the 'early' factory system and the system in 'its beginnings' Mr. Ford means the very first phases of the industrial revolution, his generalization may be just," the *Times*'s editors granted. "But for the factory system at least seventy-five years before the advent of mass production in Detroit, it certainly would not be true that it lowered the condition of the producing masses without compensation to the consuming public." Quoting extensively from "so candid a critic of the industrial system as Mr. Sidney Webb . . . in 'The Decay of Capitalist Society,'" the editors argued that the results of the factory system were positive and substantial. They concluded: "These are the precise results claimed by Mr. Ford for mass production. The advantages of mass production—increased consumption, rising standards of living, increased leisure—were claimed for the factory system after it had emerged from its dark ages. And the doubts concerning mass production which Mr. Ford meets and explains away—excessive routine, monotony, enslavement of man to the machine—are the very same doubts which the older factory system faced and answered."[10]

The editors of the *New York Times* thus identified an important question regarding mass production: What separated mass production from the factory system? In one sense, mass production was but an extension of the factory system with its characteristics of a highly divided labor force working in a centralized and mechanized facility. But as

discussed in Chapter 6, the standardization, the scale of operations, the degree of specialization both of workers and machines, the required precision, the assembly line with its
concomitant conveyor systems and time study, and the need for high wages resulted in
productivity gains that far exceeded the imaginations of those familiar with the factory
system. Mass production was, as the *Times*'s editors clearly recognized, "the super
factory system." As Edward Filene argued, mass production was to the "new" or second
industrial revolution what the factory system was to the first industrial revolution. And
mass production brought labor problems, which seemed to increase proportionately with
gains in productivity. Ford's five-dollar day temporarily relieved the symptoms of these
problems, for extraordinarily high wages did indeed ensure that workers would participate
in the consumption side of the mass production system. One could affirm Herbert Hoover's statement that "high wages [are the] . . . very essence of great production."[11]

But during the late 1920s, consumers seemed to be far more concerned with the
standardizing effects of Ford's system of mass production on the automobile itself than
with its implications for labor. The changeover to Model A served to allay consumers'
fears. The new mass production no longer meant absolute standardization. As Filene
wrote in *Atlantic* in May 1929:

> Ford, by insisting on standardizing for so long a style of car which many people thought
> none too handsome, and by allowing no deviation even in color to suit the individual
> taste, was no doubt largely to blame for the belief that mass production, involving as it
> must standardization, meant that we should have uniform ugliness thrust down our
> throats. But Ford was probably right in his insistence during the days when he was
> perfecting the methods of mass production and popularizing the automobile. To get the
> automobile widely used a very low price was necessary. Now that the automobile has
> become a necessity, and the principles of mass production have been brought to a higher
> state of perfection, Ford has redesigned his car. It is a thing of beauty, and yet it is
> standardized to the point of complete interchangeability and is produced under scientific
> mass methods.[12]

Though Filene may have been aware of the enormous and costly struggle made by the
Ford Motor Company to accommodate change, he sought to assure his readers that the
new or flexible mass production did not mean absolute standardization (as numerous
critics charged) and that it allowed for change, beauty, and aesthetic pleasure. Filene titled
his 1932 book *Successful Living in This Machine Age*. He saw mass production as the key
to achievement of this goal. To provide clarity, Filene offered on the first page of the book
a definition of mass production. His definition is a comprehensive statement about the
ideological nature of the mass production ethos in the late 1920s and early 1930s:

> Mass Production is not simply large-scale production. It is large-scale production based
> upon a clear understanding that increased production demands increased buying, and that
> the greatest total profits can be obtained only if the masses can and do enjoy a higher and
> ever higher standard of living. For selfish business reasons, therefore, genuine mass
> production industries must make prices lower and lower and wages higher and higher,
> while constantly shortening the workday and bringing to the masses not only more money
> but more time in which to use and enjoy the ever-increasing volume of industrial
> products. Mass Production, therefore, is *production for the masses*. It changes the whole
> social order. It necessitates the abandonment of all class thinking, and the substitution of
> fact-finding for tradition, not only by business men but by all who wish to live
> successfully in the Machine Age. But it is not standardizing human life. It is liberating
> the masses, rather, from the struggle for mere existence and enabling them, for the first
> time in human history, to give their attention to more distinctly human problems.[13]

Mass production—"production for the masses"—held out almost unlimited promises for positive social change. As Filene argued in an earlier article, "Mass Production makes a Better World," "Mass production holds no dangers to the common welfare, but on the contrary holds possibilities of accomplishing for mankind all of the good that theoretical reformers or irrational radicals hope to secure by revolutionary means." All that was necessary was education: "Mass production demands the education of the masses. . . . The masses must learn how to behave like human beings in a mass production world."[14]

Filene's contemporary, college president Harvey N. Davis, agreed fully with these positive views of mass production and the need to educate the masses as well as the critics of mass production. In a chapter titled "Spirit and Culture under the Machine" in Charles Beard's edition of *Toward Civilization,* Davis sought to counter the "volumes [that] have been written on the deadening effect of the Ford assembly line on the souls of noble Americans." In the same vein as Ford's *Britannica* article, Davis argued that "mass production tremendously increases the quantity of useful things in the world, and de-creases the cost of them." Moreover, mass production provided greater leisure and an ever-higher standard of living. To counter the critics, Davis added a new wrinkle to the usual argument. Having characterized mass production as "essentially a speeding up process" and "a thought-eliminating process," which "is intimately related to, if it does not inevitably breed, modern advertising," Davis argued that mass production was merely the *reflection* rather than the *cause* of twentieth-century American society. By no means affirming this society, Davis simply argued that the technology a society wants is the technology it gets. He thereby "absolve[d] mass production of the charge of having caused these unfortunate intellectual habits of our time." Criticize American society, he argued, not its system of mass production.[15]

The works of Edward Filene and Harvey Davis are merely representative of, respec-tively, the ideological expression of the ethos of mass production in the late 1920s and early 1930s and the defense against its critics. Even in the depths of the Great Depression, Filene could argue that the "masses of America have elected Henry Ford [as their exemplar]." That Filene articulated a popular view of mass production is suggested by an early annual report of the New Deal's Tennessee Valley Authority. When the TVA wanted to make its goals understandable to the American public, it chose to speak the language of the ethos of mass production. TVA, its report stated, was merely "apply[ing] to the electrical industry the principles of mass production and mass consumption which had proved successful in a number of great private industries in America."[16] The TVA report assumed that Americans knew and appreciated the great benefits of mass produc-tion: greater production, greater consumption at less cost, greater leisure, and a rising standard of living.

Industrial designers also capitalized on the ethos of mass production. During the late 1920s and early 1930s, art and architecture in the United States were infected with the rise of "machine aesthetics." This vogue appeared in response to the Exposition Interna-tionale des Arts Décoratifs et Industriels Modernes held in Paris in 1925. For some Americans, the French exhibit heralded at last the happy marriage of "art and mecha-nism." Moreover, as Jeffrey Meikle has noted, because it emphasized the minor decora-tive arts orchestrated into modernist ensemble, the Paris show "contributed to the vision of the world harmoniously made over by a supreme generalist, the industrial designer." Called into action by the American response to the Paris exhibition, artists, architects, and designers adopted a peculiarly American rhetoric of *modernique,* of machine aesthetics: the rhetoric of mass production. The art of the new age was to be industrial art or art with a machine aesthetic. Mass production of designer-created articles of everyday use would serve to uplift the unwashed American masses. Mass production of these articles would serve as a force "for alleviating suffering and relieving the dullness of the artificial environment."[17]

Initially, as Meikle points out, a striking contrast existed between the rhetoric of "social benefit of design for mass production" and the "reality of [the] custom-made luxury goods" created by those who espoused this rhetoric. These objects were products of craftsmen who were engaged in an "elite craft movement" and who displayed their goods in department store shows and museum exhibits. Though they adopted the machine aesthetic, these craftsmen "cared little for and knew nothing about the processes and materials of mass production." Soon, however, a group of designers emerged, who resolved to eliminate the contradiction between the rhetoric and the reality of machine aesthetic art. These were the industrial designers who were committed to the ideal, stated by Frederick Kiesler, that "THE NEW ART IS FOR THE MASSES." Industrial designers of the 1930s, such as Walter Dorwin Teague, Norman Bel Geddes, Henry Dreyfuss, and Raymond Loewy, radically changed the appearance of mass-produced everyday objects in America by combining machine aesthetics with the technical requirements of manufacturing. As creators of the Streamlined Decade, these industrial designers emerged from the confluence of machine aesthetics (an international movement) and the ethos of mass production (an American phenomenon).[18]

The ground swell of the ethos of mass production resulted in many attempts to bring the supposed powers of mass production to bear on a variety of problems that were exacerbated by the Great Depression. The efforts of the high priest of mass production, Henry Ford, were among the most noteworthy.

Responding to the plight of the American farmer in the era of the dust bowl, Ford offered mass production as the solution to the problem. "Machinery, chemistry, and education of the farmer toward intensive production are the best agencies of relief," Ford argued. In a vein similar to the concluding arguments in the *Britannica* article, "Mass Production," Ford pooh-poohed the idea that mass production would create overproduction of agricultural commodities: "Overproduction of foodstuffs will automatically be eliminated by development of by-products."[19]

Ford had only recently begun what his closest associates regarded as another idiosyncratic if not idiotic pursuit of finding an agricultural commodity with substantial dietary value and worth as an industrial feedstock. He settled on the soybean. As Nevins and Hill point out, Ford encouraged the farmers of Michigan "to plant [soy]beans with the assurance that the Ford Motor Company would do everything possible to provide a market." Ford set a research team to work on finding industrial uses for the bean. When they developed an oil extraction process, the Rouge plant became both a producer and a consumer of the oil. (See Figure 8.2.) The Rouge's massive foundry developed uses for soybean meal. Soybean-based fibers and plastics were developed but not used. Ford also served a complete soybean meal at a dinner at his company's Century of Progress Exhibition in Chicago. The menu included "tomato juice with soybean sauce, celery stuffed with soybean cheese, puree of soybean, soybean croquettes with green soybeans, soybean bread and butter, apple pie with soybean sauce, soybean coffee, and soybean cookies and candy." Ford ordered soymilk served for his lunches at the famous executive roundtable in Dearborn.[20]

Thus Ford's 1930 remarks were based on his experience with and expectations for the soybean. He had other crops in mind, as well as the increasing mechanization of the farm. He sought to demonstrate his view that mass production could solve agricultural problems by supporting a research team to increase yields on his Georgia plantation through mecha-

FIGURE 8.2.   Ford Soybean Processing Plant, River Rouge, 1930s. The industrial deployment of soybeans was only one of Henry Ford's ideas about how "mass production" could be applied to agriculture. (Henry Ford Museum, The Edison Institute. Neg. No. 833-74603-A.)

nization, soil study, fertilization, and crop selection—mass production methods in agriculture. When asked about the effect of mechanization on farm workers, he replied curtly, "Who's going to make the machines?"[21] Mass production in agriculture would lead to expanded mass production of agricultural equipment. Henry Ford, who had helped to spawn the ethos of mass production, continued to promote it. To him and to millions of Americans who had "elected" him as their exemplar, it hardly mattered that his solution demanded enormous capital in a severely contracted economy and that it meant, in today's parlance, agribusiness rather than agriculture. The editors of the *New York Times* took Ford's proposal severely to task. They pointed out the differences between the Ford factory and a field of arable land: "Under any circumstances, production in agriculture must be conducted through a variety of processes, applied over an area large in proportion to the anticipated output and always subject, as mass production in manufacture is not, to the vicissitudes of the seasons."[22]

Implicitly, the *Times*'s editors argued that mass production's rightful place was in manufacturing, not in agriculture. In a spirit that would have delighted Samuel Colt, the editors accepted the notion that Ford methods could be applied to any area of manufacturing. Others followed suit. In the area of housing, Foster Gunnison's work provides an excellent example of the pervasiveness of the ethos of mass production.

During the 1930s and thereafter, Foster Gunnison liked to think of himself as the "Henry Ford of housing" because he was the first to build houses on an assembly line basis. At the same time, he sought to create a "General Motors type of combine . . . with a system of plants producing [prefabricated] houses at several income levels."[23] His career as a mass production manufacturer began when the ethos of mass production—and one should add mass consumption—reached new heights in the 1930s.

From 1923 until 1932 Gunnison pursued a highly successful career as a salesman-designer of custom lighting fixtures with the New York firm of Cox, Nostrand & Gunnison, Inc. This company designed and built electric lighting fixtures for almost all of the major Art Deco buildings in New York including the Empire State Building, the Waldorf-Astoria Hotel, the Chrysler Building, and Rockefeller Center.[24] Despite the assessment by *Architectural Forum* that he was "the best . . . in New York's building field," Gunnison chose to end his career in this area of business. As the author of a confidential study of Gunnison noted, "There was no mass production or mass distribution involved in this business at all." The light fixtures designed and produced by his firm were unique for a particular building; architects always sought new designs for new buildings. "Looking back," continues the study's author, "Gunnison has remarked that this aspect of his work always bothered him. He felt that the really important contributions, the ones that were really rewarding, were those that involved mass production."[25] Like the industrial designers of the same period, Gunnison wished to create reality out of the rhetoric of mass production—to bring the machine aesthetic to the masses.

Drawing inspiration and financial support from fellow St. Lawrence University alumnus, fraternity brother, and chairman of General Electric Company Owen D. Young, Gunnison founded Houses, Inc., to stimulate research, construction, management, and financing in prefabricated housing. Under Gunnison's leadership, Houses, Inc., became an important promoter of the American Motohome, a steel-framed, asbestos cement-paneled, prefabricated house. The Motohome was the result of a cooperative venture of Houses, Inc., American Houses, Inc., General Electric, American Radiator & Standard Sanitary Corporation, and the Pierce Foundation. Although christened at a gala media event at John Wanamaker's in New York City, at which President Franklin Delano Roosevelt's mother unwrapped the Cellophane-packaged Motohome, this prefabricated house could hardly be called successful. Perhaps as many as 150 Motohomes were manufactured. But because of an internal dispute caused by the rising stars of the Gerard Swope faction at General Electric, Gunnison divested himself of Houses, Inc., in 1935 and soon formed his own prefabricated house manufacturing company in New Albany, Indiana.[26] Here Gunnison fulfilled his desire to "organize the General Motors of the homebuilding field,"[27] while at the same time becoming "the Henry Ford of housing."

Adapting the waterproof, plywood, stressed-skin panel developed by the U.S. Forest Products Laboratory, Gunnison Magic Homes, Inc., became one of the best-known manufacturers in the nascent prefabrication industry, an industry spawned by—and banking on—the ethos of mass production. Gunnison's selection of the New Albany site was not mere chance, it allowed him to rent the production facilities of Plywoods, Inc. Soon he recruited production experts from the economically depressed automobile industry, purchased the facilities, and renamed his firm Gunnison Housing Corporation. By 1937 Gunnison and his engineers had designed a standardized wall panel used to fabricate twelve different house models (twenty-four if mirror-image plans are included) and had installed a conveyor system for manufacturing the Gunnison house. (See Figures 8.3–8.6.) As William Blitzer wrote in his case study of Gunnison as an entrepreneur,

FIGURE 8.3.   Assembly Line Factory Production of Gunnison Housing Corporation, New Albany, Indiana, ca. 1937. (Courtesy of Foster Gunnison, Jr.)

FIGURE 8.4.   Automatic Paint Booths, Gunnison Housing Corporation, New Albany, Indiana, ca. 1937. (Courtesy of Foster Gunnison, Jr.)

FIGURE 8.5.   "Ford" or "Chevrolet" Equivalent of Gunnison Prefabricated House, ca. 1937. (Courtesy of Foster Gunnison, Jr.)

FIGURE 8.6.   "Buick" or "Cadillac" Equivalent of Gunnison Prefabricated House, ca. 1937. (Courtesy of Foster Gunnison, Jr.)

"This conveyor, probably the first used in the manufacture of houses, was a symbol of Gunnison's achievement, and pictures of it were reproduced very widely. It was taken as a sign that prefabrication had become a mass production affair, that it was following in the footsteps of the illustrious automobile industry and that, after many words had been exchanged on the subject, industrialization methods were at last being applied to housing production." *Architectural Forum* also stressed the importance of Gunnison's contributions, noting that he had "perfected prefabrication on a true mass-production, assembly line basis. . . . Gunnison was the first prefabricator to use a moving production line."[28]

Motivated by the ethos of mass production and seeking to capitalize on the popular imagery of the assembly line–mass production equation, Gunnison clearly demonstrated that houses could be made in a factory using a moving conveyor system. But there were problems. Mass consumption of Gunnison houses never followed from their mass production. Despite Gunnison's attempt to pursue a General Motors strategy of a car for every purse and purpose and to sell houses in the same way cars were sold, all of his houses looked very much alike, and they did not satisfy the idiosyncratic, highly personalized tastes of the American home buyer. Houses were not like the consumer durable products made on the assembly line even when they were made the same way. Houses represented lifetime or near-lifetime investments, and they were purchased on entirely different grounds than automobiles. Americans "consumed" automobiles but not houses. Even when Gunnison managed to undersell "in a few instances" conventionally constructed houses "by perhaps as much as 25%," there was no great rush to mass-consume the Gunnison magic home. Gunnison claimed to have sold forty-five hundred houses in thirty-eight states by 1941, but this figure probably included units he had under contract as part of the American war efforts. The most telling story about the scale of Gunnison's operation is that in 1948 his assembly line factory employed only three hundred workers,[29] a figure that hardly conjures up images of Ford's Highland Park and River Rouge complexes.

Foster Gunnison was only one of many entrepreneurs who sought to "Fordize" housing in America. Although prefabrication of wooden and metal structures had been practiced in a limited way since the eighteenth century in North America and the British Empire, the rise of the mass production ethos in the Depression era brought about a frenzy of prefabrication activity.[30] As the historians of prefabrication in America wrote, "The prefabrication movement was the child of depression." Mass production of houses in the "idle factories of our mass production industries" was offered as the quick solution to the serious national housing problem in the United States, just as Henry Ford had maintained that mass production could solve the severe problems of American agriculture. Despite intense promotion by individuals such as Gunnison and despite *Fortune*'s assessment that prefabrication was "the greatest commercial opportunity of the age," Fordized, mass-produced housing never caught on.[31]

As early as 1930 Lewis Mumford offered a closely reasoned analysis of why mass production of houses would not succeed and would certainly not solve American housing problems. Above all, he argued, mass production of houses was unlikely to result in significant cost reductions. Mass production technology would be applied in the fabrication of the shell of a house, which did not represent the greatest cost. Even if one reduced the cost of a shell by 50 percent, the ultimate savings would amount to only 10 percent of the total cost. A significant element of cost was represented by fixtures and mechanical items, in the making of which standardization—and mass production—already obtained in large measure. Moreover, "land, manufactured utilities, site-improvements, and finance call for a greater share of the cost than the 'building' and labor.'' Equally prob-lematical was ''the fact that mass-production brings with it the necessity for a continuous turnover.'' Although mass production worked well for items that wore out rapidly, it was inappropriate for more durable goods. ''When . . . mass-methods are applied to rela-tively durable goods like furniture or houses,'' wrote Mumford, ''there is great danger that once the original market is supplied, replacements will not have to be made with sufficient frequency to keep the original plant running.'' Both furniture manufacturers and carmakers, he continued, ''are driven desperately to invent new fashions in order to hasten the moment of obsolesence; beyond a certain point, technical improvements take second place and stylistic flourishes enter.''[32]

The other problem with the mass-produced house, Mumford noted, was what he called the ''Model T dilemma,'' by which he meant ''premature standardization.'' Mass pro-duction technology, because of its specialized machinery and ''careful interlinkage of chain processes,'' resulted in efforts ''to prolong the life of designs which should be refurbished.''[33]

Thus mass production posed a dilemma whose two horns were the ''continuance of obsolete models'' and profligate ''surface alterations of style.'' Beyond this dilemma, there was a more basic reason to view mass-produced housing as a false panacea. The real housing problem, Mumford argued, was ''housing of the lower half of our income groups, and particularly, of our unskilled workers. The manufactured house no more faces this problem than the semi-manufactured house that we know today.''[34] Mass production of houses offered no solution.

As Mumford noted, the furniture industry was subject to some of the same problems as housing. Like housing, furniture was affected—one could almost say, ''infected''—by the ethos of mass production. And, like housing, it was plagued by the same inappropriate model.

As discussed in Chapter 3, the mechanical engineering community played a central, missionary role in trying to introduce the principles of mass production, as practiced in Detroit, into the furniture and other woodworking industries. These engineers sought to bring about standardized and unchanging designs, to introduce automatic machinery in the working and handling of wood, and to improve finishing processes. Engineers were not the only ones. The great promoter of mass production, Edward Fiiene, specifically ad-dressed furniture manufacturers and urged them to ''Fordize'' their operations.[35] In language borrowed from his *The Way Out*, Filene argued that the future—and therefore the furniture industry's survival—lay in mass production. From his position as creator of a prominent chain of department stores, Filene assured furniture manufacturers that his vision of the future was sound and that mass-produced furniture would lead to greater wealth and improved life for all.

Despite the activities of the mechanical engineers, the rhetoric of Filene, and attempts by some furniture manufacturers, the furniture industry remained un-Fordized. Yet other woodworking industries, such as the manufacturers of radio cabinets, moved beyond the rhetoric of mass production to introduce successful Detroit-style production methods and to achieve comparable outputs. The very nature of furniture consumption as a deeply personal statement of a consumer's taste and personality and the relatively lengthy posses-sion of furniture by a consumer worked decisively against the introduction of mass production in the furniture industry. Furniture factories remained relatively small in numbers of employees, capital investments, and output. Although they were leading centers, Grand Rapids and High Point never became Detroits of the furniture industry.

The ethos of mass production remained largely at a rhetorical level in the furniture industry.

While mechanical engineers worked to promote standardization in furniture design and production, the critics of the ethos of mass production identified standardization as one of their central concerns about mass production. Whereas Edward Filene proclaimed that Americans had "elected" Henry Ford as their exemplar, Upton Sinclair could call the Flivver King "the most hated man" in Detroit.[36] And while millions of Americans marveled over the fast-paced efficiency of the assembly line and waxed eloquent about the beauty and performance of the Ford Model A and the V-8, others thought that "Henry Ford's process for . . . mass production [should] be shown in Museums of Unnatural History."[37]

Aldous Huxley wrote what is still a chilling view of the future in *Brave New World* (1932). To shake his readers into realizing the ultimate result of a world pervaded by the ethos of mass production, Huxley set his novel in the world of A.F. 632. In the brave new world, all time was measured after the appearance of the Ford Model T, After Ford. Henry Ford had become the lord of the brave new world; Ford was God. Rather than crossing themselves, as Roman Catholics had done before Ford, Ford worshipers made the sign of the T across their stomachs. They read the *Fordian Science Monitor,* and Ford's *My Life and Work* had replaced the *Holy Bible.* The new catechism held that "Ford's in his flivver . . . all's well with the world." The inhabitants of this brave new world had been standardized to serve single-purpose functions in an emotionless, loveless society. Thus identical Alphas were bred and brought up in large quantities to be of superior intelligence, while identical Epsilons were bred in quantities to be brutes of subhuman intelligence. Betas, Gammas, and Deltas fell between these extremes. This was "the principle of mass production at last applied to biology."[38]

Huxley used Henry Ford, his Model T, and the production system they spawned as symbols of all that was wrong with his contemporary world: standardization, specialization, uniformity of thought, the tyranny of overemphasizing scientific objectivity, the curse of bigness, and the authoritarian character of society dominated by mass production.[39]

Although less chilling, the films of René Clair and Charlie Chaplin made equally pointed statements about mass production, particularly the assembly line. Completed in 1931, Clair's *A nous la liberté* (Give us our freedom) was, as George Basalla has pointed out, "the first feature film to explore the social ramifications of mass production."[40] The film's plot revolves around two characters, Louis and Emile, who had been close friends as cellmates in prison. Louis, the cunning one, escapes and in the outside world manages to create a massive phonograph manufacturing firm. In doing so, he becomes something of a hero in his country. After serving out his sentence, Emile, who has lost touch with his old friend Louis, by chance becomes a worker in Louis's factory. There he discovers not only the true identity of the factory's owner but the secret behind his success. Louis has simply applied the techniques and management practices used in the prison workshop, where both Emile and Louis had worked on an assembly line ruthlessly bossed by prison guards. In his phonograph factory, Louis has installed a similar assembly line, and in place of prison guards he has put equally villainous foremen. The workers wear uniforms similar to those in the prison. In both prison and factory the work screams of monotony and regimentation. (See Figures 8.7 and 8.8.) Thus although Clair relied upon a device at least as old as Dickens—the comparison of prisons and factories—he focused more

FIGURE 8.7.   Shift Change at Louis's Phonograph Factory, *A nous la liberté*, 1931. (The Museum of Modern Art, Film Stills Archive.)

FIGURE 8.8.   Phonograph Manufacture on the Assembly Line, *A nous la liberté*, 1931. (The Museum of Modern Art, Film Stills Archive.)

FIGURE 8.9.   Chaplin on the Assembly Line, *Modern Times*, 1936. (The Museum of Modern Art, Film Stills Archive.)

precisely on the evils of mass production by his treatment of the assembly line and its trappings.

Clair's criticism of mass production civilization is weakened by his film's ending. Louis escapes his predicament—exposure to the police by blackmailers—by becoming a tramp with his friend Emile. His workers enjoy a similar romantic relief. Louis's phonograph factory has been fully automated, and its machinery now serves the workers rather than enslaving them. The workers enjoy their "labor" by fishing, dancing, singing, and card-playing. "Gloire au bonheur! Vive la science!"[41]

Although Charlie Chaplin's Little Tramp, like Louis, escaped the evils of mass production society by tramping down the road of endless horizons in *Modern Times* (1936), Chaplin offered no solution to the insanity-inducing assembly line as René Clair had done. After juxtaposing sheep going down a ramp to the slaughterhouse and workers entering a factory, Chaplin sets his character, the Little Tramp, down as a worker in a mass production factory. Of prominence in the factory are the scale of machinery and the assembly line. In one familiar scene, a still of which has become the poster image of *Modern Times*, the Little Tramp becomes enmeshed in the gigantic machinery—in the cogs of the symbolic Great Machine of modern civilization. Chaplin focused on the assembly line in another factory scene. (See Figure 8.9.) The Little Tramp works on an assembly line tightening bolts on a nondescript object. The pace of the line is increasingly quickened. Efficiency experts seek to improve the efficiency of the line by installing automatic feeding machines on the line so that workers no longer have to take a lunch break. When speedups on the line continue to be made, the Little Tramp eventually goes berserk, and his frame is seized by jerking motions identical to his nut-tightening operations on the line. (See Figure 8.10.) A paddy wagon carries the Little Tramp away. Later he finds

FIGURE 8.10.   Chaplin Driven Crazy on the Assembly Line, *Modern Times*, 1936. (The Museum of Modern Art, Film Stills Archive.)

peace and comfort in a prison, a radical contrast to the regimentation and depersonalization of the outside mass production culture.

The standard account of how Chaplin came to make *Modern Times*, which he had originally named *The Masses*, appeared in his autobiography. ''I remembered an interview I had had with a bright young reporter on the New York *World*,'' wrote Chaplin in 1964. ''Hearing that I was visiting Detroit, he had told me of the factory-belt system there—a harrowing story of big industry luring healthy young men off the farms who, after four or five years at the belt system, became nervous wrecks.''[42] But the factory scenes in *Modern Times* were based on more than hearsay about Detroit's ''factory-belt'' systems, and, as argued by a recent biographer, the sight of a ''miniature version of the factory-belt system in a Los Angeles restaurant.''[43] During the height of the Model T era, Chaplin actually visited Ford's Highland Park factory and was given a VIP tour by Henry and Edsel Ford. There, in addition to seeing manufacturing operations including the assembly line, Chaplin saw the power plant, which was once as much a source of pride for Ford as the line. (See Figure 8.11.) The scale and impressions of that plant would later appear in the sets of *Modern Times*. Moreover, while in Detroit Chaplin also toured the Cadillac factory, where, although he observed a different pace, he saw largely the same approach in final assembly.[44]

In Detroit during this and perhaps later visits, Chaplin also learned about some of the less than pleasant aspects of the assembly line, especially that of Henry Ford. Critics of

FIGURE 8.11.   Charles Chaplin with Edsel B. Ford and Henry Ford in the Power Plant, Highland Park Factory, 1923. (Henry Ford Museum, The Edison Institute. Neg. No. 0-4144.)

the Ford line thrived on the rumor, which had circulated since the assembly line was first introduced, that Ford workers could not leave the fast pace of the line to relieve themselves in the restroom, that lunch breaks had been reduced to a minimum, and that somebody was always watching the workman at Ford's. Such rumors had only intensified during the Model A and early V-8 eras, when Charles Sorensen became known as the ruthless big boss driver of River Rouge. When Chaplin's Little Tramp seeks relief in the washroom, he finds to his horror that he is being watched by a television camera. The factory's big boss appears on a television screen on the lavatory's wall and tells the Tramp to get back to work. Although without the futuristic television device—and feeding machines—one encounters similar accounts in works such as Upton Sinclair's *The Flivver King* (1937).

Most contemporary critics of *Modern Times* dismissed the film as naive, trivial, or poor social commentary, but the public liked it.[45] *Modern Times* was increasingly viewed as a brilliant commentary on mass production society. As James Agee said in 1948, every motion picture produced since *Modern Times* had been ''child's play.'' Most recently, George Basalla has written that despite its obvious derivation from Clair's *A nous la liberté, Modern Times* ''remains the single best film ever made for a mass audience treating technology within a social context.''[46]

Other critics of the ethos of mass production in the United States worried about its spread beyond the United States. The prolific German psychologist, Richard Müller-

Freienfels, in his 1929 work *The Mysteries of the Soul,* equated mass production, the mechanization and standardization of humans, and Americanization. He feared the Americanization of the soul, a process that bred "conformity to mass standardization [and] to convenience." "Taylorism and Fordism are the systematic accomplishment of this mechanization of the human being," Müller-Freienfels argued, and he opposed this "tyranny of technique." Writing on business in Charles Beard's edition of *Whither Mankind: A Panorama of Modern Civilization* (1930), Julius Klein expressed similar fears: "It is . . . only with the greatest caution and reserve that one can contemplate the transfer of American methods of efficiency, mass production, and rationalization to the industrial communities of Europe."[47]

In the same book, Stuart Chase wrote an article on play, which explored the impact of mass production on American leisure. Echoing other critics, Chase related the rise of mass sport in the United States to the rise of mass production. Mass sport and mass entertainment provided a degenerate way for the masses to work off the tensions caused by mass production. Chase noted that no better evidence of the negative effect of mass production on leisure was to be found than by listening to the music coming out of Detroit: "Its pounding rhythm is as simple as tightening bolts [in a Ford factory]. It gives very little scope for individual expression." The marriage of mass production and play was problematic. In contrast to Henry Ford, Chase offered no solutions. The Flivver King had said that what everybody needed was some good square dancing. But Chase closed by saying that "it will take more barn dances than Henry Ford can ever pay for, to throw off the yoke of that [mass production's] brutality."[48]

Labor union critics of mass production often spoke of the brutality of the assembly line. Among countless critiques of what has more recently become known as the blue-collar trap, one in particular stands out. In 1932, the Rock Island, Illinois, *Tri-City Labor Review* published these worker's words:

> I ran into a fellow the other day who is waiting for one of the new Fords. "Nice car," he told me.
>
> But I always think about a visit I once paid to one of Ford's assembling plants every time any one mentions a Ford car to me. Every employee seemed to be restricted to a well-defined jerk, twist, spasm, or quiver resulting in a fliver [sic]. I never thought it possible that human beings could be reduced to such perfect automats [sic].
>
> I looked constantly for the wire or belt concealed about their bodies which kept them in motion with such marvelous clock-like precision. I failed to discover how motive power is transmitted to these people and as it don't seem reasonable that human beings would willingly consent to being simplified into jerks, I assume that their wives wind them up while asleep.
>
> I shall never be able to look another Tin Lizzie in the face without shuddering at the memory of Henry's manikins [sic].[49]

Upton Sinclair's *The Flivver King* represented Ford's system of mass production as being equally brutal. Yet in this historical novel, Sinclair expressed disgust that even many of Ford's own employees believed in Henry Ford and in the benefits of mass production and mass consumption as mythologized in *My Life and Work* and other writings even though they had experienced the brutality of the Ford factory and its bosses and further dehumanization caused by massive unemployment for extended periods of time. With one exception, all members of Sinclair's Abner Shutt family in *The Flivver King* poured their lives into mass production at the Ford factory only to be robbed in the end of their homes and their humanity.

Sinclair's Detroit had become known as early as 1932 as the "beleaguered capital of mass production."[50] The Great Depression had demonstrated that mass production could bring mass unemployment or, more properly, unemployment of the masses. While many such as Filene continued to argue that now more than ever mass production was the salvation of modern society, Paul Mazur noted that the Great Depression had demonstrated that "mass production has not proved itself to be an unmixed blessing; in the course of its onward march lie overproduction and the disastrous discontinuity of industry that comes as a consequence." Mazur pleaded with his *New York Times* readers to see mass production as "an alluring but false doctrine." Moreover, Mazur argued, "It is essential for business to realize that unquestioning devotion to mass production can [only] bring disaster."[51]

Mazur's critique of mass production appeared on the heels of a previous *Times Magazine* article, "Gandhi Dissects the Ford Idea." In his article, Harold Callender established a radical dichotomy between Ford's and Gandhi's respective views of happiness. While Gandhi offered widespread handicrafts as the solution to global problems of unemployment and hunger, Ford offered his doctrine of mass production. The *Times Magazine* juxtaposed a photograph of an assembly line with one of a group of Indian hand spinners. Captions under the two photographs read: "The Ford Formula for Happiness—A Mass-Production Line" and "The Gandhi Formula for Happiness—A Group of Handcraft Spinners."[52]

Few Americans would have resorted to the Gandhi formula, but it seemed obvious to most of them that developments in mass production had not been matched by the development of mass consumption. As Mazur put it, "The power of production . . . has been so great that its products have multiplied at geometric rates . . . at the same time the power of consumption—even under the influence of stimuli damned as unsocial and tending toward profligacy [for example, advertising and built-in obsolescence]—has expanded only at a comparatively slow arithmetic rate."[53] Curtailing production and allowing consumption to match production was one solution—a sensible one in the minds of the critics of mass production. But alongside what Mazur called the "industrial cult" of mass production there arose a cult of mass consumption.

The problem, argued its high priests, was not mass production but mass consumption. Depression times demanded a new breed of engineer: the "consumption engineer."[54] The new consumption engineer would complement the production engineer through aggressive marketing activities, especially advertising. He would manufacture customers. The 1930s witnessed the publication of a phenomenal amount of literature on the "economics of consumption," much of which dealt in one form or another with marketing and advertising.[55] Despite the fears of many, such as the banker Mazur, that consumption engineering tended toward profligacy, advertising grew in its "slickness."

As Stuart Ewen points out in his *Captains of Consciousness,* advertisers saw their role as helping society to overcome a "puritanism in consumption."[56] They did so by avoiding what the leading advertising copy writer, Helen Woodward, called the "real, inner truth." In a revealing piece, she shared the secret of her success: "If you are advertising any product never see the factory in which it was made. . . . Don't watch the people at work. . . . Because, you see, when you know the truth about anything, the real, inner truth—it is very hard to write the surface fluff which sells it." It was precisely this posture that also led many critics of mass production to call into question the institution of advertising, which the historian David Potter later called the "institution of abundance.''[57] By its very own needs, mass production had created a profligacy that was "damned as unsocial," just as was mass production's transformation of workers into automata.[58]

Of the critics of the ethos of mass production who dared to see the "real, inner truth" of a mass production factory, Diego Rivera was in many ways the most pointed and dramatic. In 1932–33, Rivera painted an impressive series of frescoes in the Garden Court of the Detroit Institute of Art portraying mass production at Ford's River Rouge complex as well as the larger operations of the company.[59] (See Figures 8.12–8.14.) The frescoes, which took eight months to complete, were the result of Rivera's intense, month-long study of the Rouge and its operations (supported by the documentary photography of Ford Motor Company photographer W. J. Stettler). And, of course, the context of Rivera's work was Detroit in the Depression, Detroit the "beleaguered capital of mass production."

The men who commissioned Rivera's murals, Edsel B. Ford, Albert Kahn, Charles T. Fischer, and Julius H. Haas, had never imagined that the Mexican artist would focus on mass production at the Rouge. In his initial letter of commission, the director of the Detroit Institute of Arts informed Rivera that these arts commissioners had originally said they "would be pleased if you could find something out of the history of Detroit, or some motif suggesting the development of industry in the town; but at the end they decided to leave it entirely to you."[60]

The Rouge captivated Rivera from the moment he arrived in Detroit in April 1932. Edsel Ford had offered to open the doors of any firm or factory in Detroit, but Rivera found this unnecessary. Although perhaps in extremes, the Rouge spoke of the times and of the place—of an era in which, despite a severe economic depression, an ethos of mass production prevailed and of the firm that had created mass production. The Rouge provided Rivera the means to comment on the era and to interpret mass production. He was originally commissioned to produce frescoes for only two panels of the Garden Court. When he first learned of the sum of the commission, before he arrived in Detroit, Rivera said that he would do only parts of those panels. But the month he spent at the Rouge compelled him to want to paint the entire Garden Court of twenty-seven panels for a fraction over twice the figure of the original two-panel commission.

In two major panels on the north and south walls (each roughly 45 × 17.5 feet), their twelve predella panels, and two vertical panels on the west wall, Rivera captured the motion and pace, the power and scale, the total environment of the mass production factory. By his inclusion of a gallery of tourists and the very dynamism conveyed in these paintings, he stated that mass production was a spectacle to be heard and felt as well as seen.[61] Rivera, a Marxist, chose to focus on workers whose pace and motions were controlled by the massive machinery surrounding them. A number of workers are placed in the foreground, but they are dominated by the machines behind them, machines which, as in the case of a fender stamping press, seem to be anthropomorphic.[62] Highly animated and distorted by the machine while working, the weary laborers sit hunched over during their lunch break and file sheepishly out of the factory after collecting their wages at day's end.

Although providing more information about the product being made than Chaplin did in *Modern Times*, Rivera, like the filmmaker, showed a far greater interest in the process and pace of mass production than in the product. Just as Chaplin chose not to specify objects being made on the line in *Modern Times*, Rivera elected not to paint completely manufactured and assembled Ford automobiles. The "real, inner truth" of mass produc-

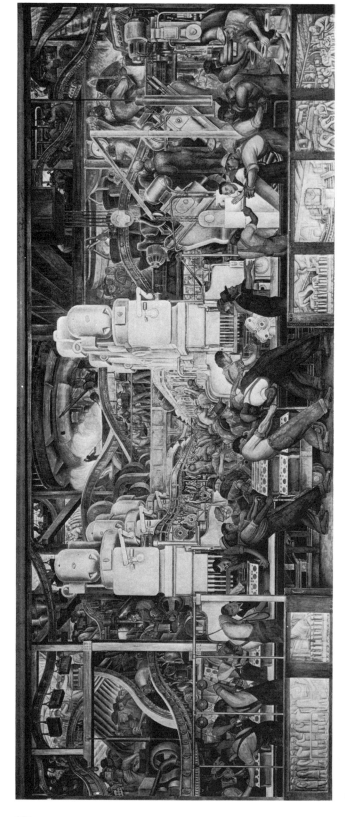

FIGURE 8.12. North Wall, *Detroit Industry*, Diego Rivera, Detroit Institute of Arts, 1932. (Founders Society Purchase, Edsel B. Ford Fund & Gift of Edsel B. Ford. Courtesy of the Detroit Institute of Arts.)

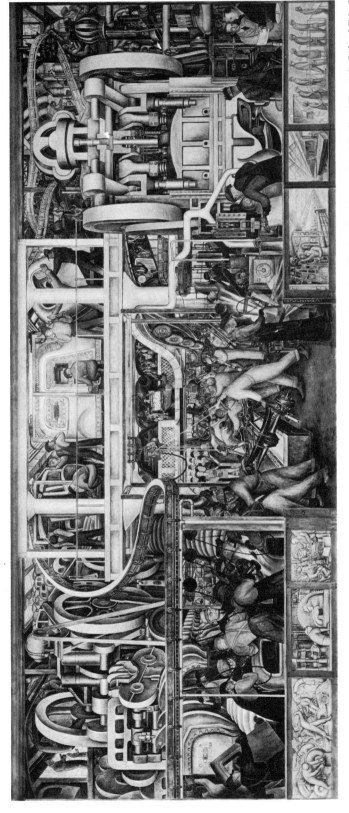

FIGURE 8.13. South Wall, *Detroit Industry*, Diego Rivera, Detroit Institute of Arts, 1932–33. (Founders Society Purchase, Edsel B. Ford Fund & Gift of Edsel B. Ford. Courtesy of the Detroit Institute of Arts.)

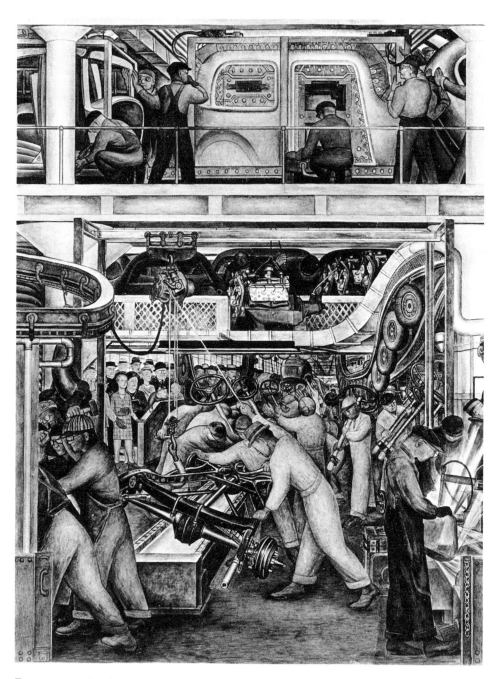

FIGURE 8.14. Detail, South Wall, *Detroit Industry,* Diego Rivera, Detroit Institute of Arts, 1932–33. Note the gallery of spectators. (Founders Society Purchase, Edsel B. Ford Fund & Gift of Edsel B. Ford. Courtesy of the Detroit Institute of Arts.)

tion was what took place in the factory, not its product. Thus Rivera stood Helen Woodward, the advertiser who had admonished her peers never to go near their clients' factories, on her head.

When the Rivera frescoes were opened for public viewing, they created an intense outcry. Some people labeled them blatantly communistic; others simply said they should be whitewashed (a portent of what would later happen to Rivera's Rockefeller Center mural). Rivera said in *Art Digest* that "I paint what I see. Some society ladies have told me they found the murals cold and hard. I answer that their subject is steel, and steel is both cold and hard."[63] But as Rivera well knew, the subject of the frescoes went far beyond steel; it went to the very heart of mass production—to the mechanized, conveyor-laced, assembly-line-dominated factory and its human appendages.

Ford's River Rouge plant, captured by Rivera as a major symbol of mass production, was the result of more than a century and a quarter of development since the United States Ordnance Department first committed the nation's resources to achieving uniformity in muskets and other small arms. The road from the Springfield and Harpers Ferry armories to Detroit was neither short nor direct. Though the advantages of the "uniformity principle" seemed obvious to the Enlightened, French-influenced officers of the Ordnance Department, and though they could draw extensively upon the public treasury, the actual attainment of uniform or interchangeable parts manufacture eluded almost two generations of the nation's most talented mechanics.

The production of interchangeable parts entailed far more than its early proponents visualized, and it demanded more of them than anyone first expected. Not only did the armory mechanics have to figure out how their goal could be realized in theory but also in practice. This achievement required developing a system of jigs, fixtures, and gauges that were based on a "perfect" model and establishing procedures to maintain a constant vigil over the accuracy of these special devices and the machine tools with which they were used. Armory mechanics also pursued the development of machine tools. Their efforts paralleled those of other American inventors and mechanics in that the focus of much of their work was in the construction of special- or single-purpose machine tools.

For the noted English machine tool builder Joseph Whitworth, these special-purpose machine tools distinguished what later became known as the "American system" of manufactures. As he noted in his special report on American manufacturing, Americans "call in the aid of machinery in almost every department of industry. Wherever it can be introduced as a substitute for manual labour, it is universally and willingly resorted to."[64] Pinmaking, clockmaking, barrelmaking, and other traditional handicrafts had been mechanized in the United States by the early 1850s.

Armory mechanics brought this stream of development of special-purpose machines and tools together with their quest for interchangeability. In his small-scale Rifle Works at Harper's Ferry, John H. Hall was perhaps the first mechanic to succeed, but it was at the Springfield Armory in the 1840s and early 1850s that small arms were manufactured with interchangeable parts by special machines and tools on a "large" scale. It was to these developments at Springfield that latter-day historians alluded when they wrote of the American system of manufactures. But to the mechanics of the nineteenth century, the developments at Springfield and other armories of special-purpose machines and tools; a model-based jig, fixture, and gauging system; and even the system of shop management were known simply as "armory practice." The mechanic who had been schooled in armory practice in the antebellum period became a fundamental agent in the diffusion of this technology to other manufacturing industries. Wherever armory practice was adopted, it was invariably the work of a mechanic or group of mechanics who had worked in the federal or the prominent private armories.

The sewing machine industry was perhaps the first American industry to adopt *in toto* the techniques used in small arms manufacture. Yet not every manufacturer of sewing machines relied upon armory practice. Some sewing machine makers initially or quickly employed armory mechanics and their techniques, but at least one did not do so. Singer Manufacturing Company demonstrated that success in the sewing machine industry did not necessarily depend on the adoption of this new manufacturing technology. Using what contemporaries called the European approach to manufacture—skilled workers using general-purpose machine tools and extensive hand labor—Singer succeeded in the business because its leaders built a finely tuned international marketing organization for the Singer sewing machine.

Eventually, when Singer began to dominate the industry and when its sales had grown to tens of thousands and soon hundreds of thousands of units, the company's leaders pushed for the gradual adoption of armory practice. When annual production reached almost half a million sewing machines in the early 1880s, it seemed imperative to Singer executives to achieve the manufacture of interchangeable parts for the Singer sewing machine. But to the dismay of many, Singer's production experts found that such a goal was not easy to achieve. Indeed, manuscript records end before there is clear evidence that Singer realized its goal. The Ford Motor Company may have been the first manufacturer to have produced as impressive a number of units as Singer with parts that interchanged.

Paradoxically, though the Singer Company demonstrated that an American manufacturer did not have to adopt the "universally resorted to" special machinery to succeed in the sewing machine industry, it showed that woodwork could be turned out in large quantities with specialized machinery in comparatively large factories. The case or cabinet manufacturing operations of Singer's South Bend, Indiana, and Cairo, Illinois, factories offer a stark contrast to most woodworking factories of the late nineteenth century. In particular, the example of Singer demonstrates that it was at least technically possible to "mass-produce" furniture, something that many have thought difficult if not altogether impossible. The case of Singer suggests that other factors—primarily style changes and matters of personal taste—have mediated against the rise of a Ford in the manufacture of furniture.

The history of the McCormick reaper company closely parallels that of Singer in the manufacture of the sewing machine. A careful examination of McCormick production technology suggests that, like Singer, McCormick relied for a long time upon general machines and skilled workers. And like Singer, McCormick succeeded in the reaper business not because of its production technology but because of the company's emphasis on marketing its product. Eventually, McCormick paid closer attention to its manufacturing operations. Cyrus McCormick finally moved the company away from older manufacturing techniques toward armory practice in the early 1880s, when he hired an armory mechanic well-versed in its methods. McCormick's move occurred thirty years after he is commonly believed to have adopted the American system of manufactures.

Whereas the cases of Singer and McCormick suggest the relative slowness of the diffusion of techniques developed in antebellum armories, the history of the bicycle industry clearly demonstrates that different manufacturing approaches were developed that would later be of major importance in the development of mass production at the Ford Motor Company. Bicycle makers located primarily in New England brought to perfection the techniques of armory practice. Indeed, contemporaries argued that with the bicycle came the first widespread diffusion of armory practice. Abandoning time-honored metal-working techniques, however, midwestern manufacturers pioneered in the production of metal bicycle parts from sheet steel. Though viewed by armory mechanics as ''cheap and sorry,'' pressed steel parts would become an essential technique in the mass production of the automobile.

In less than a decade, the Ford Motor Company brought together the best of armory practice with the rapidly developing techniques of pressed steel work. To these approaches, Ford production men added fresh thinking about work sequences, tool design, and controlling the pace of work in the Ford factory. When challenged to produce more of the standardized Model T's at lower cost and when allowed virtually free rein to experiment with new methods, the Ford team brought about a revolution in manufacture: mass production. The development of the moving assembly line was a key element in this revolution. But in addition, one must add Ford's ability to manufacture tremendous quantities of interchangeable parts, the installation of unprecedented numbers of highly specialized machine tools, and conveyor systems that were applied everywhere. To ensure that workers would continue to labor in the increasingly mechanized and faster driven factory, Ford initiated the five-dollar day, an unprecedented incentive that proved to be the final element in the development of mass production at Ford.

Thirteen years and fifteen million Model T's later, however, Ford's system of mass production reached a dead end. The single-purpose manufacture of a single car model with single-purpose machinery resulted in saturation of the market for Model T's. During the 1920s Ford saw its share of the market drop from more than half to less than a sixth. Such a decline can be comprehended only by understanding how entirely committed Henry Ford was to his company's system of mass production. While Ford's company was losing its share of the market, Henry Ford was enlarging and integrating the operations of the Rouge with the implicit idea that the Highland Park factory, the birthplace of mass production, was not a satisfactory home for mass production to be developed as fully as possible.

Though Henry Ford continued to enlarge the Rouge, the precipitous decline of the Model T led him finally to introduce a new car, the Model A. The extent of the overhaul required to manufacture the Model A and the problems encountered make clear how far the Ford Motor Company had traveled down the dead-end road of single-purpose or mass production. With the Model A and its successor the V-8, Ford adopted ''flexible mass production,'' a development in which General Motors—especially its Chevrolet Division—had pioneered. The lessons of changing over from one model to another were painfully learned by Ford production experts, and industry analysts questioned as late as 1933 whether they had really mastered this fundamental part of the new mass production technology.

As pointed out above, however, Ford's long-term success with the Model T and its changeover to the Model A led many Americans to believe that a new era of style and beauty, economy, and abundance was at hand. These were the Americans who in the late 1920s and early 1930s helped to create an ethos of mass production.

As the decade of the 1930s wore on, many of those who had espoused the ethos of mass production clung tenaciously to its precepts. The critics of mass production also continued to fire salvos at the machine, while industrial designers worked to reshape almost every part of the era's material culture, thus aiding the increasingly powerful new consumption engineers of Madison Avenue.[65] But the Depression did not go away. In spite of its benefits, even when coupled with the new emphasis on stylized consumption, mass production failed to fulfill the vision of its prophets.

An external force finally brought an end to the Depression and renewed America's faith in its mass production system. That force was World War II. Lewis Mumford had tied all the pieces together in his classic work of 1934, *Technics and Civilization,* in which he pointed to the military as the source, and indeed the salvation, of mass production. Growing out of the standardization and quantity production of military weapons in the late eighteenth and early nineteenth centuries, mass production was at once the means of support for and an extension of the regimentation and quantity consumption of soldiers outfitted in standardized uniforms consuming standardized goods. The army was, in Mumford's estimation, the perfect "pattern not only of production but of ideal consumption under the machine system." Mumford concluded: "Quantity production must rely for its success upon quantity consumption; and nothing ensures replacement like organized destruction. . . . War . . . is the health of the machine."[66] The "organized destruction" of World War II, made possible by the mass production of war materiel, served not only to bring an end to the Great Depression but to solidify the position of mass production in American society. Nevertheless, the war brought an end to the era dominated by an ethos of mass production, and, of course, its critics. The specter of the atomic bomb removed mass production from its central place in American consciousness and forced the nation and the world into the nuclear age. Like the era dominated by the ethos of mass production, the nuclear age held out unlimited promises to some Americans while for others it heralded the ultimate destruction of humanity.

The Evolution of the Expression *The American System of Manufactures*

T he *Official Descriptive and Illustrated Catalogue* of the Crystal Palace Exhibition included entries for Colt, McCormick, Singer, and Robbins & Lawrence but noted only that the latter's rifles were manufactured such that "the various parts [were] made to interchange."[1] Samuel Colt's display of pistols interested Englishmen enough to win for Colt an invitation to deliver an address before the Institution of Civil Engineers. In his paper, "On the Application of Machinery to the Manufacture of Rotating Chambered-Breech Fire-Arms and the Peculiarities of Those Arms," delivered November 25, 1851, Colt never referred to his method of manufacture as the American system. During the discussion session that followed, however, a Mr. Hodge alluded to "Colonel Colt's system of manufacturing."[2] Subsequently, the *Civil Engineer and Architect's Journal* summarized Colt's paper. The editor who wrote the summary mentioned "the system [of manufacture] adopted in England and on the Continent, of making firearms almost entirely by manual labor." By contrast, one can infer either Colt's system of manufacture or the American system of manufactures.[3]

1851

When Charles Tomlinson published the 1853 edition of his *Rudimentary Treatise on the Construction of Locks,* he contrasted the *"handicraft"* system with the *"factory* system." A critical dimension of the factory system was the "mode of producing . . . articles" by machinery. Tomlinson noted that the "system of manufacturing on a large scale [was] more nearly universal in the United States than in England." He implied, however, that what was taking place in the United States was simply a part of the development of the factory system.[4]

1853

Upon their return to England from their tour of duty as commissioners to the New York Crystal Palace Exhibition (1853), Joseph Whitworth and George Wallis wrote special reports, which were published in 1854. Neither report contained the expression *American system of manufactures,* yet both stressed that in the United States special-purpose machines and machine tools were very common. As Whitworth wrote, Americans "call in the aid of machinery in almost every department of industry. Wherever it can be introduced as a substitute for manual labor, it is universally and willingly resorted to."[5]

1854

Following the publication of these reports, Parliament's Select Committee on Small Arms conducted its investigation of the "existing system" of arms procurement in England and the "advantages of producing muskets by machinery." As stressed above in Chapter 1, throughout the proceedings and testimony of this committee, the developments in American small arms production were referred to variously as the "American plan," the "Springfield principle," the "Springfield system," the "Russian Plan," and the "almost perfect system of Samuel Colt." While on the witness stand, John Anderson was asked how long it would take him and a committee of others "to learn the whole of the American system" of arms production. Although this is without question the clearest early use of the expression that historians later wrote about, it appears that neither the select committee nor Anderson's committee endowed the term with any great significance.[6]

1855   The lack of significance in the select committee's casual use of the term *American system* is underlined by the fact when Anderson's committee was formed, it was called the Committee on the Machinery of the United States of America rather than the Committee on the American System of Manufacture.[7] Moreover, as noted in Chapter 1, the committee's report of 1855 referred to American manufacturing methods in a number of ways. Only once did the principal author, Anderson, use the words "American system."[8] Without question, Anderson was alluding to methods of small arms production used at the Springfield Armory and other arms factories, but he employed the term casually, without emphasis, and without any transcendent meaning.

Not long before Colt's London armory closed, the Charles Dickens–run *Household Words* featured an article on it.[9] The writer described the manufacture of the pistol in depth but did not identify these procedures as the American system of manufactures.

1857   The author of a similar article on Colt's Hartford armory, published in the *United States Magazine,* also failed to label Colt's production methods with any special term. This fact deserves emphasis because the article described Colt's display at the London Crystal Palace, his lecture before the Institution of Civil Engineers, his testimony before the select committee, and the visit of the Anderson committee to the Colt armory in 1854.[10]

When Anderson had completed the task of establishing the American-equipped Enfield Arsenal, he gave a paper to the Royal Society of Arts, "On the Application of Machinery in the War Department."[11] Anderson's paper is filled with the word *system,* but in no instance did he employ the expression *American system* or *American system of manufactures.*

1859   In a subsequent paper, "On Some Applications of the Copying or Transfer Principle in the Production of Wooden Articles," Anderson described the Blanchard-type gunstocking machinery at Enfield as well as the "American system of turning long poles."[12] In his writings at least, Anderson did not restrict his use of *American system* to any particular set of techniques or methods. Nor did the author of an article on the American Alfred C. Hobbs's English lock factory. The writer described the "English system—or rather, no system—of lock-making" and then contrasted this with "Mr. Hobbs' system of manufacturing." Hobbs, like Colt, employed a "system of working from standard gauges."[13]

But the *Engineer's* illustrated description of the Enfield Arsenal was more precise in linking the techniques used at Enfield with American practice. "This arrangement," reported the *Engineer,* "is on the general system in use in the United States."[14] *American system of manufactures* still had not been used nor had it yet become a meaning-laden phrase.

When John Anderson delivered a paper, "On the Application of the Copying Principle **1862** in the Manufacture and Rifling of Guns," to the Institution of Mechanical Engineers in 1862, he made no mention of the American system.[15]

Charles Hutton Gregory devoted part of his 1868 presidential address before the **1868** Institution of Civil Engineers to the first ten years of the Enfield Arsenal. Gregory noted that the work of Samuel Colt, John Anderson, and others had led the parliamentary select committee to recommend "a new system" of military arms production in England. The method adopted was "the system of concentration and of copying-machinery."[16] Nowhere did Gregory use the expression *American system,* although he pointed out that Enfield had been "stocked with improved machinery, founded on that already in use in the United States Arsenals at Springfield and Harper's Ferry."[17]

A decade later, in 1878, Sir Thomas Brassey delivered a series published as *Lectures* **1878** *on the Labour Question* in which he talked about the diffusion of interchangeable parts arms production technology to other areas of manufacture. Brassey cited the case of the Peabody Rifle Company of Providence, Rhode Island, which had initiated manufacture of sewing machines.[18] In no instance, however, did he employ the expression *American system of manufactures.*

In the United States, Thomas Edison invented the phonograph and projected its widespread manufacture. In a letter to his agent George S. Nottage, he stressed that "this machine can only be built on the American principle of interchangeability of parts like a gun or a sewing machine."[19]

With the publication of Charles H. Fitch's Tenth U.S. Census "Report on the Manufac- **1881** tures of Interchangeable Mechanism" the term *American system* began to take on a specific and definite meaning. In his initial section on the manufacture of firearms, Fitch equated the "development of the interchangeable system" with the "development of the American system." He also traced the "adoption of the American system in foreign countries." "The interchangeable system," Fitch wrote, "with its astonishing results and its ingenious plants of machinery, was distinctively American."[20]

Sir Edmund Beckett's article "Lock" appeared in the ninth edition of the *En-* **1882** *cyclopaedia Britannica* in 1882. It concluded with the following statement:

> It should be added that Mr. Hobbs introduced into England in 1851 the American system of manufacturing every part of a lock by machinery, so that all similar locks of a given size are exactly alike, except the keys, and the gating in the tumblers, which is cut when they are lifted by the key; and even those are done by machinery adjustable to secure what may be called an infinite number of variations. The same system has been adopted in the Government rifle manufactories, and for clocks and watches; and no hand-work can compete with it.[21]

This is the first unambiguous, definitional statement published about the American system of manufactures by an Englishman of which I am aware. Since the first volume of both the English and the American editions of the ninth edition appeared in 1875, Beckett probably wrote his article in the 1870s—some twenty years after Englishmen are supposed to have commonly used the expression *American system of manufactures*.

1883   The publication of the autobiography of James Nasmyth, edited by Samuel Smiles, in New York in 1883 solidified the term *American system*, at least for latter-day historians. Nasmyth recounted the select committee's deliberations of 1854 and the Board of Ordnance's plans to "introduce the American system, by which arms might be produced much more perfectly, and at a great diminution of cost."[22]

Although it is fair to assume that Nasmyth was alluding to the manufacture of interchangeable parts firearms, a contemporary article, "Cutlery at Sheffield," stated that "Sheffield is now alive to the American system," implying only that English cutlery manufacturers were adopting special-purpose machinery.[23]

Writing under the pseudonym of Chordal, James W. See touched on American manufacturing methods in several of his letters to the *American Machinist* in 1883. In May he wrote, "Our American systems of labor-saving machinery and devices are probably, in a general way, the most marvelous and perfect in the world."[24] A month later he used the sewing machine industry as an example of the "Scientific System of Production" or "Yankee manufacture." For Chordal, "the manufacturing system" implied interchangeable parts manufacture.[25]

Chordal's letters generated several "Personal Recollections" of S. W. Goodyear, also published in *American Machinist*. Goodyear recalled some interesting episodes in the early sewing machine industry, but his major emphasis was in making the distinction between building and manufacturing. He also discussed interchangeable manufacture and the demands it placed on its creators. Even in this discussion, however, Goodyear did not employ the magic words *American system*.[26]

1884   Charles H. Fitch returned to the history of interchangeable manufacture in 1884, when he published a clever article in the *Magazine of American History* entitled "The Rise of a Mechanical Ideal." Of course, that mechanical ideal was the "interchangeable system of manufacture," which was pursued and achieved in America.[27]

1886   John Anderson's death in 1886 occasioned his obituary writer to note, "In 1854, [Anderson] with two artillery officers, visited America and minutely examined the American system of small-arms manufacture."[28] The term had acquired a specific meaning.

1890   Writing in *American Machinist* in 1890, B. F. Spalding endowed the expression *American system of manufacture* with its more modern meaning and its importance even though he used the term variously: "The 'American System' of Manufacture," "The 'American' System of Manufacture," "The 'American' system," and "the American system." Interchangeable manufacture and the American system were intimately linked, Spalding argued. In detailing the gunmaking machinery that was exported all over the world, Spalding wrote that "it was for such doings as these that the term 'American' was bestowed upon the system, and it was a title earned, won, and granted in the open field." Spalding also wrote that the "feature of interchangeability is not all there is to the American system; this alone falls far short of it." Mechanized production—use of special-purpose machinery—was essential. He made this point by contrasting the "European and American systems."[29]

At the Columbian Exhibition, the American Society of Mechanical Engineers heard **1893** William F. Durfee deliver a paper, "The History and Modern Development of the Art of Interchangeable Construction in Mechanism." Durfee argued strenuously that the idea of interchangeability was not American. Yet because Americans had transformed this idea into a reality, he could speak of the "American interchangeable system" or the "American system of interchangeable construction."[30]

An Englishman, John Rigby, read a paper, "The Manufacture of Small Arms," which *Engineering* printed. In a section titled "Origin of the Interchangeable System," Rigby argued that the "credit of originating the interchangeable system, and of applying machine work in general to the production of military arms, belongs undoubtedly to the United States."[31] The "new system of manufacturing" the Enfield rifle, therefore, was the American system.

Writing a six-part series of detailed articles, "The Revolution in Machine-Shop Prac- **1894** tice," Henry Roland (a pseudonym of Horace L. Arnold) discussed the history of British and American machine tool making.[32] Roland also included details about the production of sewing machines, small arms, bicycles, and watches, yet he did not employ the words *American system of manufactures*. Arnold's long experience in American manufacturing, however, must have supplied him with plenty of knowledge of the Americanness of interchangeable manufacture.

Frederick A. McKenzie's title *The American Invaders* referred to the invasion of **1901** American products into English markets, but he did not allude to an American system of manufactures. The invading goods, he recognized, had been made in highly mechanized American factories.[33]

For Joseph V. Woodworth, author of *American Tool Making and Interchangeable* **1905** *Manufacturing,* the "development of the modern system of manufacturing" was not only that of interchangeable manufacturing; it was also American. Nevertheless, he did not highlight the American system of manufactures.[34]

Joseph W. Roe, a Yale professor of mechanical engineering, provided modern histo- **1914** rians with their understanding of the meaning and origins of the American system when he published "Development of Interchangeable Manufacture" in *American Machinist.* Though of French origins, interchangeability was taken over by gunmakers in the United States. Therefore, wrote Roe, the "system of interchangeable manufacture is generally considered to be of American origin. In fact, for many years it was known in Europe as the 'American System' of manufacture." In the same article, Roe repeated that the interchangeable system "was known everywhere as 'the American system.' "[35]

Two years later, in 1916, Roe reprinted his *American Machinist* article on interchange- **1916** able manufacture as a chapter in his book, *English and American Machine Tool Builders.* But Roe made a minor addition to the original article by including a section from James Nasmyth's autobiography, quoted above, to buttress his argument that interchangeable manufacture "was known everywhere as 'the 'American system.' " Roe even put

Nasmyth's words, "introduce the American system," in italics.[36] The term was now complete with meaning, authority, and definition, as evidenced by subsequent citations of and work by Roe.

1923   Guy Hubbard's series, "Development of Machine Tools in New England," published in the *American Machinist,* relied heavily upon Roe's work. As Hubbard concluded in his initial article, "New England . . . gave to the world the interchangeable, or American system of manufacture." Eli Whitney's work in particular "became known as the American system."[37]

1924   L. P. Alford also drew directly on Joseph Roe's book for his short history of the "so-called 'American' System" of manufacture that he included in his article, "Duplicate and Interchangeable Manufacture." Alford also laid out and extended the principles of the American system of manufactures.[38]

1937   Joseph W. Roe laid the capstone to his earlier treatment of the American system when he delivered a Newcomen Society of North America address on interchangeable manufacture, which was first published by *Mechanical Engineering.* Upon the recommendation of the Board of Ordnance in 1854, Roe argued, "the British government resolved to introduce 'the American System.'" (At this point, he cited the Nasmyth autobiography.) He also concluded that interchangeable manufacture "constitutes one of the greatest contributions of the machine age, and we can be proud that when the British government introduced it into England they called it 'the American System.'"[39]

Singer Sewing Machine Artifactual Analysis

W hile I was a predoctoral fellow of the National Museum of History and Technology (now the National Museum of American History), Smithsonian Institution, I took the opportunity to inspect visually a number of Singer New Family sewing machines made as early as 1865 and as late as 1884–85. Specifically, I studied machines with the following serial numbers: 87104 (ca. 1865); 106092 (ca. 1865); 459,834 (ca. 1870); 1038977 (ca. 1872–73); 1081835 (ca. 1872–73); and 5235877 (ca. 1884–85).*

Simple, visual comparison of the two machines made about 1865 revealed profound and easily recognizable differences in their parts, suggesting that standardized manufacturing techniques had not yet been adopted at Singer. The parts of one machine are finely finished, whereas those of the other are more coarsely made. Some of the individual castings are different. These variations are clearly documented in Figures A.1 through A.8.

One other major observation about these 1865 machines is significant. Many of their parts were stamped with the serial number of the machine in which they were found, suggesting that they had been custom-fitted into the machine. These numbers are evident in Figures A.1 through A.8.

Aided by William Henson of the Museum's Technical Laboratory, I attempted to swap parts of the two machines made about 1872 or 1873 (I was given permission to do so because these sewing machines had never been accessioned or cataloged). Generally, the parts would not interchange. Neither the vertical nor the horizontal shafts would interchange, nor would the beveled gears that run on these shafts. As with the 1865 Singer machines, the 1872–73 artifacts revealed how the gears and thrust bearings were originally fitted into the machine. (See Figure A.9.) The workman fitted the gear correctly on the shaft, tightened down a set screw, drilled a tapered hole through the gear and shaft, stamped the correct serial number on the gear so that after finishing it, it could be put back in the machine to which it had been fitted, and finally took the gear off the shaft and permanently removed the set screw. The head of the machine was then taken to the japanning room for painting and ornamentation. After these processes had been completed, the head and the parts were reunited.

The shuttle carriers of the 1872–73 sewing machines would not interchange. More-

---

*Dates for these machines are based on the estimates in Grace Rogers Cooper, *The Sewing Machine,* p. 114.

FIGURE A.1. Underside of Singer New Family Sewing Machine, Serial Number 87104, ca. 1865. The overall workmanship of this machine lacks uniformity and careful regulation. For example, the slot in the head of one of the four screws which holds the arm and base of the sewing machine together was cut both unstraight and off center. Timing marks on the counterbalanced crank are also visible.

FIGURE A.2. Detail of Figure A.1. The serial number has been stamped on an adjustment plate for the four-motion feed mechanism. This is one of several places on the machine where the number appears.

FIGURE A.3. Needle Bar and Presser Foot Bar of New Family Machine Number 87104. The photograph shows the serial number stamped in the presser foot bar.

FIGURE A.4. Underside of Singer New Family Sewing Machine, Serial Number 106092, ca. 1865. Compare the overall workmanship, especially on the counterbalanced crank, with that in Figure A.1. Also, compare the manner in which the cloth feed mechanism is recessed into the bed with that in Figure A.1.

FIGURE A.5.   Detail of Figure A.4. Serial numbers are stamped on several of the parts.

FIGURE A.6.   Detail of Needle Bar, Needle Bar Cam, Presser Foot Bar, and Faceplate, Machine Number 106092. Note the evidence of rough machining on the cam and on the face plate. The serial number on the needle bar is also evident.

FIGURE A.7. Detail of Bevel Gears, Machine Number 106092. This photograph illustrates that the serial number of the machine was also stamped on the bevel gears of the New Family sewing machine. (See Figure A.9 for a detailed view of how these gears were fitted into the machine.) The grooved cam on the right was not originally part of the machine; it was installed to drive an attachment, whose function I am not sure of.

FIGURE A.8. Detail of Needle Bar, Needle Bar Cam, Presser Foot Bar, and Faceplate, Serial Number 360834, ca. 1870. (This is a sub-serial number of the main serial number 459834. See the reference in note 1 for details on Singer's two-number system.) This photograph is included for the sake of comparison with the 1865 machine illustrated in Figure A.6. Although this mechanism has been altered slightly for some sort of attachment, it still suggests the level of workmanship in 1870.

FIGURE A.9.   Detail of Horizontal Shaft and Bevel Gear of New Family Sewing Machine Number 103977, ca. 1872–73. This photograph illustrates the way the bevel gears and thrust bearings were fitted into the sewing machine. Two holes are visible in the bevel gear. When the gear was originally fitted onto the shaft, only the top hole existed, which is threaded. This hole held a set screw, which, after the shaft had been turned down to the correct size for the particular gear, was tightened down to hold the gear (the slight indentation mark of the set screw on the shaft is visible). When the gear was temporarily secured, the workman drilled a tapered hole through both the gear and the shaft. The machine was then disassembled so that the head could be painted and decorated. When the machine was reassembled, the gear was secured on the shaft only by means of the tapered pin.

FIGURE A.10.   Detail of Crank or Cam Mechanism, Which Operates the Grooved Cam of the Needle Bar, Machine Number 1038977. This was one of the many parts of the two Singer sewing machines made about 1872 or 1873 which did not interchange. It was clear from the attempt to interchange these parts that originally they were custom-fitted to the shaft using a lathe to turn down the shaft to the "snug fit" size.

FIGURE A.11.   Underside of Singer New Family Sewing Machine, Serial Number 5235877, ca. 1884 or 1885. Compare the details of construction and the general level of workmanship with those of Figures A.1 and A.4. Adjustment of the four-motion cloth feed is different. Overall machine work is superior. No serial number is present on any parts.

FIGURE A.12.   Detail of Bevel Gears of Machine 5235877. This photograph shows that as late as 1884–85 Singer was still fitting its bevel gears into its machine with the same methods used in the 1860s.

over, the crank devices that operated the needle bar cams would not come close to interchanging. (See Figure A.10.) And, finally, each of the needle bar faceplates, although interchangeable, fitted onto the other machine only in the crudest manner. When switched, the plates did not properly align with the arm of the machine. Each was off by about one-thirty-second of an inch, resulting in a visually incorrect fit. In comparison with the two 1865 machines, these early 1870s machines possessed greater visual uniformity in overall appearance, suggesting that production techniques had become more standardized at the Singer factory. Still, however, the critical fitting parts of each machine all bear the same serial number.

The sewing machine made in 1884 or 1885 possessed no serial numbers except on the designated plate located on the base of the machine. (See Figures A.11 and A.12.) Gears, linkages, shafts, the shuttle race, and the pitman all lack numbers, and their finish is remarkably uniform. Filing marks and tool marks are not as apparent as with the earlier machines. The lack of numbers suggests that the parts of the Singer New Family machines were perhaps more fully interchangeable by 1884 or 1885, but certainly the manuscript evidence discussed in Chapter 2 suggests otherwise. The critical test would be to try to interchange parts of two New Family machines made during 1884–85, but at the Smithsonian Institution this was not possible because the museum did not have a second sewing machine of the same vintage.

Notes

Introduction

1. Henry Ford, "Mass Production." Unfortunately, the letter to Ford requesting the article is not in the Ford Archives, Dearborn, Michigan. Related correspondence survives, however. See especially C. A. Zahnow to W. J. Cameron, October 29, 1925, Acc. 285, Henry Ford Office, Box 359; W. J. Cameron to L. J. Thompson, September 28, 1953, Acc. 1, Fair Lane Papers, Box 89. The latter file contains much of the supporting contemporary documentation for this article.

2. Cameron to Thompson, September 28, 1953.

3. *New York Times*, September 19, 1926, sec. 10, p. 1; editorial, "The Super-Factory System," ibid., sec. 2, p. 1.

4. I have found the term *mass production* in print before 1925 in W. E. Freeland, "Mass Production at the Winchester Shops," *Iron Age* 101 (1918): 616–21; "Mass Motor Car Works," *Engineer* 128 (1919): 627–28; and Henry Obermeyer and Arthur L. Greene, "Mass Production in British Motor Industries," *American Machinist* 57 (1922): 524–26. In 1925 a British journal entitled *Mass Production* began publication, but its title was soon changed to *Fuel Economist*. Carter Goodrich used the term in his 1925 work, *The Miner's Freedom*, pp. 3, 12, 105, 169, and mentioned "mass production as efficient as that at Ford's." In *The Way Out* (1925), Edward A. Filene equated "Fordism" or "Fordizing" with "mass production" and argued that this was indeed "the way out," the way of the future. (See Chapter 8 below.) I suspect but cannot prove that Filene's book stimulated the *Britannica*'s editor to request Ford's article on mass production. The only standard work on the English and American language to deal with *mass production*, Raymond Williams, *Keywords* (pp. 161–62) identifies the term as from the United States in the 1920s, thus supporting the notion of the importance of the Ford article. I am indebted to George Basalla for this reference.

5. Roger Burlingame, *Engines of Democracy*, pp. 391, 395. For a more recent restatement of this thesis, see Edwin A. Battison, *From Muskets to Mass Production*, p. 3.

6. An introduction to the American system may be found in Nathan Rosenberg, ed., *The American System of Manufactures;* Merritt Roe Smith, *Harpers Ferry Armory and the New Technology;* and Otto Mayr and Robert C. Post, eds., *Yankee Enterprise.*

7. Battison, "Eli Whitney and the Milling Machine"; Ferguson, *Bibliography of the History of Technology*, p. 299.

8. Merritt Roe Smith, "Eli Whitney and the American System of Manufacturing." See also Robert A. Howard, "Interchangeable Parts Reexamined," and Chapter 1 below. As will be noted below, Samuel Colt's work can be seen in much the same light as that of Eli Whitney.

9. *Harpers Ferry Armory;* esp. chaps. 7 and 8.

10. See Merritt Roe Smith, "Military Entrepreneurship," in Mayr and Post, eds., *Yankee Enterprise*, pp. 63–102.

11. See Chapter 1 below.

12. *Report of the Committee on the Machinery of the United States of America*, in Rosenberg, ed., *American System of Manufactures*, pp. 121–22.

13. Major James Dalliba, quoted in Smith, *Harpers Ferry Armory*, p. 109. On the relative costs of small arms, see Felicia Johnson Deyrup, *Arms Makers of the Connecticut Valley*, p. 132; Deyrup discusses costs, pp. 52, 118–19, and Table 1, Appendix B, pp. 229–32. The figures for Springfield Armory's production of muskets, rifles, and carbines, 1795–1870, are given in ibid., p. 233. Anderson is quoted in Rosenberg, *American System of Manufactures*, pp. 65–66.

14. Rosenberg, "Technological Change in the Machine Tool Industry, 1840–1910."

15. Ottilie M. Leland, *Master of Precision*.

16. See Chapters 2 and 4 below. One notable exception to the prevailing interpretation of Singer's success is Andrew B. Jack, "The Channels of Distribution for an Innovation." See also Elizabeth M. Bacon, "Marketing Sewing Machines in the Post–Civil War Years."

17. I. M. Singer & Co. [Edward Clark] to William F. Proctor, July 16, 1855, Papers of the Singer Manufacturing Company, Box 189.

18. On Singer's marketing strategy see Robert Bruce Davies, *Peacefully Working to Conquer the World*, and Alfred D. Chandler, Jr., *The Visible Hand*, pp. 303–5, 402–5.

19. Spalding, "The 'American System' of Manufacture," p. 11.

20. Minutes of a meeting held at Elizabethport Factory, March 26, 1883, Singer Papers, Box 239.

21. A. D. Pentz to G. R. McKenzie, August 17, 1884, ibid., Box 198.

22. "Mass Production," p. 822.

23. David A. Hounshell, "Public Relations or Public Understanding?" and Chapter 4 below.

24. Chandler, *Visible Hand*, pp. 406–8.

25. David A. Hounshell, "The Bicycle and Technology in Late Nineteenth Century America," and Chapter 5 below.

26. Joseph V. Woodworth, *American Tool Making and Interchangeable Manufacturing*, p. 516.

27. George Pope to David Post, January 12, 1891, Papers of the [Pope] Hartford Cycle Company.

28. See Henry Ford to the editor, *Automobile* 14 (January 11, 1906): 107–9, quoted in John B. Rae, ed., *Henry Ford*, pp. 18–19.

29. *Oxford English Dictionary*.

30. Drucker, *Concept of the Corporation*, pp. 219–20. See Chapter 6 below.

31. See Horace L. Arnold and Fay L. Faurote, *Ford Methods and the Ford Shops*, for figures on productivity gains; Allan Nevins and Frank Ernest Hill, *Ford: The Times, the Man, the Company*, p. 447.

32. Reminiscences, Ford Archives.

33. This is the general interpretation of Eugene S. Ferguson, *Oliver Evans*, pp. 13–28.

34. Henry Ford with Samuel Crowther, *My Life and Work*, pp. 129–30.

35. Stephen Meyer III, *The Five Dollar Day*.

36. See the article on Gunnison in the *Dictionary of American Biography*, 7th Suppl. (1981), and Chapter 8 below. For earlier efforts at prefabrication, see Gilbert Herbert, *Pioneers of Prefabrication*.

37. This mission may be followed in the pages of *Mechanical Engineering* between 1919 and 1925, the date when the Wood Industries Division was established.

38. See, for example, *New York Times*, March 7, 1930, p. 48.

39. See Chapter 3 below.

40. *A Crack in the Rear View Mirror: A View of Industrialized Building* (New York: Van Nostrand, 1973), quoted in William J. Abernathy, *The Productivity Dilemma*, p. 8.

41. See Chapter 7 below.

42. Anne O'Hare McCormick, "The Future of the Ford Idea," p. 1.

43. Several dozen books on "economics of consumption" were published in the 1930s, each suggesting in one way or another that getting the consumer straightened out would lead to the return of good times. One can begin with the following: E.E. Cal-

kins, "The New Consumption Engineer and the Artist"; Harry Tippen, *The New Challenge of Distribution: The Paramount Industrial Problem* (New York: Harper & Brothers, 1932); John B. Cheadle, et al., *No More Unemployed* (Norman, Okla.: University of Oklahoma Press, 1934); Lewis Corey, *The Decline of American Capitalism* (New York: Covici, 1934); Maurice Leven et al., *America's Capacity to Consume* (Washington, D.C.: Brookings Institution, 1934); William H. Lough, *High Level Consumption: Its Behavior; Its Consequences* (New York: McGraw-Hill, 1935); Carle C. Zimmerman, *Consumption and Standards of Living* (New York: D. Van Nostrand Co., 1936); Charles S. Wyand, *The Economics of Consumption* (New York: Macmillan, 1937); Elizabeth Ellis Hoyt, *Consumption in Our Society* (New York: H. Holt, 1938); Roland Snow Vaile, *Income and Consumption* (New York: H. Holt, 1938); and Alfred P. Sloan, Jr., *The Creation of Abundance*, pamphlet, March 11, 1939, Eleutherian Mills Historical Library.

44. See Abernathy, *Productivity Dilemma*.

Chapter 1

1. Arthur C. Cole, "American System."

2. Eugene S. Ferguson, *Bibliography of the History of Technology*, p. 298. Ferguson later gave a much broader definition in his Dibner symposium paper, "History and Historiography," in Otto Mayr and Robert C. Post, eds., *Yankee Enterprise*, pp. 1–23, esp. pp. 14–15.

3. Nathan Rosenberg, ed., *The American System of Manufactures*, p. 5. Anderson's wording in *The Report of the Committee on the Machinery of the United States* in ibid. is on p. 143 and as follows: p. 89 ("system" the committee used in writing its report); p. 100 ("new system of casting" metal); p. 104 ("same system of special machinery"); p. 113 ("system" of testing quality of castings); p. 114 (James's rifle hand-made "system" of production); p. 119 (study of "the system, machinery, and apparatus by which" small arms were made in the United States); p. 128 ("System, Machinery, and Apparatus by means of which Small Arms are produced"); p. 128 ("an almost perfect system of manufacture"); p. 129 ("the thorough manufacturing system"); p. 129 ("system of forging gun barrels"); p. 135 ("clamp milling system"); p. 136 (interchangeable methods cannot be understood "by those who have been engaged on a ruder system"); p. 143 ("American system"); p. 159 ("system of moulding and casting brass guns"); p. 159 ("our system"); p. 162 ("system of levers"); p. 160 ("ordinary system of boring"); p. 169 (tub- and pail-making—"a very perfect sys-

tem''); p. 170 (''A similar system'' used for furniture); p. 172 (''system of combining a pendulum saw with a planing machine''); p. 172 (''system of smoothing and polishing wood''); p. 172 (''system of bending timber''); p. 173 (''system of clamp milling''); p. 175 (''Dahlgren's book on 'System of Armament for Boats' ''); p. 175 (''whole system'' of clock manufacture); p. 193 (''systematic arrangement in the manufacture''); p. 193 (''admirable system everywhere adopted''); p. 194 (''system of paying by the piece''); p. 195 (labor ''affecting the system of the whole manufactory); p. 195 (''This [piece-work] system''); p. 195 (''system of contracting for machinery in ordinary use''); p. 196 (information in this report being ''arranged, systematized, and copied out hurriedly'').

4. Joseph W. Roe, *English and American Tool Builders*, pp. 129, 140–41. There is little doubt that Roe obtained his views from Charles H. Fitch's undocumented ''Report on the Manufactures of Interchangeable Mechanism.'' Fitch equates the ''American system'' and the ''interchangeable system'' and implies that the term came into use around the time of the 1851 London Crystal Palace Exhibition (pp. 618–20). Cf. Roe's later account, ''Interchangeable Manufacture,'' p. 758.

5. John E. Sawyer, ''The Social Basis of the American System of Manufacturing''; H. J. Habakkuk, *American and British Technology in the Nineteenth Century*, pp. 104, 120; Rosenberg, ed., *American System of Manufactures;* and Paul Uselding, ''Studies of Technology in Economic History,'' pp. 160–61, 168–78. In his prefatory discussion of the American system Uselding states: ''The story of American prowess at the Crystal Palace, the displays of the Colt Revolver, the McCormick Reaper, the Singer Sewing Machine—all embodying the principle of interchangeability—is now a familiar tale'' (pp. 160–61). This statement is indicative of the legendary character of the term. In this regard, it is significant that D. L. Burn, whose 1931 article inspired Sawyer's seminal article, did not use the expression *American system of manufactures*, nor did a 1900–1901 *Times* (London) study of American engineering competition (Burn, ''The Genesis of American Engineering Competition, 1850–1870''; *Times* (London), *American Engineering Competition*).

6. Great Britain, Parliament, *Report from the Select Committee on Small Arms*, p. iii. See also the Introduction in Rosenberg, ed., *American System of Manufactures*, pp. 29–72.

7. The contrast is painted in Charles Tomlinson, ed., *Rudimentary Treatise on the Construction of Locks*, pp. 154–63. See Rosenberg's description of the British contract system in his introduction to *American System of Manufactures*, p. 39.

8. *Report on Small Arms*, p. iv.

9. The entire list of witnesses appears in ibid., p. xxxvi.

10. Anderson's report on U.S. machinery clearly illustrates his advocacy; see *The Report of the Committee on the Machinery of the United States of America*, in Rosenberg, ed., *American System of Manufactures*, pp. 87–197.

11. *Report on Small Arms*, Q. 1010, p. 79; Q. 341, p. 27. In a different report written for the Board of Ordnance's Committee of Small Arms, a special committee pointed out in January 1854: ''The principle of substituting machinery for labour is established beyond a doubt, and its application to the manufacture of small arms can no longer be disputed; *but if any proof were wanting in this respect, it will be found by a visit to Colonel Colt's manufacture of the American revolving pistol*'' (testimony of Lieutenant-Colonel James S. Tulloh, *Report of the Select Committee on Small Arms*, Q.443, p. 42; italics added). Sir Thomas Hastings, a member of the Board of Ordnance and its comptroller of stores, testified before the Select Committee that he had opposed the idea of a government-owned, mechanized armory. Then he saw Colt's London armory and ''became a convert'' to this system (Q. 205, p. 15). Anderson's testimony is in Q. 1266, p. 99, and Q. 1404–08, p. 110. Nasmyth's testimony is on pp. 107–24; Whitworth's, pp. 138–52; Prosser's, pp. 172–84.

12. Ibid., pp. 122–24; Nathan Rosenberg, ''Technological Change in the Machine Tool Industry, 1840–1910.''

13. *Report on Small Arms*, Q. 927, p. 73. The committee sought information on manufacturing costs at the Springfield and Harpers Ferry armories but found none conclusive. There was almost universal sentiment that labor was ''more dear'' in the United States than in England, although Colt testified that English labor was costing him almost as much as American labor (ibid., Q. 1093, p. 85). See Russell I. Fries, ''British Response to the American System.''

14. *Report on Small Arms*, Q. 932, p. 74; Q. 7415, p. 420.

15. Report on the Committee on Patents to accompany H.R. 59, December 21, 1853, as quoted in ibid., p. 419.

16. Ibid., Q. 7418, p. 420.

17. Nasmyth quoted in Whitworth's testimony, ibid., Q. 2205, p. 150. Colt testified that he made his pistol ''entirely'' by machinery with some minor exceptions (ibid., Q. 1084, p. 85).

18. Ibid., Q. 2780, p. 178. The London gunmaker Charles Clark argued with Prosser, saying that at least 50 percent hand labor would be required and that labor would have to be first-rate (Q. 7173, p. 400). Board of Ordnance report quoted, Q. 443, p. 42.

19. Ibid., Q. 7432, Q. 7433, Q. 7439, and Q.

7440, p. 421; Q. 7511, pp. 424–25. On Stickney's qualifications, see Q. 7419–26, p. 421. When questioned closely, Colt had said that his machines "save me more than one-half the entire cost of the arm. I would say that it enables me to produce almost three times as many arms for the same money" (Q. 1166–67, p. 92). For the extent that the Board of Ordnance planned to displace labor, see Anderson's testimony, Q. 933–37, p. 74.

20. Ibid., Q. 7723 and Q. 7724, p. 436. Charles Clark agreed with Richards (Q. 7168, p. 400).

21. Ibid., Q. 2131, p. 147, Q. 2148, p. 148; Q. 1570, p. 118. The Board of Ordnance's position, essentially one of "good enough," was expressed by member Sir Thomas Hastings, Q. 1348–52, p. 149.

22. Ibid., Q. 1089, p. 85; Q. 1114 and Q. 1115, p. 87.

23. Colonel Boldero, ibid., Q. 1198, p. 94.

24. Ibid., Q. 1057 and Q. 1058, p. 82; Q. 1357 and Q. 1359, p. 107; Q. 1361, Q. 1362, and Q. 1365, p. 108.

25. Ibid., Q. 1944 and Q. 1946, p. 138. In its final report, the select committee included Whitworth's views on the impossibility of interchangeability (ibid., pp. v–vi).

26. Ibid., Q. 2036, p. 143. Richard Prosser agreed on precisely the same grounds (Q. 2788, and Q. 2804, p. 179). Ibid., Q. 2037, p. 179; Q. 2095, p. 146. See also, Q. 2096, p. 146.

27. Ibid., Q. 2121, p. 147; Q. 2124, p. 147. See George Wallis's conflicting testimony on the issue of the interchangeability of Springfield muskets, Q. 1700–1704, Q. 1727, pp. 125–26, 127.

28. Ibid., Q. 1525, p. 116. See other of Nasmyth's answers to questions about uniformity: Q. 1514–27, Q. 1565–70, and Q. 1638, pp. 116, 118–20, 122.

29. Q. 1120, p. 87; Q. 1121, p. 87; Q. 1116, p. 87; Q. 1166, p. 92; Q. 7725, p. 436. Charles Clark had also purchased the "highest class" of Colt's pistols and concluded that "they will not reverse or interchange; you cannot get two to do so" (Q. 7175, p. 400).

30. Ibid., Q. 7446, Q. 7453, Q. 7454, pp. 421–22. Not long after the select committee finished its investigation a writer for Charles Dickens's weekly journal, *Household Words,* wrote that Colt's London-made revolvers were distinguished by their "uniformity": "No one part belongs, as a matter of course, to any other part of one pistol; but each piece may be taken at random from a heap, and fixed to and with the other pieces until a complete weapon is formed" ("Revolvers," *Household Words* 9 (1855), as reprinted in Charles T. Haven and Frank A. Belden, *A History of the Colt Revolver,* p. 348. The select committee's final report is inconclusive on the issue (ibid., pp. iii–xi).

31. Ibid., Q. 1367, p. 108; Q. 2074–75, p. 145. Compare Whitworth's testimony to his *Special Report* in Rosenberg, ed., *American System of Manufactures,* p. 343.

32. *Report on Small Arms,* Q. 2646, p. 172; Q. 2790, p. 179. The book Prosser showed the committee was Joseph Gamel, *Description of the Tula Weapon Factory in Regard to Historical and Technical Aspects,* which has been translated for the Smithsonian Institution and the National Science Foundation by Franklin Book Programs, Inc., 1975. Edwin A. Battison has edited the translation, which will be published by the Smithsonian Institution. For more information, see Edwin A. Battison, "Searches for Better Manufacturing Methods, Section Two." James Nasmyth had earlier pointed out what he had done for the Russian factory at Tula (*Report on Small Arms,* Q. 1377, Q. 1404–8, pp. 109–10).

33. *Report on Small Arms,* Q. 2853, p. 181. On the centrality of the Springfield stocking machinery, see Anderson's testimony in which he quotes his report of December 23, 1853, to the Board of Ordnance Committee on Small Arms, ibid., p. 27; Whitworth's testimony, Q. 1969, p. 140; and Wallis's testimony, Q. 1777 and Q. 1778, pp. 127, 129.

34. Ibid., p. x.

35. Discussion of John Anderson, "On the Application of Machinery in the War Department," p. 164.

36. For the economic approach see H. J. Habakkuk, *American and British Technology,* which spawned an entire body of literature on the subject of American labor-saving technology. This literature is reviewed in Uselding, "Studies of Technology." The social approach may be found in Sawyer, "The Social Basis of the American System of Manufacturing"; see also, Eugene S. Ferguson, "The American-ness of American Technology."

37. Merritt Roe Smith, "Military Entrepreneurship," in Mayr and Post, eds., *Yankee Enterprise,* pp. 66–67; Selma Thomas, "The Greatest Economy and the Most Exact Precision"; see also articles on Gribeauval in *La grande encyclopédie* and *Biographie universelle (Michaud) ancienne et modern.*

38. Smith, "Military Entrepreneurship," pp. 63–102, is the outstanding example of recent scholarship.

39. August 30, 1785, *Papers of Thomas Jefferson,* ed. Julian P. Boyd, 8:455. Jefferson wrote virtually the same account in a letter to the governor of Virginia, January 24, 1786, ibid., p. 214.

40. Thomas, "The Greatest Economy."

41. Jefferson to Knox, September 12, 1789, *Papers of Thomas Jefferson,* 15:422; Jefferson to Knox, November 24, 1790, ibid., 18:69. In France, subsequent to Blanc's *Mémoire,* the Académie Royale des Sciences was requested by the minister of war to re-

view Blanc's work. A committee conducted the study and issued a report to the academy March 19, 1791. This report has been translated and published in William F. Durfee, ''The First Systematic Attempt at Interchangeability in Firearms.'' See also Durfee, ''The History and Modern Development of the Art of Interchangeable Construction in Mechanism.''

42. Other French military engineers who served in the United States during the early national period are discussed in Norman B. Wilkinson, ''The Forgotten 'Founder' of West Point.''

43. Wilkinson, ''The Forgotten 'Founder' of West Point.'' A different view of West Point's establishment (which also emphasizes the French influence) is presented in Peter M. Molloy, ''Technical Education and the Young Republic.''

44. Quoted from Smith, ''Military Entrepreneurship,'' pp. 66, 97.

45. Ibid.

46. In my discussion of Simeon North, I have drawn upon S.N.D. North and Ralph H. North, *Simeon North.*

47. Deyrup, *Arms Makers of the Connecticut Valley,* p. 46; the quotation is in North and North, *Simeon North,* p. 64.

48. North and North, *Simeon North,* p. 81.

49. Ibid., p. 106.

50. Edwin Battison, recently retired curator at the Smithsonian Institution, who knows a good deal about North pistols, has told me that the parts of the first hundred or so pistols that North made under this system would indeed interchange.

51. Merritt Roe Smith, ''John H. Hall, Simeon North, and the Milling Machine,'' pp. 577, 574. North had also developed gun barrel turning machinery by 1816; see Smith, *Harpers Ferry Armory,* p. 117.

52. Whitney quoted in Robert S. Woodbury, ''The Legend of Eli Whitney and Interchangeable Parts,'' p. 238, and in Constance M. Green, *Eli Whitney and the Birth of American Technology,* p. 110; Smith, ''Eli Whitney and the American System,'' p. 50. Although lacking footnotes, Smith's essay is the best source for the current perspective on Whitney.

53. Quoted in Edwin A. Battison, ''Eli Whitney and the Milling Machine,'' p. 20.

54. For only a small sampling of the work of early American mechanicians, see Ferguson, ''The American-ness of American Technology.''

55. Durfee, ''First Systematic Attempt at Interchangeability.''

56. Russell I. Fries, ''A Comparative Study of the British and American Arms Industries, 1790–1890,'' p. 19. See also Gene S. Cesari, ''American Arms-Making Tool Development, 1789–1855,'' p. 79.

57. Whitney delivered his contract weapons as follows: September 1801, 500 muskets; June 1802, 500; March 1803, 500; July–October 1803, 1,000; 1804, 1,000; 1805, 1,000; 1806, 1,500; 1807, 2,000; February 1808–January 1809, 1,500. See Green, *Eli Whitney,* pp. 131–35.

58. Cesari, ''American Arms-Making,'' pp. 36–37, 47–48; Woodbury, ''Legend of Eli Whitney,'' pp. 241–42. Edwin Battison, who is working on a full study of Whitney, agrees with this assessment (Report to the Department of Science and Technology, National Museum of History and Technology, Smithsonian Institution, Spring 1977).

59. Smith, ''Eli Whitney and the American System,'' p. 51; Jefferson quoted in Green, *Eli Whitney,* pp. 133–34.

60. Smith, ''Eli Whitney and the American System,'' p. 58.

61. See Green, *Eli Whitney,* pp. 135–36.

62. On Blanc's methods, see Durfee, ''First Systematic Attempt at Interchangeability.'' On Whitney's methods and the lack of interchangeability, see Battison, ''Eli Whitney and the Milling Machine.'' Cesari, ''American Arms-Making,'' p. 40.

63. Smith, *Harpers Ferry Armory,* pp. 28–29.

64. Output at Springfield, 1795 to 1802, was as follows: 1795, 245; 1796, 838; 1797, 1,028; 1798, 1,044; 1799, 4,595; 1800, 4,862; 1801, 3,205; and 1802, 4,358 (Deyrup, *Arms Makers of the Connecticut Valley,* p. 233).

65. Springfield Armory's early history is treated in Fries, ''Comparative Study of British and American Arms,'' pp. 31–34. Output during Morgan's tenure was as follows: 1803, 4,775; 1804, 3,566; and 1805, 3,535 (Deyrup, *Arms Makers of the Connecticut Valley,* p. 233).

66. Fries, ''Comparative Study of British and American Arms,'' pp. 34–37. Springfield's output statistics in Deyrup, *Arms Makers of the Connecticut Valley,* give an indication of its checkered history, 1806–15.

67. Charles Babbage, *On the Economy of Machinery and Manufactures,* pp. 119–21, 173–74.

68. Smith points out that as early as 1813 Wadsworth had espoused ideas about system and uniformity ''in language reminiscent of Tousard'' (''Military Entrepreneurship,'' p. 68); Smith, *Harpers Ferry Armory,* p. 107. Information in this paragraph has been taken from ibid., pp. 104–7, and from ''Military Entrepreneurship,'' pp. 68–76.

69. Smith, *Harpers Ferry Armory,* p. 107; see also Charles H. Fitch, ''The Rise of a Mechanical Ideal,'' pp. 518–19.

70. The first and last quotes are in Edward Hartwell White, Jr., ''The Development of Interchangeable Mass Manufacturing in Selected American Industries from 1795 to 1825,'' pp. 109, 111; Lee's

complaint quoted in Smith, *Harpers Ferry Armory*, p. 108.

71. For more on inspection of arms, see Smith, "Military Entrepreneurship," pp. 74–76. Dalliba quoted in Smith, *Harpers Ferry Armory*, pp. 109–10. In "Military Entrepreneurship," p. 75, Smith identifies master armorer Adonijah Foot as the innovator of this gauging system.

72. Quoted in Smith, "Military Entrepreneurship," p. 75.

73. Smith, *Harpers Ferry Armory*, p. 109.

74. Formal orders to the master armorer at Springfield, March 24, 1818, quoted in White, "Development of Interchangeable Mass Manufacturing," pp. 113, 115.

75. Smith's discussion of Thomas Blanchard is the best that exists, and I have drawn heavily from it. See *Harpers Ferry Armory*, pp. 124–38. Certainly the British in the 1850s believed that the Blanchard stocking machinery exemplified American manufacturing technology. See the testimony of John Anderson, Thomas Hastings, Lieutenant-Colonel James S. Tulloh, George Wallis, Joseph Whitworth, and Richard Prosser in the *Report on Small Arms*, the *Special Reports* of Joseph Whitworth and George Wallis reprinted in Rosenberg, ed., *American System of Manufactures*, and Anderson, "On the Application of Machinery."

76. "Elegant" in the sense that Eugene S. Ferguson used the expression in "Elegant Inventions."

77. This system was described by Lieutenant-Colonel Tulloh in his testimony before the Select Committee on Small Arms: "In fact, it was something so different from anything I expected, and so beautiful, that I, like everyone else who has seen it, was highly gratified. The consecutive arrangements of the machinery were such as I have never seen in any department before. There were, I think, about 150 machines at work, and those machines were all placed in a kind of consecutive arrangement to produce the pistol of which Colonel Colt is the manufacturer. It seemed to be a kind of stream of work flowing through the manufactory in consecutive order" (*Report on Small Arms*, Q. 413, p. 40).

78. Quoted in Smith, *Harpers Ferry Armory*, p. 125.

79. Ibid.

80. The Brunel system is covered admirably in K. R. Gilbert, *The Portsmouth Block-making Machinery*. Carolyn C. Cooper has given us a more comprehensive view of the systematic arrangement and operation of Brunel's machinery in "The Production Line at Portsmouth Block Mill."

81. The critical importance of Blanchard's invention is recognized in Smith, *Harpers Ferry Armory*, p. 127, and Cesari, "American Arms-Making," p. 247.

82. Smith covers the history of Blanchard's relationship with the national armories in *Harpers Ferry Armory*, pp. 124–38. On inside contracting see John Buttrick, "The Inside Contract System."

83. My discussion of Hall is based almost completely on Smith's book. See esp. pp. 184–251.

84. Ibid., pp. 190–92.

85. Ibid., p. 192.

86. Ibid., p. 143.

87. Ibid., p. 144. The committee's report is reprinted in Claud E. Fuller, *The Breech-Loader in the Service*, pp. 29–32.

88. Quoted in Fitch, "Rise of a Mechanical Ideal," p. 520. The figure of sixty-three gauges appears on p. 523.

89. Hall quoted in Smith, *Harpers Ferry Armory*, p. 227. Samuel Colt emphasized the use of bearing points in his 1851 address before the Institution of Civil Engineers, "On the Application of Machinery to the Manufacture of Rotating Chambered-Breech Fire-Arms, and the Peculiarities of those Arms," p. 45.

90. Smith, *Harpers Ferry Armory*, p. 208.

91. Ibid., pp. 210–11.

92. Ibid., pp. 214–15.

93. Fuller, *The Breech-Loader*, p. 32.

94. *Report on Small Arms*, Q. 413, p. 40.

95. Smith, *Harpers Ferry Armory*, p. 221.

96. Compare Hall's attitudes with those of Samuel Colt (and those of Colt as expressed through Tulloh's testimony) in his testimony before the Select Committee on Small Arms (*Report on Small Arms*, pp. 84–99 [Colt], pp. 39–52 [Tulloh]).

97. Smith, *Harpers Ferry Armory*, p. 249.

98. Ibid., pp. 241–51.

99. These years are covered in Charles Fitch, "Rise of a Mechanical Ideal," and Felicia Deyrup, *Arms Makers of the Connecticut Valley*, pp. 98–159, see esp. p. 132.

100. "Rise of a Mechanical Ideal," pp. 521, 525.

101. Ibid., pp. 526–27. See also the *Report of the Committee on the Machinery of the United States* in Rosenberg, ed., *American System of Manufactures*, pp. 102–3, 117.

102. Deyrup, *Arms Makers of the Connecticut Valley*, p. 119.

103. Ibid., pp. 117–21.

104. Rosenberg, ed., *American System of Manufactures*, p. 66. Colt's claim was corroborated by Anderson's committee in 1854 (ibid., p. 195).

105. Colt tells the history of his early work in "On the Application of Machinery." See also Deyrup, *Arms Makers of the Connecticut Valley*, pp. 144–59; Fries, "Comparative Study of British and American Arms," pp. 57–150; and Cesari, "American Arms-Making," pp. 165–229, 284–90.

106. Colt, "On the Application of Machinery,"

p. 44. Howard argues that the Paterson arms made by Colt were of "high quality workmanship" ("Interchangeable Parts Reexamined," p. 643).

107. B. F. Spalding, "The 'American' System of Manufacture," p. 2.

108. Jack Rohan, *Yankee Arms Maker*, p. 165; Cesari, "American Arms-Making," p. 184. Colt's arrangement with Warner provides the primary documentation for Gage Stickney's claim in his testimony before the Select Committee on Small Arms that "Colonel Colt has copied his [machinery] from Springfield" (*Report on Small Arms*, Q. 7462, p. 422).

109. Cesari, "American Arms-Making," pp. 186–87, 192.

110. See Howard L. Blackmore, "Colt's London Armoury," pp. 171–95, and Rosenberg, ed., *American System of Manufactures*, pp. 44–47.

111. Rohan, *Yankee Arms Maker*, p. 223, and Cesari, "American Arms-Making," p. 197.

112. Paul Uselding, "Elisha K. Root, Forging and the 'American System'." John Anderson regarded the Colt/Root die-forging machinery as "the most perfect thing that I have ever seen," and he celebrated the "almost perfect system" of special-purpose machine tools used sequentially at Colt's London armory (*Report on Small Arms*, pp. 32, 27). Cesari, "American Arms-Making," pp. 215–20. In 1854, Tulloh estimated that Colt's London armory had some 150 machines at work (*Report on Small Arms*, Q. 413, p. 40).

113. On the Colt gauging system, see Cesari, "American Arms-Making," p. 217. James Nasmyth was struck by the Colt gauging methods used at his London armory: "[H]e had a testing machine, to test in every case, whether the part so produced by machinery was brought to the exact size that it ought to be; that was a remarkable part of the system, those criticising machines: little instruments that were in every department, and the part, whatever it was, was placed in different directions, with parts that came over it, and they could detect the one-thousandth of an inch difference; and in order to keep this machine perfectly correct, he had a duplicate machine kept in store as a standard to refer to, in case the workmanship of the shop standard getting out of order" (*Report on Small Arms*, Q. 1527, pp. 116–17).

114. Throughout his "Comparative Study of British and American Arms," Russell Fries implies that interchangeability lowered cost. Fries explicitly argues that interchangeability reduced costs in the antebellum period in his article, "British Response to the American System." Howard, "Interchangeable Parts Reexamined," pp. 633–47, provides evidence that absolute interchangeability raised costs, not lowered them.

115. "Mass Production," p. 822.

116. Testimony of Gage Stickney, *Report on Small Arms*, Q. 7447–51, p. 442, and Cesari, "American Arms-Making," p. 204. For a recent analysis of the entire concept of interchangeability with emphasis on Colt, see Howard, "Interchangeable Parts Reexamined," pp. 633–47.

117. Colt, "On the Application of Machinery," p. 45.

118. Howard concludes: "Looking at the total record of American arms production, it becomes evident that the use of machine tools was of paramount importance, but the interchangeable part was not achieved to any great measure" ("Interchangeable Parts Reexamined," p. 649).

119. Roe, *English and American Tool Builders*, p. 170. The inside contracting system has been touched on by many authors; the best account is Buttrick, "Inside Contract System." Other connections among these mechanics are represented graphically by Guy Hubbard, "Development of Machine Tools in New England," pp. 2–3, who demonstrates that many of Colt's inside contractors had been apprentices and workmen of the Robbins & Lawrence Co. See, for example ibid., pp. 617–20.

120. Buttrick, "Inside Contract System," pp. 205–7.

121. An expression used by Robert James Walker in commenting upon Samuel Colt's 1851 address before the Institution of Civil Engineers, "On the Application of Machinery," p. 65.

122. See Ferguson, "The American-ness of American Technology."

123. This conclusion is based on a perusal of *Scientific American*, 1844–50. See also Peter Haddon Smith, "The Industrial Archeology of the Wood Wheel Industry in America," pp. 54–57.

124. Babbage, *On the Economy of Machinery*, pp. 176–90.

125. Chauncey Jerome, *History of the American Clock Business for the Past Sixty Years*, p. 92. For the early American clock industry, I have relied principally upon John Joseph Murphy, "Entrepreneurship in the Establishment of the American Clock Industry," and Murphy, "The Establishment of the American Clock Industry." White, "Development of Interchangeable Mass Manufacturing," also includes a chapter on the American clock industry. See also the thorough study of Terry by Kenneth D. Roberts, *Eli Terry and the Connecticut Shelf Clock*, and Roberts, *Some Observations Concerning Connecticut Clockmaking, 1790–1850*.

126. Murphy, "Establishment of the American Clock Industry," pp. 32–42.

127. Fitch, "Report on the Manufactures of Interchangeable Mechanism," p. 684; Murphy, "Establishment of the American Clock Industry," pp. 62–63, 60. Henry Terry noted in his *American Clock*

Making that Yankee clockmakers were "fitting wheels and different parts to their proper place in each clock and putting one at a time in running order" (quoted in Murphy, "Establishment of the American Clock Industry," p. 89).

128. See Murphy's admirable chapter, "Capturing a Rural Market," in "Establishment of the American Clock Industry," pp. 114–74.

129. Ibid., pp. 96, 98, 104–6, 112.

130. Murphy relies primarily on Terry, *American Clock Making*, and Hiram Camp, *Sketch of the Clock Making Business, 1792–1892*. On gauges, see Murphy's long quotation from Camp in "Establishment of the American Clock Industry," pp. 89–90.

131. David Pye, *The Nature and Art of Workmanship*, pp. 4–8.

132. W. G. Lathrop, "The Development of the Brass Industry in Connecticut," p. 12; Murphy, "Establishment of the American Clock Industry," pp. 178–79. On Ives, see also Kenneth D. Roberts, *The Contributions of Joseph Ives to Connecticut Clock Technology, 1810–1862*.

133. *Special Report of Joseph Whitworth*, in Rosenberg, ed., *American System of Manufactures*, p. 341.

134. Eugene Ferguson has often noted that clocks require only very rough uniformity compared to firearms. Perhaps it is this difference in the requirement of uniformity or precision that determined the difference of approaches to production between firearms makers and clock producers. Robert Howard concurs, arguing that "in the making of clocks . . . the required level of precision was generally very low." ("Interchangeable Parts Reexamined," p. 633).

135. Murphy, "Establishment of the American Clock Industry," p. 197, 201–2, 219.

136. Quoted in ibid., p. 210.

137. Ibid., pp. 184–85.

138. Both Whitworth and the Anderson committee visited Connecticut clock factories, notably that of Jerome. See *Special Report of Joseph Whitworth*, in Rosenberg, ed., *American System of Manufactures*, pp. 341–42, and *Report of the Committee on the Machinery of the United States*, in ibid., pp. 104, 161, 175; testimony of John Anderson, *Report on Small Arms*, Q. 1064, p. 82.

139. *Report of the Committee on the Machinery of the United States*, in Rosenberg, ed., *American System of Manufactures*, p. 92; see "Journal of the Committee" in ibid., pp. 98–118.

140. Rosenberg, ed., *American System of Manufactures*, pp. 1–86.

141. *Special Report of Joseph Whitworth*, in ibid., p. 387.

142. *Report on Small Arms*, Q. 2043, p. 144. American opinions on the issue of American versus British machine tools differed. George Escol Sellers believed that as early as 1832 American machine tools

excelled those of the British (Eugene S. Ferguson, ed., *Early Engineering Reminiscences (1815–40) of George Escol Sellers* [Washington, D.C.: Smithsonian Institution, 1965], pp. 109–12). The armsmaker Joshua Stevens argued that as late as 1849, however, "we were obliged to buy most of our machinery in England, as comparatively little progress had been made in this country in the manufacture of fine tools of that character" ("Sixty Years a Mechanic," *Machinery* 1 [October 1894]: 4).

143. *Special Report of Joseph Whitworth*, in Rosenberg, ed., *American System of Manufactures*, p. 343.

144. Ibid., p. 365. Whitworth's report, with its extended discussion of special-purpose machine tools in America and scant attention to interchangeability, led me to overstate that the so-called "American system of manufactures" referred to special-purpose machinery rather than machine-made, interchangeable parts manufacture. See David A. Hounshell, "The System: Theory and Practice," in Mayr and Post, eds., *Yankee Enterprise*, and "From the American System to Mass Production," chap. 1.

145. *Special Report of George Wallis* in Rosenberg, ed., *American System of Manufactures*, p. 261.

146. *Report of the Committee on the Machinery of the United States*, in Rosenberg, ed., *American System of Manufactures*, pp. 121–22.

147. *Report on Small Arms*, Q. 341, pp. 27–28.

148. *Report of the Committee on the Machinery of the United States*, in Rosenberg, ed., *American System of Manufactures*, pp. 143, 187, 136.

149. Ibid., p. 193

150. Ibid., pp. 104, 193–94.

151. John Anderson, *General Statement of the Past and Present Condition of the Several Branches of the War Department* (London: HMSO, 1857), quoted in Rosenberg, ed., *American System of Manufactures*, pp. 65–66.

152. John Anderson, "On the Application of Machinery," p. 157.

153. *Report of the Committee on the Machinery of the United States*, in Rosenberg, ed., *American System of Manufactures*, p. 193.

154. Ibid., p. 107.

155. The phrase "high price of labour" appears in ibid., p. 91. Throughout the reports of Whitworth, Wallis, the Committee on the Machinery of the United States, and the Select Committee on Small Arms, there is universal agreement that labor was "more dear" in the United States than in England and that for this reason mechanization had arisen in the United States.

Chapter 2

1. The best history of this early period is Frederick G. Bourne, "American Sewing-Machines." See

also Grace Rogers Cooper, *The Sewing Machine,* pp. 3–42.

2. See production figures for the Singer and Wheeler and Wilson companies in Bourne, "American Sewing-Machines," p. 530. General company history of Wheeler and Wilson is based on Bourne, "American Sewing-Machines"; Cooper, *The Sewing Machine;* [Wheeler and Wilson Manufacturing Company], *The Sewing Machine; Dictionary of American Biography,* s.v. "Wheeler, Nathaniel"; *National Cyclopaedia of American Biography,* s.v. "Wheeler, Nathaniel" and "Wilson, Allan Benjamin"; and "Wheeler and Wilson Manufacturing Co.: Nathaniel Wheeler and A. B. Wilson," in J. D. Van Slyck, *New England Manufacturers and Manufactories,* 2: 672–82. I have found no manuscripts surviving from the Wheeler and Wilson Manufacturing Company.

3. Cooper, *The Sewing Machine,* p. 29; "Wilson Improved Patent Sewing Machine."

4. Of the capital, $100,000 was assigned to the Wilson patent and $60,000 to the factory (Van Slyck, *New England Manufacturers,* 2: 678, 682). For Woodruff's later career, see Robert Bruce Davies, *Peacefully Working to Conquer the World,* p. 39.

5. Van Slyck, *New England Manufacturers,* 2: 682.

6. Ibid., p. 679.

7. Guy Hubbard, "Development of Machine Tools in New England," p. 877; and B. F. Spalding, "The 'American' System of Manufacture," p. 11.

8. Hubbard, "Development of Machine Tools," p. 877.

9. [Wheeler and Wilson], *The Sewing Machine,* pp. 10–16.

10. Ibid., p. 21.

11. Ibid., p. 24.

12. Ibid., p. 14; "American Industries—No. 10."

13. [Wheeler and Wilson], *The Sewing Machine,* pp. 15–16.

14. "American Industries—No. 10," p. 274.

15. Correspondence between Singer Manufacturing Company's production expert, E. H. Bennett, and Franklin Park, the works manager at Singer's Kilbowie, Scotland, factory, May 20, July 8, 1909, vol. 124, Papers of the Singer Manufacturing Company, State Historical Society of Wisconsin, Madison.

16. Brown & Sharpe Manufacturing Company's Historical Data Files, Patent Library, North Kingston, R.I., list Willcox as a hardware merchant (all subsequently cited Brown & Sharpe correspondence is at this repository). Cooper, *The Sewing Machine,* p. 46, calls him a patent model builder. See also *National Cyclopaedia of American Biography,* s.v. "Willcox, Charles Henry."

17. Henry Dexter Sharpe, "Joseph R. Brown, Mechanic, and the Beginnings of Brown & Sharpe";

see also Howard Francis Brown, "The Saga of Brown and Sharpe (1833–1968)."

18. See J. R. Brown & Sharpe to Friend [William G.] Angell, March 11, 1858, Historical Data Files (Bound Volumes); J. A. Willcox to J. R. Brown & Sharpe, February 16, 1858, Historical (Misc.) Drawer. On Angell, see *Dictionary of American Biography,* s.v. "Angell, William Graham."

19. These tools are listed by Sharpe, "Joseph R. Brown."

20. J. R. Brown & Sharpe (by Lucian Sharpe) to James Willcox, n.d. [June or July 1858], Historical Data Files (Bound Volumes).

21. J. A. Willcox to J. R. Brown & Sharpe, March 19, 1858, ibid.

22. J. R. Brown & Sharpe to J. A. Willcox, March 22, 1858, and n.d. [June or July 1858], ibid.

23. Both parties had been considering increasing the number of machines to one hundred since early March 1858. See J. R. Brown & Sharpe to Friend [William G.] Angell, March 15, 1858, to J. A. Willcox, n.d. [June or July 1858], July 21, 1858, ibid.

24. J. R. Brown & Sharpe to James Willcox, August 24, September 7, 1858, ibid.

25. J. A. Willcox to J. R. Brown & Sharpe, March 19, 1858; J. R. Brown & Sharpe to James Willcox, September 21, 1858, ibid.

26. J. R. Brown & Sharpe to James Willcox, September 25, n.d. [ca. October 1, 1858], ibid.

27. J. R. Brown & Sharpe to James Willcox, October 30, 1858, ibid. I do not know the identity of the Brown & Sharpe workman who had been employed at the Sharps rifle works.

28. Ibid.

29. Brown, "Saga of Brown and Sharpe," p. 26, states that 1861 marks the first date of machine tool sales, in this case, a turret lathe.

30. For sales of Willcox & Gibbs sewing machines, see Table 2.1. An article in the *Providence Daily Journal* (December 19, 1863) states that "nearly twenty thousand [Willcox & Gibbs sewing machines] have already been made, while the continued supply of seven hundred per month scarcely suffices to meet the regular demand for this valuable household companion."

31. The company noted in its *Catalogue and Price List,* October 1, 1885, that the tools advertised "are manufactured with the intention of having combined in each, respectively, all those qualities best adapted to serve the uses for which they are designed. They are mostly the outgrowth of our own wants in the business of manufacturing, and therefore have been proved by experience, to be well adapted to their several uses."

32. Biographical information on Howe can be found in Robert S. Woodbury, *History of the Milling Machine,* pp. 29–76; Hubbard, "Development of

Machine Tools,'' pp. 437–41; and Lucian Sharpe to Wm. G. Angell, July 15, 1861, Brown & Sharpe, Historical Data Files (Bound Volumes). This letter contains a five-page account of Howe's prior work and his contact with Joseph R. Brown & Sharpe.

33. This interpretation, although contrary to Woodbury's, is conjectural, but the extended communication between Brown & Sharpe and Howe, documented in Sharpe's letter cited in n. 32, strongly suggests collaboration.

34. Luther D. Burlingame to Joseph W. Roe, September 10, 1914, Brown & Sharpe, Historical Data Files (Bound Volumes) (Burlingame was Brown & Sharpe's ex-officio historian); Henry M. Leland to John R. Freeman, February 11, 1927, ibid.

35. Biographical material on Leland may be found in the letter cited in n. 34; Henry M. Leland to Brown & Sharpe Manufacturing Company, April 15, 1914, and "Henry M. Leland Conference with L. D. Burlingame, August 24, 1929," both in Brown & Sharpe, Historical Data Files (Bound Volumes); and Ottilie M. Leland, *Master of Precision.*

36. "Henry M. Leland Conference with L. D. Burlingame," pp. 6, 11.

37. Leland to Freeman, February 11, 1927, p. 3.

38. "Henry M. Leland Conference with L. D. Burlingame," p. 11.

39. According to Robert S. Woodbury, Brown & Sharpe had begun grinding experiments before 1860 (*History of the Grinding Machine*, p. 61). Leland to Freeman, February 11, 1927, pp. 3–4.

40. Leland to Freeman, February 11, 1927, p. 4. For more information on the development of grinding techniques at Brown & Sharpe, see Woodbury, *History of the Grinding Machine*, pp. 58–71.

41. "Reminiscences of W. A. Viall," Brown & Sharpe, Sewing Machine and Small Tool Drawer.

42. Gene Cesari makes the distinction between process and product orientation of manufacturerers and machine tool makers in the antebellum period in "American Arms-Making Machine Tool Development."

43. Singer Manufacturing Company's early history is found in Bourne, "American Sewing-Machines"; Cooper, *The Sewing Machine;* and Davies, *Peacefully Working.* For biographical information on Singer and Clark see *Dictionary of American Biography,* s.v. "Singer, Isaac"; *National Cyclopaedia of American Biography,* s.v. "Singer, Isaac"; Peter Lyon, "Isaac Singer and His Wonderful Sewing Machine"; and "Edward Clark."

44. Both the *Boston Almanac,* 1854–56, and the *Boston Directory,* 1853, list Phelps as a maker of "philosophical" (scientific) instruments.

45. See Box 235; Cash Account Book, 1851–52, Box 233; "Bills Receivable, Bills Payable," Box 184; other bills from 1851 and 1852, Box 234; J. M.

Emerson to I. M. Singer & Co., March 21, 1851, Box 189, all in Singer Papers. Singer purchased its upright drill from David Chamberlain of Boston for $95.

46. J. M. Emerson to I. M. Singer & Co., March 24, 1851, Box 189, ibid.

47. J. M. Emerson to I. M. Singer & Co., March 28, 1851, Box 189, ibid.

48. This conclusion is based on the hundreds of bills found in Box 204 of the Singer Papers.

49. See Lyon, "Isaac Singer."

50. Contracts for the room and power are in Box 155, Singer Papers. *Sewing Machine Times,* n.s., 18 (October 25, 1908):8, notes that the size of the room was 30 × 30 feet and that the company rented additional rooms of this size in the same building. A Singer Sewing Machine Co. publication, *Mechanics of the Sewing Machine* p. 45, claims that the room was 25 × 50 feet. For information leading to identification of Ezra Gould, I am indebted to Carlene Evans Stephens of the National Museum of American History, Smithsonian Institution. For the purchases, see Bills, Boxes 231 and 234, Singer Papers. I. M. Singer & Co. [Edward Clark's handwriting] to William F. Proctor, July 16, 1855, Box 189, ibid.

51. *United States Magazine* 1 (September 15, 1854), opposite p. 161. The Singer company claimed in 1914 that this illustration appeared in August 1853 (*Mechanics of the Sewing Machine,* p. 45).

52. *Sewing Machine Times,* n.s., 18 (October 25, 1908): 8, acknowledged the crude production technology of the Singer company. So did the Singer company publication of 1914, *Mechanics of the Sewing Machine,* which stated on p. 45: "It is apparent that the greater part of the sewing machine construction at that time was produced by hand at the bench, so that no two machines or the parts composing them were precisely alike, either in shape or fitting."

53. Singer Sewing Machine Co., *Mechanics of the Sewing Machine,* p. 45; Singer & Co. [Clark] to Proctor, July 16, 1855.

54. Even the most casual inspection of Singer sewing machines made from 1851 through the 1870s reveals differences among them. These differences—even in machines of the same model and made in the same year—are products of the idiosyncrasies of individual workmen and are to be expected from the European approach to manufacture. See Appendix 2 for an account of variations in Singer machines.

55. See "Edward Clark."

56. Singer & Co. [Clark] to Proctor, July 16, 1855.

57. Bourne, "American Sewing-Machines," attributes this innovation to Clark, but, as Robert B. Davies points out, Clark credited the Wheeler and Wilson Company with this marketing device (*Peacefully Working,* p. 20).

58. Singer's marketing strategy is discussed in

Davies, *Peacefully Working,* and in Andrew B. Jack, "The Channels of Distribution for an Innovation." Alfred D. Chandler draws upon and clarifies the work of Davies and Jack in *The Visible Hand,* pp. 303–8, 402–5. See also Elizabeth M. Bacon, "Marketing Sewing Machines in the Post-Civil War Years." In an important article on American law and corporate marketing strategy, Charles W. McCurdy suggests that Singer's successful marketing strategy owed a great deal to the company's work in the courts breaking down legal trade barriers to the establishment of a national market ("American Law and the Marketing Structure of the Large Corporation, 1875–1890").

59. Cooper, *The Sewing Machine,* p. 35, claims that the first Singer family machine proved to be too light for effective work.

60. "Sewing Machines," pp. 421–23.

61. Spalding, "The 'American' System of Manufacture," p. 11. I cannot precisely identify Spalding. I have identified a Frank Spalding, who about 1859 learned the machinist's trade at the Cheney Bros. Silk Mills at Manchester, Connecticut, and who worked for various sewing machine manufacturers including Stedman of Meriden, Connecticut, the Finckle & Lyon Sewing Machine at Middletown, Connecticut, and its successor the Victor Sewing Machine. In 1883 he went to work for Brown & Sharpe as a toolmaker. Before he retired in 1919 he had received thirty patents on tools. Frank Spalding published articles in the technical press and could have written the series on the American system. See Luther D. Burlingame, "Frank Spalding," Brown & Sharpe, Historical Data Files (Bound Volumes).

62. Horace L. Arnold [Henry Roland], "Six Examples of Successful Shop Management—V," p. 997. I am indebted to Daniel Nelson for this reference.

63. Ibid.

64. Copy of responses to questionnaire attached to a letter from L. B. Miller (who supplied the information) to the Singer Mfg. Co., August 3, 1885, Box 200, Singer Papers. See also the company's publication by John Scott, *Genius Rewarded; or the Story of the Sewing Machine,* pp. 29–32; and George R. McKenzie's remarks at the groundbreaking ceremony for the new Scotland factory in Singer Manufacturing Co., *Report of the Proceedings on the Occasion of Breaking Ground for the Singer Manufacturing Company's New Factory,* p. 18, Box 233, Singer Papers.

65. The many complaints are evident from a cursory reading of the company's incoming correspondence, which is bound in several hundred letterbooks that form a part of the Singer Papers, Boxes 1 through 53 and 1975 accession (unprocessed: twenty-two large moving boxes full of letterbooks). See esp. M. B. Heine to I. M. Singer & Co., January 21, 1863; and Edwin Dean to I. M. Singer & Co., February 2, 1863,

Box 6, Singer Papers. Problems abroad are reflected in correspondence from Singer's British agent, W. E. Broderick; see, for example, Broderick to I. M. Singer & Co., February 25, 1862, Box 1, ibid. Needlemaking problems and Carter's recruitment are documented in George B. Woodruff to I. M. Singer & Co., March 3, 14 (two letters), May 13, 20, 1862, Box 2; Jerome Carter to I. M. Singer & Co., September 29, 1862, Box 5; and power of attorney form, Mrs. Lydia M. Carter, Administratix of Jerome Carter to Singer Mfg. Co., Sept. 19, 1876, Box 190, ibid.

66. 44 (1922): 927–28. The hiring of Miller may have been the occasion recalled by Isaac Singer, who still worked around the factory in 1863, that "his foreman nearly ruined the business by his first attempt to produce parts on the interchangeable plan without the exactness of duplication that was later obtained" ("Looking Back: A Bird's Eye View of Our Trade," *Sewing Machine Times,* n.s., 18 (October 25, 1908): 8.

67. Waldo E. Nutter, *Manhattan Firearms;* Arnold's letter to Root is reprinted on p. 129.

68. Arnold [Roland], "Six Examples of Successful Shop Management," p. 997.

69. Singer Manufacturing Co., "The Singer Gauge System," *Catalogue of Singer Sewing Machines, Illustrating Their Construction, Their Variety, and Their Special Uses by Manufactures,* p. 51. In this and other advertising literature the Singer company stressed the assembly system in which "the working parts are 'assembled' . . . and each placed in its proper working position. Each of these parts has been so accurately made that all are perfectly interchangeable and require no adjustment, each fitting properly to its intended position, and resulting in a complete and harmonious whole" (ibid., p. 41).

70. "Sewing Machines," pp. 421–23. The dropforged parts probably included the thread tension discs, tension spring, presser foot, presser foot lever, shuttle race slides, shuttle carrier, pitmans (or connecting rods) for the shuttle race carrier, two parts for the cloth feed, and several washers and pins. The majority of the parts, however, were cast. My source for these statements is the sewing machines themselves. I have inspected the following New Family machines which were made in the years indicated in parenthesis: Serial Numbers 87104 (ca. 1865), 106092 (ca. 1865), 360834 (ca. 1869), 459,834 (ca. 1870), 1038977 (ca. 1873), 1081835 (ca. 1873), 5235877 (ca. 1884/85). These machines are preserved by the Division of Textiles, National Museum of American History, Smithsonian Institution. See Appendix 2 for a report of inspection and analysis procedures.

71. The published account of the Singer factory describes the operation of these gear cutters: "One of these [beveled gear blanks] is placed upon a machine

about the size of a man's hat, when a circular saw, in circumference of the size of a silver dollar, approaches and cuts a cog in solid metal with as much apparent ease as it is in soft wood. It then recedes, when the cog wheel turning on its axis, presents a new front which is instantly cut. The saw-dust falling into a little tin gutter, is conducted to a box for preservation. In a few minutes all the cogs being cut, the machine stops itself, when the operator replaces the finished wheel by a fresh one, and the machine proceeds as before. The wheel thus cut is however not perfect. The cogs being rough and square, would not mesh with those of another wheel. It is therefore put in another machine similar to the first, with the exception that instead of a circular saw, there revolves against the cogs a wheel having on its circumference a groove which rounds off and polishes the cogs, stopping itself when it has completed its work'' (''Sewing Machines,'' p. 423).

72. ''Inventory, Glasgow Factory, 31st Dec. 1868,'' p. 21, Box 189, Singer Papers. See also the letter from John MacDonald to Singer Mfg. Co., May 19, 1870, requesting one thousand pounds of ''grinding stones for grinding gears'' (Foreign Letter File 2, 1975 accession, ibid.).

73. That this system was in use is evident from later correspondence and from the Time Book folio or Payroll Ledger, vol. 277, ibid. Since this ledger begins in 1868, precise dating of the adoption of the inside contract system is impossible. Horace L. Arnold [Roland] says that inside contracting ''became general in 1866–7'' (''Six Examples of Successful Shop Management,'' p. 997).

74. ''Sewing Machines,'' p. 422; production figures for 1863, however, indicate that the factory turned out more than four hundred machines. ''Our Manufacturing Industries, no. III,'' *New York Times,* July 9, 1865, p. 5.

75. Quoted from the May 28, 1867, Minutes of the Board of Directors Meetings of the Singer Manufacturing Company by Davies, *Peacefully Working,* p. 44. These minutes are held today by the Singer company in New York, and I was denied access to them. Because of personal friendship with a former board chairman of Singer, Davies, while a graduate student at the University of Wisconsin in 1960, was allowed to examine the minutes as well as other Singer records. Upon the chairman's retirement, however, the company refused further access to Davies (ibid., Acknowledgments).

76. Ibid., pp. 44–45.

77. The production figure is given in ibid., p. 46, in a letter from John MacDonald to John Anderson, December 14, 1867, Box 36, Singer Papers, which I evidently missed. The correspondence between MacDonald and the New York offices in 1868 is contained in letterbook volumes 92 through 96 of the 1975 Singer Papers accession (uncataloged). See, for example,

letters of September 16, 1868, vol. 92, and November 28, 1868, vol. 96. See also the 1868 Glasgow factory inventory, Box 189 of the papers accessioned earlier and cataloged.

78. The figure of seven hundred is not absolutely certain. The new factory at Glasgow had a weekly output of six hundred machines, and I have assumed that the original factory produced one hundred machines per week. Robert B. Davies, in ''International Operations of the Singer Manufacturing Co., 1854–1895,'' p. 122, gives a figure of 792. The figure of six hundred is taken from the Singer company's *Report of the Proceedings of Breaking Ground,* p. 34–35, in which Alexander Anderson states that ''we were receiving the parts from America, in a partly finished state.'' *Scientific American* 24 (1871): 406, reprints a small article taken from *Engineering.* Its output figure of one thousand four hundred machines per week is suspect, although by 1879 the Glasgow factory was turning out 189,085 machines, or roughly four thousand machines per week (Alex Anderson to George R. McKenzie, December 26, 1879, Box 191, Singer Papers).

79. Davies, *Peacefully Working,* p. 47, citing minutes of August 30, 1869; the directors had been unable to find an adequate site in New York City. The evidence on the two-thousand-machine limit is conflicting. Davies also wrote that ''in January 1869, McKenzie spoke in terms of a weekly production of 1,500 family machines for United States domestic sales, and semi-finished parts of 300 machines weekly for the Glasgow plant. To meet *this goal* [my italics] by May 1, he was authorized to buy $12,264.00 of new and used tools to increase output'' (''International Operations,'' p. 120, citing Directors' Minutes, January 1, 1869, and February 17, 1869).

80. Davies, ''International Operations,'' p. 48, citing minutes of December 15, 1869.

81. ''Inventory, Glasgow Factory, 31 December 1868,'' Box 189, Singer Papers.

82. The wage differential is based on average wage figures for factory employees in the two countries and an exchange rate of $5 per pound sterling. The American factory built tools for the Glasgow factory well into the 1880s, as evidenced by scores of letters from John MacDonald to Singer Mfg. Co. and from L. B. Miller to George Ross McKenzie from 1868 to 1885, Singer Papers.

83. Evidence for this outcry is contained in the verbose letters from George B. Woodruff, who headed Singer's European sales in London, dating from 1867 and 1868. Letters from 1867 are in Boxes 30–36; from 1868, 1975 accession, vols. 92–96, Singer Papers. One year earlier, however, Woodruff had complained about the ''exceedingly bad workmanship in all the machines sent from New York [in which] the cam rollers and crank pins are so badly

fitted'' (Woodruff to Singer Mfg. Co., April 12, 1865, Box 16, ibid.). Similar protests were recorded in 1883, when the company established a factory in Montreal, Canada. This time the company sent circulars around to its branch agents to give them ''an opportunity of comparing them with those which we have been importing for some years. We should like to have your opinion regarding same. The machine is the same in every respect, nevertheless we would like to have your opinion upon it as compared with the Glasgow machine'' Box 196, ibid.).

84. ''Report'' (1872) in Archives Nationales, F12-6907, ouvrages en matières diverses; machines à coudre; admission en franchise, 1844–72. I am indebted to James Edmonson for pointing out the existence of this report.

85. Davies, *Peacefully Working*, p. 50, citing Board Minutes of January 1, 1869; Singer Factory Journal, 1868–73, p. 689, 1975 accession, Singer Papers.

86. Davies, ''International Operations,'' p. 124, citing Singer Board Minutes, February 18, October 19, 31, 1870, and January 25, 1871. Davies does not indicate whether the expert was hired. See also his *Peacefully Working*, p. 48.

87. See Memorandum of the Providence Tool Company, December 19, 1870, for details of the contract, Box 190; see also John B. Anthony, Providence Tool Co., to Inslee A. Hopper, Singer Mfg. Co., December 28, 31, 1870, 1975 accession, vol. 135, Singer Papers. A letter written by Anthony, president of Providence Tool Co., to T. B. Terry & Co., October 27, 1881, published in the *Sewing Machine Advance* 4 (January 1882): 3, confirms the terms of the contract.

88. Proposed agreement between Domestic Sewing Machine Co. and Providence Tool Co., October 17, 1873, Box 190; see also notes of a meeting between Singer Mfg. Co. and Providence Tool Co., November 22, 1872, Box 191; Inslee A. Hopper to Providence Tool Co., December 21, 1872, for the new contract, Box 190, Singer Papers; Anthony to Terry & Co., October 27, 1881.

89. Articles of agreement between Domestic Sewing Machine Company and Singer Mfg. Co. (draft) [ca. May or June], 1873, Box 228. Providence Tool terminated the contract in a letter from Anthony to Singer Mfg. Co., June 24, 1873, Box 190. Singer took the initiative, as shown in Anthony to Singer Mfg. Co., July 3, 1873, Box 190, and D. Blake, Domestic Sewing Machine Co. to Singer Mfg. Co., June 25, 1873, which refers to a termination notice of June 23, 1873, Box 190, ibid.

90. D. Blake, Domestic Sewing Machine Company, to Inslee A. Hopper, Singer Mfg. Co., Jan. 27, 1873; John B. Anthony, Providence Tool Co., to Domestic Sewing Machine Co., May 10, 22, 1873, Box 190, ibid. Anthony's letter published in *Sewing Machine Advance* suggests that the entire one hundred thousand machines were delivered. But the Providence Tool Company never delivered more than five thousand machines in one month. See 1873 correspondence between Providence Tool and Singer as well as ''Settlement with and result of contracts with Domestic Sewing Machine Co., Oct. 28, 1873,'' Box 190, Singer Papers.

91. Draft, Articles of Agreement [creating the Domestic Manufacturing Company], [May or June] 1873, p. 7, Box 228, Singer Papers.

92. John B. Anthony, Providence Tool Company, to I. A. Hopper, Singer Mfg. Co., November 27, 1872, Box 190, ibid.

93. See Appendix 2 for a technical description of inspection of Singer sewing machines, ca. 1873.

94. Spalding, ''The 'American' System of Manufacture,'' p. 11.

95. Eugene S. Ferguson, ''History and Development of Statistical Quality Control,'' unpublished manuscript, April 15, 1964; Daniel J. Boorstin, *The Americans*, pp. 193–200.

96. L. B. Miller to G. R. McKenzie, April 12, 1881, Box 193, Singer Papers.

97. Captain Charles Chapman, *The Ocean Waves*, and Scott, *Genius Rewarded*.

98. *Elizabeth Daily Journal*, October 9, 1907, Box 231, Singer Papers; Davies, *Peacefully Working*, p. 48; Chapman, *Ocean Waves*, p. 143. Because Singer was so defensive about the sewing machine combination, particularly under the editorial fire of journals such as *Scientific American*, it never published articles about its factory in trade journals, a common practice in the nineteenth century.

99. Chapman, *Ocean Waves*, pp. 143–51. Regarding the skill of Singer's molders, L. B. Miller wrote in a Bureau of Labor inquiry about ''those who make our castings. . . . Although they make molds, they cannot be called molders who understand the business throughout, and can make molds of any kind'' (Miller to Singer Mfg. Co., August 3, 1885, Box 200, Singer Papers).

100. E. H. Bennett to G. R. McKenzie, July 30, 1884, Box 198, Singer Papers. Bennett was complaining about castings made at Montreal.

101. Chapman, *Ocean Waves*, pp. 158–60.

102. See Appendix 2 for documentation.

103. Chapman, *Ocean Waves*, pp. 158–59.

104. Ibid.

105. See Appendix 2.

106. On marketing, see Bacon, ''Marketing Sewing Machines.''

107. ''The 'American' System of Manufacture,'' p. 11.

108. L. B. Miller to Singer Mfg. Co., Feb. 11, 1875, Letterbook 226, 1975 accession. Singer Papers.

109. Scott, *Genius Rewarded,* pp. 48, 50; Singer Manufacturing Co., *Catalogue,* and Singer Sewing Machine Co., *Mechanics of the Sewing Machine;* L. B. Miller to Singer Mfg. Co., August 3, 1885, Box 200, Singer Papers.

110. Scott, *Genius Rewarded,* p. 50.

111. Ibid., pp. 55–56. On John Scott as McKenzie's attorney, see S. A. Bennett to G. R. McKenzie, March 30, 1881, Box 193, Singer Papers.

112. Singer Sewing Machine Co., *Mechanics of the Sewing Machine,* p. 50.

113. Miller to McKenzie, March 30, 1881, Box 193, Singer Papers.

114. A. D. Pentz to McKenzie, December 9, 1881; Clark to McKenzie, April 13, 1881, ibid.

115. Pentz to McKenzie, April 2, 1881, ibid.

116. Miller to McKenzie, April 5, 1881, ibid.

117. Clark to McKenzie, April 13, 1881, and Sydney A. Bennett to McKenzie, April 15, 1881, ibid.

118. Miller to McKenzie, April 12, 1881; Pentz to McKenzie, April 14, 1881, ibid.

119. Sydney A. Bennett to McKenzie, April 21, 1881; Miller to McKenzie, April 20, 1881, ibid.

120. Miller to McKenzie, April 27, 1881; Pentz to McKenzie, May 3, 1881, ibid.

121. Sydney A. Bennett to McKenzie, April 27, May 12, 1881; William F. Proctor to McKenzie, June 2, 1881, ibid.

122. Pentz to McKenzie, May 19, 1881; Miller to McKenzie, May 11, 24, 1881; Sydney A. Bennett to McKenzie, May 20, 1881, ibid.

123. Pentz to McKenzie, June 1, 1881, ibid.

124. Ibid.

125. Sydney A. Bennett to McKenzie, June 1, 1881, ibid.

126. Pentz to McKenzie, June 1, May 19, 1881, ibid.

127. Pentz to McKenzie, June 13, 29, 1881, ibid.

128. Miller to McKenzie, July 26, 1881; Pentz to McKenzie, June 13, 1881, ibid.

129. Miller to McKenzie, June 8, 28, 1881, ibid. Total production that week was 8,379 machines: 7,000 New Family, 205 Medium Manufacturing, 435 Number 4 Manufacturing, 175 Improved Family, 550 Portable Hand, and 14 Number 3 Standard Manufacturing, in addition to 3 Button Hole machines.

130. Sydney A. Bennett to McKenzie, June 2, 1881; Pentz to McKenzie, June 29, 1881, Miller to McKenzie, August 10, 1881, ibid.

131. Sydney A. Bennett to McKenzie, June 15, July 1, 19, 1881; Pentz to McKenzie, July 8, 1881, ibid.

132. Edward Clark to McKenzie, September 9, 19, 1881, ibid.

133. Miller to McKenzie, May 3, 1882. McKenzie had "consider[ed] it very important" to have Pentz

work with him at the Glasgow factory. See G. R. McKenzie to Edward Clark, March 28, 1882, both letters in Box 195, ibid.

134. McKenzie to Sydney A. Bennett, May 29, 1882, ibid. Earlier Pentz had flatly told McKenzie that he was "an oponent [*sic*] of the principle of night work. . . . Two men cannot, in my opinion, use the same tools intermittently with the greatest economy to either the tools or the work produced" (Pentz to McKenzie, May 19, 1881, Box 193, ibid.).

135. McKenzie to Sydney A. Bennett, June 6, May 29, 1882, Box 195, ibid. In September, Miller snapped back at the New York office for making what seemed to him impossible demands on the Elizabethport factory. The central office had asked Miller to prepare eight automatic "gearing machines" for the factory in Scotland. He replied, "We have so much work that is legitimately called for, in the way of special tools to meet our necessary demand for machines, that it would be impossible for us to fill an order for Glasgow for a long time to come; if we did, it would be at a serious loss to this factory" (Miller to Singer Mfg. Co., September 30, 1882, ibid.).

136. Obituary of E. H. Bennett, *Transactions of the American Society of Mechanical Engineers* 19 (1897–98): 979–80.

137. Diehl's presence at these meetings also indicates important changes. A German immigrant, he had been associated with Singer off and on since 1868, when he worked at the Blees Factory, which was making sewing machines for Singer. Between 1868 and 1871 he worked for the Remington Company in Ilion, perhaps making sewing machines. Singer hired Diehl again in 1874 to work as a designer and inventor at the Elizabethport factory. Sometime—perhaps in 1883, when changes began to take place at the Singer factory—he set up a gauge department at Elizabethport. See his obituary in *Sewing Machine Times,* n.s., 23 (April 25, 1913): 1.

138. Minutes of a meeting held at Elizabethport factory, March 26, 1883, Box 239, Singer Papers.

139. Ibid.; minutes of March 27, 1883 meeting, Box 239; Hugh Wallace to McKenzie, June 28, 1883, Box 197, ibid.; obituary of Philip Diehl. See also Miller to McKenzie, March 29, 1884, Box 198, Singer Papers, which discusses the progress of the gauge department.

140. Minutes of meetings held at Elizabethport factory, March 26, 27, 1883, Box 239, Singer Papers.

141. Pentz to McKenzie, April 19, 1883, Box 197; Miller to McKenzie, July 25, 1883, Box 196, ibid.

142. Miller to McKenzie, May 21, 1884, Box 198, ibid.

143. F. Lander to McKenzie, May 17, 26, 1884, Box 199; E. H. Bennett to McKenzie, May 24, July 7, 1884, Box 198, ibid.

144. E. H. Bennett to McKenzie, July 23, 30, 1884, Box 198, ibid. McKenzie had great confidence in Bennett's work. About the Montreal situation, he wrote to Sydney Bennett: "I have rec'd Ed Bennett's report & . . . he cd. not have done better. . . . His report lays bare a deplorable state of affairs & is just what I suspected all along. I don't know anyone better qualified to organize the factory out of chaos than Ed" (June 7, 1884, Box 199, ibid.).

145. E. H. Bennett to McKenzie, June 24, 1885, Box 200, ibid.

146. Weekly production figures appear in letters from Miller to McKenzie, 1883–85; for example, May 21, 1884, 7,552; May 30, 1884, 7,656; July 23, 1884, 7,361; August 19, 1884, 7,111; July 24, 1885, 7,238; July 31, 1885, 6,567; August 21, 1885, 5,588; September 10, 1885, 6,432; September 18, 1885, 6,315.

147. S. W. Goodyear, "Personal Recollections," *American Machinist* 7 (January 12, 1884): 7.

148. E. H. Bennett to McKenzie, August 17, 1884, Box 198, Singer Papers.

149. Pentz to McKenzie, December 12, 1884, ibid.

150. L. B. Miller to Singer Mfg. Co., August 3, 1885, responding to a questionnaire by the U.S. Bureau of Labor, Box 200, ibid.

151. See Miller's and Bennett's obituaries in *Transactions of the American Society of Mechanical Engineers* 44 (1922): 927–28 and 19 (1897–98): 979–80.

152. Arnold [Roland], "Six Examples of Successful Shop Management," pp. 994–1000. The date of 1883 for this change from the inside contracting system to paid foremen is not absolutely sure. Arnold said he drew upon "precise information." In one place, he said the date was 1883 but later wrote "about 1883." The establishment in 1883 of an Elizabethport Managing Committee supports Roland's argument. In 1885, however, Miller and McKenzie entertained the idea of returning to the contract system. See Miller to McKenzie, July 25, 31, August 11, 21, 1885, Box 200, Singer Papers.

153. Minutes of meeting held at Kilbowie, June 26, 1886, Box 204, Singer Papers. Unfortunately, the blue book does not exist, but some details about it survive in the contextual information in these minutes.

154. *Report of the Proceedings of Breaking Ground,* p. 4; E. H. Bennett to McKenzie, June 28, 1886, Box 204, Singer Papers. For more information on Singer's Kilbowie factory, see S. B. Saul, "The Market and the Development of the Mechanical Engineering Industries in Britain, 1860–1914," pp. 160–61, and Davies, *Peacefully Working,* pp. 55–91.

155. Minutes of meeting held at Kilbowie, June 26, 1886.

156. An expression, used by Robert B. Davies, *Peacefully Working,* from the Kilbowie ground-breaking ceremony booklet, p. 22.

Chapter 3

1. Great Britain, Parliament, House of Commons, *Report of the Select Committee on Small Arms,* Q. 2043, p. 144. For Whitworth's general views on American woodworking, see *Special Report of Joseph Whitworth* reprinted in Nathan Rosenberg, ed., *The American System of Manufactures,* pp. 343–48.

2. *Report of the Committee on the Machinery of the United States of America* in Rosenberg, ed., *American System of Manufactures,* pp. 89–197.

3. John Anderson, "On the Application of Machinery in the War Department," p. 157. For a significantly different view of American woodworking machinery at Enfield, see Thomas Greenwood, "On Machinery for the Manufacture of Gunstocks."

4. "The Engineer and the Woodworking Industry," p. 448.

5. "Engineering in Furniture Factories," p. 85.

6. Thomas D. Perry, "The Wood Industries," p. 434. Between 1919 and 1925, when the Wood Industries Division was created, the ASME published many papers concerning woodworking under the name of the Forest Products Section.

7. Nathan Rosenberg, "America's Rise to Woodworking Leadership"; Polly Anne Earl, "Craftsmen and Machines"; Edward Duggan, "Machines, Markets, and Labor."

8. Alfred D. Chandler, *The Visible Hand,* pp. 247–49.

9. The recent study by Robert Bishop and Patricia Coblentz, *The World of Antiques, Art, and Architecture in Victorian America,* includes sewing machine cabinets as a part of Victorian furnishings, particularly for women. I am indebted to Maureen Quimby for this reference. Why sewing machine manufacturers decided to encase their machines in cabinets is a subject beyond the scope of this study but is provocatively addressed in Diane Douglas, "The Machine in the Parlor."

10. P. 199.

11. J. A. Fay and C. B. Rogers. See the numerous woodworking machines displayed at the Crystal Palace in the *Official Catalogue of the New York Exhibition of the Industry of all Nations,* Classes 5 & 6, "Machines for Direct Use; Machinery, Tools, etc.," pp. 32–44.

12. Ettema, "Technological Innovation and Design Economics," and James Lindsey Hallock, "Woodworking Machinery in Nineteenth Century America." Hallock argues that "American inventors were interested in general purpose machines, and

these they developed in western isolation until American woodworking machinery became the most advanced in the world'' (p. 10).

13. *Official Catalogue of the New York Exhibition*, pp. 32–44.

14. *Special Report of Joseph Whitworth*, in Rosenberg, ed., *American System of Manufactures*, p. 343.

15. For example, Anderson described at length the equipment and processes used to make wooden tubs and pails, rocking chairs, chairs, and bedsteads. For each, manufacturers had designed and constructed special tools for the various operations which significantly cut down on labor input. Anderson's committee report concluded that carriage wheels were made entirely by special machines, yet these special machines were ''common all over the country'' (*Report of the Committee of the Machinery of the United States of America*, p. 168).

16. A cursory reading of recent secondary literature on American woodworking technology in the nineteenth century demonstrates that little material on this subject exists beyond Richards's work. Consequently, the strong views and the abundance of information included in Richards's *Treatise on the Construction and Operation of Wood-Working Machines* (and his later books) have shaped both perceptions and assumptions about woodworking technology. For the purpose of this chapter, the most serious distortion resulting from surviving literature on woodworking technology is the view, stated implicitly in Richards, that only woodworking machinery companies made woodworking machines. The major innovators in the development of mass production of metal goods from rifles to sewing machines to bicycles to automobiles, however, often built their own highly specialized machine tools, which the machine tool industry did not market. Sometimes these companies designed machine tools and then relied upon the machine tool industry to build them. There is little reason to doubt that the same was true with woodworking, particularly in firms that also worked metal.

17. David Bigelow, *History of Prominent Merchants and Manufacturing Firms*, p. 250.

18. [Wheeler and Wilson Manufacturing Company], *The Sewing Machine*, p. 18.

19. Chauncey Jerome, *History of the American Clock Business for the Past Sixty Years*, p. 92. See also John Joseph Murphy, ''The Establishment of the American Clock Industry,'' pp. 65–66, 185–86.

20. The evidence on the precise nature of this factory is conflicting. A *Scientific American* article on Wheeler and Wilson (40 [1879]: 271, 274–75) stated that ''the raw material is cut to dimension at the company's mill in Indianapolis, and transported here [Bridgeport] to be worked up into the desired forms.'' Yet Thomas D. Perry wrote in *Modern Plywood* that the ''Sewing Machine Cabinet Company of Indi-

anapolis was started about 1867, by the Wheeler & Wilson Sewing Machine Company, to make plywood parts for various sewing-machine cabinets. At first they are reported to have used only sliced veneer, but after acquiring a veneer lathe about 1870, they commenced to use rotary-cut crossbands or crossings'' (p. 29). See also the article on the Sewing Machine Cabinet Company in *Asher & Adams' Pictorial Album of American Industry*, p. 10.

21. Before it made fancy cases and cabinets, Singer sold large, stout boxes to hold the first heavy, cast-iron machines, which weighed over one hundred pounds. These boxes were used both for shipping and as stands for sewing machines.

22. Singer Manufacturing Co., *Catalogue of Singer Sewing Machines, Illustrating Their Construction, Their Variety, and Their Special Uses by Manufactures*, preface.

23. On Singer's Russian factory, see Robert B. Davies, *Peacefully Working to Conquer the World*, pp. 275–305.

24. These figures are from ''Cabinet Factory and Western Foundry—South Bend, Ind.,'' a brief Singer company manuscript history, compiled in 1920 by David Pollock, manager of Singer's South Bend cabinet works, and now located in the Northern Indiana Historical Society, South Bend, Indiana.

25. The following newspaper articles, maintained by the South Bend Public Library, contain information about Pine and the early days of the South Bend case factory: C. N. Fassett, ''Leighton Pine: A Human Dynamo'' (date and source not identified); ''Obituary'' (date and source not identified); ''Leighton Pine—Sudden Death,'' November 5, 1905 (source not identified); Mary Roberts, ''City's Claim to World Fame Rests in Part on Work of Leighton Pine,'' *News Times* (South Bend), November 23, 1928; E. Fred Grether, ''How the Singer Works Came to South Bend,'' *Sun News*, October 16, 1900; ''Mammoth New Plant of Singer Manufacturing Company,'' *Sun News*, October 15, 1901; ''Carloads of Sewing Machines Shipped Daily to all Parts of World from Singer's Great Plant; Has 30 Acres of Floor Space,'' *News Times*, February 23, 1919. See also ''Singer Manufacturing Company—South Bend, Ind. Works'' (1957), pamphlet at South Bend Public Library.

26. E. Fred Grether, ''How the Singer Works Came to South Bend,'' *Sun News*, October 16, 1900.

27. ''Singer Manufacturing Company—South Bend, Ind. Works,'' p. 2.

28. John A. Liebert to Inslee A. Hopper, October 27, 1869; J. C. Derr to John A. Liebert, February 27, 1870; and for the friction between the three cabinet-makers, see also Leighton Pine to Inslee A. Hopper, September 16, 1869; James Bolton to Inslee A. Hopper, October 19, 1869; ''Your Servant'' to the Singer Mfg. Co., April 19, 1870; and an undated [1870]

memorandum [in Pine's handwriting] outlining policy and managerial changes and rules made at South Bend, all in Box 155, Papers of the Singer Manufacturing Company, State Historical Society of Wisconsin, Madison.

29. Leighton Pine to Singer Mfg. Co., November 20, 1869, vol. 112 (1975 accession), p. 306; Leighton Pine to G. R. McKenzie, July 13, 1870, vol. 126 (1975 accession), p. 139; Leighton Pine to Singer Mfg. Co., December 8, 1869, vol. 113 (1975 accession), Singer Papers. For information on the various styles of woodwork, see "Cabinet Work, June 1, 1870," Box 155, ibid.

30. Pine to Hopper, September 16, 1869.

31. George R. McKenzie wrote to S. A. Bennett in 1882 that "I anticipate there will be clashing between Mr. Pine & Mr. Russell, in the management of the business but I am doing my very best to keep it down & things may go on smoothly—I hope so" (June 6, 1882, ibid.).

32. Leighton Pine to Singer Mfg. Co., December 24, 1870, Box 37, ibid. South Bend's use of veneers supports the view about Wheeler and Wilson's Indianapolis works expressed by Perry and quoted in n. 20 above.

33. The following letters support this statement: Leighton Pine to Singer Mfg. Co., August 13, 1870, January 20, February 2, 10, 1871, Box 37, Singer Papers. There are dozens of letters dated in the 1880s which make clear that Singer's machine factory built South Bend's and Cairo's woodworking machinery.

34. Leighton Pine to Singer Mfg., Co., January 28, 1871, Box 37, ibid.

35. See biographical material on Pine, cited in n. 25.

36. Indianapolis Cabinet Co. to Singer Mfg. Co., March 5, 1881, Box 193; Indianapolis Cabinet Co. to Singer Mfg. Co., April 6, 1881, Box 194; S. A. Bennett to G. R. McKenzie, April 21, June 1, 24, August 3, 1881, Box 193; Jacob Fox to G. R. McKenzie, December 20, 1881, Box 193; numerous letters from Indiana Mfg. Co., Peru, Indiana, to Singer Mfg. Co. Box 194; numerous letters from Mr. Kelsall, Cincinnati, to Singer Mfg. Co., Box 194; Leighton Pine to G. R. McKenzie, June 17, 1881, Box 193, Singer Papers.

37. Sligh Furniture Co. to Singer Mfg. Co., March 2, 1881; W. R. Clark to Singer Mfg. Co., March 3, 1881, Box 193, ibid.

38. Numerous letters in Box 194 of the Singer Papers testify to these complaints. See also Leighton Pine to G. R. McKenzie, July 19, 1882, Box 195, ibid.

39. Pine to McKenzie, February 7, 1881, Box 193. But compare G. R. McKenzie to S. A. Bennett, June 6, 1882, Box 195, ibid.

40. A broad reading of South Bend correspondence to Singer executive offices suggests that style-consciousness and the products of competitors played a crucial role in Singer's adoption of veneer and built-up work. The letter from Leighton Pine to G. R. McKenzie, July 19, 1882, suggests, however, that "when I was put in chge. at So. Bd., Nov. 1879, and our Wal[nut] for tables was about exhausted, . . . we *had* to find an immediate substitute." Yet timber inventories of the period which survive from the South Bend factory do not suggest the exhaustion of walnut supplies. See, for example, Pine to McKenzie, July 26, 1884, Box 198, ibid.

41. Perry, *Modern Plywood,* p. 29. A built-up cabinet, made by Van Dyke & Downs of New York was illustrated in 1876 in *Asher & Adams' Pictorial Album of American Industry,* p. 124. The Van Dyke could have been James Van Dyke, who later helped Singer established its veneer mill in Cairo, Illinois. The sewing machine industry's move toward veneer products fits in nicely with Charles Lock Eastlake's *Hints on Household Taste.* See Mary Jean Smith Madigan, "Influence of Charles Lock Eastlake on American Furniture Manufacture, 1870–1890," pp. 4–5.

42. Pine to McKenzie, March 5, 1881; see also Pine to McKenzie, January 6, 1881, Box 193, Singer Papers.

43. Pine to McKenzie, March 1, May 28, 1881, ibid.

44. On plate drying see Pine to McKenzie, March 5, 1881; James Van Dyke to McKenzie, May 14, 1881, ibid. On box covers see Pine to McKenzie, March 5, 1881.

45. See Pine to McKenzie, February 2, 1881, ibid.: "I am acquainted with a party who has the run of the Domestic Cabinet shops, and can get us any information we require as to their built up work and apparatus. Will have a sketch of their gluing machine in a short time."

46. See the following letters concerning the Indianapolis operation: Indianapolis Cabinet Co. to Singer Mfg. Co., March 5, April 6, 1881; James Fox to McKenzie, March 11, 1881; James Van Dyke to McKenzie, May 14, 1881; Pine to McKenzie, June 27, 1881, ibid.

47. Van Dyke to McKenzie, May 14, 1881.

48. Van Dyke to McKenzie, June 25, 1881, ibid.

49. Ibid.

50. Van Dyke to McKenzie, May 14, 1881.

51. Pine to McKenzie, May 28, 1881; Van Dyke to McKenzie, May 14, 1881, ibid.

52. See, for example, William Noyes, *Wood and Forest,* pp. 180–81, and C. S. Sargent, *Woods of the United States,* pp. 50–51. But compare the views of E. Vernon Knight and Meinrad Wulpi, eds., *Veneers and Plywood,* pp. 168–70.

53. Pine to McKenzie, May 28, 1881.

54. Van Dyke to McKenzie, May 14, 1881; Pine to McKenzie, May 28, 1881. Pine's enthusiasm for

built-up work is revealed in McKenzie to S. A. Bennett, June 6, 1882: "You are perfectly correct about Pine. I have attempted to guide him myself but of late I cannot do it. I cannot tell what he & the others have done at Cairo, but I trust they have not made preparations to do a great deal of made up work, because, by this time, you know the agents on this side demand solid gum wood cabinetwork & that will facilitate & simplify the whole affair in time" (Box 195, ibid.).

55. McKenzie to S. A. Bennett, June 2, 1881, Box 194; Pine to McKenzie, June 17, July 1, 18, 1881; Van Dyke to McKenzie, June 25, July 12, 1881, Box 193, ibid.

56. Pine to McKenzie, July 1, 1881.

57. Van Dyke to McKenzie, July 12, 1881.

58. Pine to McKenzie, July 26, 1881, ibid.

59. Production figures in Pine to McKenzie, February 7, 1883, Box 196, ibid. Blow-by-blow details on the problems may be found in correspondence in Box 195, ibid.

60. Pine to McKenzie, January 16, 1882, Box 195, ibid.

61. See, e.g., Pine to McKenzie, July 19, 1882, ibid.

62. Edward Skillman to McKenzie, June 20, 1882; W. B. Russell to McKenzie, March 9, 1882, and Pine to McKenzie, July 2, 1882 (see also n. 31); Pine to McKenzie, October 7, 1882, ibid.

63. Pine to McKenzie, December 5, 1882, ibid.

64. Knight and Wulpi, eds., *Veneers and Plywood*, p. 169.

65. Pine to McKenzie, February 7, 1883.

66. Box 196, ibid., contains several dozen letters concerning these matters.

67. Pine to McKenzie, May 9, 1883 (see also Pine to McKenzie, March 16, 1883); Pine to McKenzie, January 15, 1884, Box 198 (see also Pine to McKenzie, December 14, 1883, Box 196), ibid.

68. Boxes 197 and 198, ibid., contain letters from 1884.

69. S. A. Bennett to Pine, June 3, 1884; Pine to Singer Mfg. Co., June 6, 1884 (see also S. A. Bennett to McKenzie, June 7, 1884); Pine to McKenzie, June 26, July 24, August 18, 1884, Box 198, ibid.

70. Pine to McKenzie, August 25, 1885, Box 200; see also Pine to Singer Mfg. Co., February 15, 1886, and Pine to Henry Calver, March 15, 1886, Box 203, ibid.

71. See, for example, Pine to Singer Mfg. Co., December 23, 1892, Box 114, ibid.

72. Pine to Singer Mfg. Co., May 5, 1888, Box 209, ibid.

73. Pine to McKenzie, January 17, 1887, Box 203, see also Pine to Singer Mfg. Co., January 1, 1887, Box 205, ibid.

74. U.S. Census Office, *Tenth Census, 1880, Census Reports* (Washington, D.C.: U.S. Government Printing Office, 1883), 2: xvii.

75. Frank Edward Ransom, *The City Built on Wood*, p. 13.

76. "Mammoth New Plant of Singer Manufacturing Company," *Sun News* (South Bend), October 15, 1901.

77. Ransom, *City Built on Wood*, pp. 56, 13.

78. U.S. Bureau of the Census, *Manufactures 1905, Part I, United States by Industries*, p. 9.

79. See Richards, *Treatise*, pp. 32–39, for his discussion of these general themes.

80. Veneered woodworking, especially that of peeled veneer, necessarily made the woodworking industry a heat-using industry. With this method of construction, logs must be boiled for long periods to introduce water into the wood, the veneer must be dried with heat to precise moisture content, and when the veneer is laid on a large scale, heat must be used to expedite the glue-drying process. Bentwood design further increased heat inputs as did the increasingly important adoption of kilns for controlling moisture content in all woods used in construction of products. Potentially, at least, woodworking could be interpreted within Chandler's general framework (see *The Visible Hand*) rather than being cast as an industry in which all change occurred in the antebellum period.

81. U.S. Bureau of the Census, *Manufactures 1905*, pp. clx–clxi, 9.

82. B. A. Parks, "Engineering in Furniture Factories," *Transactions of the American Society of Mechanical Engineers* 42 (1920): 881–82; italics added.

83. *Encyclopaedia Britannica*, 15th ed., s.v. "Furniture Industry."

84. Archer W. Richards, "Mass Production of Radio Cabinets," p. 65.

85. "Carriage and Coaches," pp. 362–63.

86. A brief history of this early period can be found in Albert Russel Erskine, *History of the Studebaker Corporation*. See also Kathleen A. Smallzried and Dorothy J. Roberts, *More Than You Promise*. The expression derives from an unidentified article, probably from a Utah newspaper in 1880, entitled "Studebakers," which I found in a clipping book of the McCormick Reaper Company. This book is now in the McCormick Collection, Accession 2C, vol. 28, State Historical Society of Wisconsin, Madison, Wisconsin. The article contains a lengthy history of the Studebaker company taken from the *Louisville Courier-Journal*.

87. "Studebakers."

88. In the 1850s, both Joseph Whitworth and John Anderson had mentioned wheelmaking machinery in the United States (see note 15 for Anderson's remarks). These observations are well documented for the nineteenth century in Peter Haddon Smith, "The Industrial Archeology of the Wood Wheel Industry in America," pp. 21–66.

89. Erskine, *History of the Studebaker Corporation*, p. 19.

90. Smallzried and Roberts, *More than You Promise*, p. 66, and Studebaker Corporation, *100 Years on the Road*, p. 5.

91. Studebaker Corporation, *100 Years*, p. 8. Much of this history also appears in "Studebakers," 2C, vol. 28, McCormick Collection.

92. "Studebakers."

93. Smith, "Industrial Archeology," pp. 51–54.

94. "Studebakers."

95. Ibid.

96. Studebaker Brothers Manufacturing Company, *Illustrated Catalogue of Studebaker Brothers Manufacturing Company Wagons, Buggies, and Carriages*, pp. 4, 15, and Studebaker Brothers Manufacturing Company, *Illustrated Souvenir of the Studebaker Brothers Manufacturing Company, South Bend, Ind., U.S.A.* The latter is a photographic study of the Studebaker factory; it contains almost no text.

97. For the history of resistance welding, see John B. Schmitt, "The Invention of Electric Resistance Welding." Additional titles in William B. Gamble, "List of Works in the New York Public Library Relating to Electric Welding."

98. "Electric Welding."

99. Frederic A. C. Perrine, "Practical Aspects of Electric Welding."

100. Studebaker Brothers, *Illustrated Catalogue*, p. 15.

101. These photographs appeared in Studebaker Corporation, *100 Years*, p. 6.

102. Studebaker's *Illustrated Souvenir* of 1893, however, showed some automatic hub mortising machinery in the South Bend factory.

103. Erskine, *History of the Studebaker Corporation*, pp. 25–27, 14, 23.

Chapter 4

1. This story has been told many times. The best works to consult on the long history of the reaper are Robert L. Ardrey, *American Agricultural Implements*, pp. 40–52, and William T. Hutchinson, *Cyrus Hall McCormick: Seedtime, 1809–1856*, chaps. 3, 4, 5, 7, 8.

2. Hutchinson, *McCormick: Seedtime*, p. 182. I have relied on Hutchinson, particularly chapters 8 and 13, for the history of McCormick's early reaper production.

3. C. H. McCormick to Wm. S. McCormick, November 15, 1846, Series M/I, Box 18, McCormick Collection, State Historical Society of Wisconsin, Madison. All manuscripts, unless otherwise noted, will be cited like this one. The series M/I identifies the C. H. McCormick Estate Papers. Other series cited below are 1A (C. H. McCormick), 2A (C. H. McCormick), X (McCormick Harvesting Machine Company), and 4C (C. H. McCormick, Jr.).

4. Hutchinson, *McCormick: Seedtime*, p. 198; see Hutchinson's chapter "Advertising the Reaper and Mower," pp. 327–49.

5. C. H. McCormick to Wm. Maguire, December 24, 1842, 1A, Box 2; C. H. McCormick to Wm. S. McCormick, August 6, 1845, 1A, Box 3.

6. C. H. McCormick to Wm. S. McCormick, December 11, 1846, 1A, Box 3.

7. C. H. McCormick to Leander McCormick, January 8, 1847; C. H. McCormick to Seymour, Chappell & Co., March 19, 1847, ibid.; Hutchinson, *McCormick: Seedtime*, p. 307.

8. Hutchinson, *McCormick: Seedtime*, pp. 259–62. The very complicated story of the dispute is detailed in ibid., pp. 256–66.

9. Ibid., pp. 265–68.

10. January 18, 1849, quoted in ibid., pp. 268–69.

11. Ibid., p. 269.

12. *Chicago Weekly Democrat*, March 29, 1851, quoted in Hutchinson, *McCormick: Seedtime*, p. 270.

13. Wm. S. McCormick to Jas. D. Davidson, March 24, 1851, 1A, Box 4; Hutchinson, *McCormick: Seedtime*, p. 271.

14. These contracts are in 1A, Box 3.

15. Contract with Magnus Norbo and Tobias Jackson, January 20, 1851, 1A, Box 4; Hutchinson, *McCormick: Seedtime*, p. 325.

16. Hutchinson, *McCormick: Seedtime*, p. 317.

17. Hutchinson cites an article from the *Chicago Tribune* of 1859 which describes this process (ibid., p. 308).

18. L. J. to C. H. McCormick, July 8, 1859, 1A, Box 15; for a request for a wooden part, see C. H. McCormick (by Hanna) to H. H. Tarbell, May 16, 1856, 1A, Box 8.

19. McCormick's dependence on skilled workers in all of the factory departments is reflected not only in contemporary accounts of the factory but in the solid unionization of the shops during and after the Civil War. James Campbell carried on an extensive correspondence with Leander about the acquisition of machine tools—lathes, presses, and planers—for the McCormick works. All tools mentioned are general machine tools. See L. J. McCormick folder, 1A, Box 6.

20. The duties of the brothers emerge from an extensive survey of their correspondence with each other. For an agent's complaint, see Misc. Ms. 1A, Box 4.

21. C. H. McCormick (by Wilson) to Alfred Johnson, May 2, 1856, 1X, vol. 1, p. 259; Wm. S. to Cyrus McCormick, May 16, 21, 23, 1856, 1A, Box 9.

22. W. J. Hanna to Wm. S. McCormick, July 17, 1856, 1A, Box 8.

23. C. H. to Wm. S. McCormick, September 2, 1856, ibid.

24. C. H. to L. J. [or Wm. S.] McCormick, August 13, 1856, ibid.

25. Ibid.; C. H. to Wm. S. McCormick, September 6, 1856, ibid.; Wm. S. McCormick to James Campbell, November 12, 1856, 1A, Box 9; see also Wm. S. to C. H. McCormick, September 12, 1856, ibid.

26. L. J. to C. H. McCormick, March 14, 1857, Wm. S. to C. H. McCormick, May 30, 1857, 1A, Box 11; C. H. to Wm. S. & L. J. McCormick, December 10, 1857, 1A, Box 10; C. H. to Wm. S. McCormick, October 30, 1858, 1A, Box 14.

27. C. H. to Wm. S. McCormick, May 2, 21, 1857, 1A, Box 10; arrangements with Campbell discussed in C. H. to Wm. S. McCormick, May 30, 1857, ibid.; "Statement 1849–1857," November 6, 1858, 1A, Box 13.

28. Wm. S. to C. H. McCormick, February 1, 1859, 1A, Box 15.

29. William T. Hutchinson, *Cyrus Hall McCormick: Harvest, 1856–1884*, pp. 109–10, gives the complete details of this partnership.

30. Ibid., pp. 127, 130–33.

31. Stock Bond (Mfg.), 1860–79, 3X [Jallings no. 55]. This volume contains a "Memorandum of Machinery in Reaper Factory when put in hands of McCormick & Bros. Nov. 1, 1859 still being property of Cyrus H. McCormick," pp. 381–83, and an "Inventory of Tools, etc. in Reaper Factory Nov. 1, 1859, charged up to C. H. McCormick & Bros.," pp. 374–80. The enumeration of tools below is derived from this stock book.

32. For an enumeration of the machinery ordered by the Anderson Committee see *Report of the Committee on the Machinery of the United States of America*, in Nathan Rosenberg, ed., *The American System of Manufactures*, pp. 180–92.

33. Hutchinson, *McCormick: Harvest*, is sympathetic to the McCormicks' problems during the Civil War.

34. There is a remarkable contrast between Leander's nonresponse to strikers during the Civil War and Cyrus McCormick, Jr.'s, actions taken after a long strike by foundrymen in 1885. As soon as the molders struck in 1885, McCormick investigated and soon adopted pneumatic molding machinery, eliminating the need for skilled molders. This chapter attempts to explain the differences in attitude between Cyrus McCormick's brother and his son.

35. Wm. S. to C. H. and L. J. McCormick, September 27, 1862, 1A, Box 17.

36. Ibid.; L. J. to C. H. McCormick, April 7, 1863, ibid.

37. L. J. to C. H. McCormick, undated (probably January–March 1863); Wm. S. to C. H. McCormick, March 1, 1863, ibid.

38. L. J. to C. H. McCormick, undated (probably March or April 1863), ibid. I have been unable to determine the nature of this new machinery and of the fixtures.

39. L. J. to C. H. McCormick, April 7, 1863; Wm. S. to C. H. McCormick, April 8, 15, 1863, ibid. (see also Wm. S. to Cyrus McCormick, April 11, 1863, ibid.); Wm. S. to C. H. McCormick, June 7, 1863, ibid.

40. Wm. S. to C. H. McCormick, July 12, 1863, ibid.; the clerk is quoted by Hutchinson, *McCormick: Harvest*, p. 89; Wm. S. to C. H. McCormick, August 19, 1863, 1A, Box 17.

41. L. J. to C. H. McCormick, August 20, 1863; Wm. S. to C. H. McCormick, September 27, 1863, 1A, Box 17.

42. Cyrus had unjustly accused Leander of "cruel treatment" in not finishing an adequate number of machines for his European sales. This criticism greatly angered Leander. See L. J. to C. H. McCormick, August 8, 1863, ibid.

43. Wm. S. to C. H. McCormick, October 4, 1863, ibid.

44. L. J. to C. H. McCormick, November 22, December 9, 1863, ibid.

45. L. J. to C. H. McCormick, January 15, 1864; Wm. S. to C. H. McCormick, January 24, February 7, 21, December 21, 1864, 1A, Box 18.

46. L. J. to C. H. McCormick, October 7, November 29, 1865, 1A, Box 19.

47. "New Machinery, in a/c with C. H. McCormick," 2A, Box 45.

48. C. H. McCormick & Bro. to Crosby Bros., June 5, 1866, 1X, vol. 90, p. 617.

49. C. A. Spring to C. H. McCormick, August 21, 1886, 1A, Box 20.

50. L. J. to C. H. McCormick, October 7, 1867, 1A, Box 24.

51. In its sales catalog for 1868, the McCormick company described its factory and falsely alleged the uniformity of its reaper parts: "[Our factory] is filled with the newest and most approved machinery, directed by skillful mechanics, under our own personal care and control, and our manufacturing facilities are such that we produce during the season at the rate of one complete machine every ten minutes. The result is, that if one machine works well, all must do so, for all the different parts are so accurately made by machinery, that one piece will fill its allotted place in any one of ten thousand machines, one machine being an exact counterpart of all of the same style in one year" (*McCormick's Prize Harvester* . . . [1868], p. 5, 2A, Box 15).

52. L. J. to C. H. McCormick, August 8, 1869, 1A, Box 33.

53. L. J. to C. H. McCormick, September 10, 11, October 12, 1869, ibid.

54. L. J. to C. H. McCormick, October 12, 1869.

55. L. J. to C. H. McCormick, August 30, September 1, 1870, 1A, Box 36.

56. L. J. to C. H. McCormick, September 13, 1870, ibid. Unfortunately, I have been unable to secure any more information about these McCormick-built lathes. Leander reported two months later that they worked "well" (L. J. to C. H. McCormick, November 11, 1870, ibid.).

57. L. J. to C. H. McCormick, August 3, 1870, ibid.; L. J. to C. H. McCormick, March 2, 1871, 1A, Box 39.

58. C. H. to L. J. McCormick, March 13, 1871, 4A, Box 2, vol. 1, 2d ser., pp. 433–35.

59. L. J. to C. H. McCormick, July 12, August 10, 1871, 1A, Box 39.

60. L. J. to C. H. McCormick, August 10, 1871.

61. L. J. to C. H. McCormick, August 22, 1871, ibid.

62. The actual number of machines burned was 1,969 ("Machs Sold, Burnt & on hand," February 24, 1872, 1A, Box 45).

63. "Estimate of Loss by Fire Oct. 9, 1871," ibid.

64. C. H. McCormick & Bro. to L. W. Pond, March 5, 1871, 1X, L.P.C.B. 132, p. 524; Pond to C. H. McCormick & Bro., March 9, 1872, 2X, Box 154; C. H. McCormick & Bro. to William Sellers, December 21, 1871, 1X, L.P.C.B. 131, p. 273; C. H. McCormick & Bro. to Stiles & Parker, April 18, 1872, 1X, L.P.C.B. 133, pp. 401–2; N. C. Stiles, Stiles & Parker, to C. H. McCormick & Bro., April 29, 1872, 2X, Box 156. Argument about the price was waged between the two firms until the end of August 1872, when Stiles offered to accept $400 for them so that he could be done with the matter (Stiles to C. H. McCormick & Bro., August 29, 1872, ibid.).

65. Nowhere in the records of the McCormick company in this period are there references to turret lathes, milling machines, or grinding machines. In his efforts to reequip the factory, Leander carried on correspondence with the following machine tool companies: L. W. Pond, New York Tool Company, Brown & Sharpe, Corliss, Sellers, Plumb & Burdict (Buffalo Bolt & Nut Works), Putnam Machine Co., RS & J Gear & Co., Simonds Manufacturing Co., Lane & Bodley, Providence Bolt & Screw Co., Rollstone Machine Works, Lowell Machine Works, Fitchburg Machine Co., New York Steam Engine Co., Ball & Co., and Stiles & Parker. All the tools Leander sought were general-purpose machine tools. These inquiries and purchases may be found in 1X, L.P.C.B. 130–40.

66. C. A. Spring to C. H. McCormick, July 10, 1872, 1A, Box 47; see letters from L. J. to Cyrus McCormick, July 18, 21, 25, 1872, 1A, Box 45, and from C. A. Spring to L. J. McCormick, July 26, 1872, 1A, Box 47.

67. L. J. to C. H. McCormick, May 17, July 15, 1873, 1A, Box 50; italics added.

68. C. H. to L. J. McCormick, August 1, 1873, ibid.

69. C. H. to L. J. McCormick, November 17, 1873, ibid.

70. "McCormicks' New Reaper Works," January 10, 1873, 2A, Box 15. This flyer described the "great works": "The buildings, five stories high, occupy three sides of a square, each side 360 feet long, with the engine room and a three-story middle building between the two wings. The floor surface of the works would cover an area of six acres."

71. See C. H. McCormick to L. J. McCormick, March 19, 1877, 1A, Box 68.

72. Chas. Colahan to C. H. McCormick, June 14, 1877, 1A, Box 67. The self-binder was not a McCormick invention. Its history is treated in Ardrey, *American Agricultural Implements*, pp. 64–77. The McCormick version of this invention is well illustrated in McCormick Harvesting Machine Co., *Triumph Throughout All Nations*, pp. 10–13.

73. Chas. Colahan to C. H. McCormick, June 30, 1877, 1A, Box 67.

74. "Estimate of machines Sold & on Hand in 1877," 1A, Box 68, and "Estimate of machines Sold & on Hand 1878," 1A, Box 72.

75. Chas. Colahan to C. H. McCormick, June 15, 1878, 2A, Box 51. Colahan would later write from Brockport, New York, that, unlike the McCormick reaper works, "the Manufacturers here are progressive" (Chas. Colahan to C. H. McCormick, July 8, 1879, 2A, Box 52).

76. See C. H. to L. J. McCormick, June 17, 1878, 1A, Box 72.

77. See, for example, L. J. to C. H. McCormick, August 5, 1878, 2A, Box 51; *McCormick: Seedtime*, chap. 5, "The McCormick Reaper Controversy," and *McCormick: Harvest*, passim.

78. Geo. B. Averill [a foreman at the McCormick works] to C. H. McCormick, July 17, 1879, 2A, Box 52.

79. Articles of association, August 11, 1879, 1A, Box 74.

80. See Charles Spring's account of this meeting to Cyrus McCormick (who was absent) in a letter of September 9, 1879, 2A, Box 52.

81. C. A. Spring to C. H. McCormick, October 2, 5, 18, 1879, ibid.

82. J. P. Whedon to C. H. McCormick, Jr., October 20, 1879, 2A, Box 31.

83. C. A. Spring to C. H. McCormick, September 17, 1879; Chas. Colahan to C. H. McCormick, October 1, November 10, 1879; C. A. Spring to C. H. McCormick, November 15, 1879; Chas. Colahan to C. H. McCormick, November 30, 1879, 2A, Box 52.

84. I have not seen this statement, although

William Hutchinson notes it in *McCormick: Harvest*, p. 638.

85. On April 6, 1880, Hall and Leander submitted a printed pamphlet to the board calling Cyrus's statement "paltry" and "of a personal nature." The same day, Hall submitted his letter of resignation and Leander his letter to advise the board of his six-month trip to Europe. These communications are noted in the Original Minute Book of Board of Directors of the McCormick Harvesting Machine Company, September 9, 1879–April 6, 1886, M/I, vol. 35, Box 24. Leander's letter is in M/I, Box 3. Cyrus submitted to the board a lengthy statement to preface his proposal to fire Leander, which appears in the above-cited minute book. In this document, Cyrus included statements signed by various foremen at the factory and office personnel which documented Leander's and Hall's "want of interest in the business at the Works" and lack of attendance. See pp. 41–43 of the minute book.

86. Production figures from 1880 to 1902 may be found in M/I, Box 18.

87. See C. H. McCormick, Jr.'s, pocket diary accounts about the hiring of Wilkinson, April 30, May 1, 6, 1880, 4C, vol. 8, Box 1, and his main diary entry, May 6, 1880, 4C, vol. 7, Box 1.

88. C. H. McCormick, Jr., diary entry, May 13, 1880, 4C, vol. 7, Box 1.

89. Draft of letter, C. H. McCormick to Henry Day, December 27, 1880, 1A, Box 76; italics added.

90. See Cyrus McCormick, Jr.'s, pocket diary for 1881, 4C, vol. 13, Box 1.

91. C. A. Spring to C. H. McCormick, November 2, 1880, 2A, Box 53.

92. See Cyrus McCormick, Jr.'s, pocket diary for 1881, 4C, vol. 13, Box 1.

93. "American Industries—No. 73, Harvesting Machines," p. 307. See Cyrus McCormick, Jr.'s, pocket diary of 1881 for the background of the *Scientific American* article. The outline of the article and its manuscript appear in 4A, vol. 5, Box 2, pp. 408–26. In his outline (dated April 12, 1881), under a section on the ironworking department, the young McCormick planned a subsection on "special machines & jigs." For more information on this article, see David A. Hounshell, "Public Relations or Public Understanding?"

94. See the McCormick exposition booklet, *La Compagnie MacCormick de Chicago a l'Exposition de 1900*, pp. 60–63, a copy of which may be found in 6X, Box 2.

95. See Cyrus McCormick, Jr.'s, diary entry of September 7, 1881, 4C, vol. 12, Box 1, which describes the way he announced his superintendency and his hopes for the factory's operation. See also his letter to his father, September 8, 1881, 1A, Box 80.

96. C. A. Spring to C. H. McCormick, September 26, 1881; C. A. Spring to C. H. McCormick, Jr., November 14, 1881, 2A, Box 53.

97. C. H. McCormick to Henry Day, February 13, May 5, 1882, 1A, Box 85; C. H. McCormick, Jr., to C. H. McCormick, August 1, 1882, 2A, Box 53; C. A. Spring to C. H. McCormick, March 30, 1883, 2A, Box 54.

98. June 6, 1883, 4C, vol. 15, Box 1.

99. June 30, 1884, 4C, vol. 18, Box 18; C. H. McCormick & Bro. to Crosby Bros., June 5, 1866.

100. See R. L. Ardrey, "The Harvesting Machine Industry." Cyrus McCormick, Jr.'s, diaries reveal his continued interest in the factory's operations. See, for example, his entry of April 3, 1899, when he noted, "at works: taking up with the foremen the question of output" (4C, vol. 41, Box 3).

101. Robert Ozanne, "Prelude to Haymarket," chap. 1 of *A Century of Labor-Management Relations at McCormick and International Harvester*, pp. 3–28. For a brilliant discussion of mechanization and control, see Harry Braverman, *Labor and Monopoly Capital;* also see David Montgomery, "Whose Standards? Workers and the Reorganization of Production in the United States, 1900–20," in *Workers' Control in America*, pp. 113–38.

102. William J. Abernathy, *The Productivity Dilemma*.

Chapter 5

1. John Higham, "The Reorientation of American Culture in the 1890s," offers a profound interpretation of this period.

2. Reprinted in A. Hunter Dupree, ed., *Science and the Emergence of Modern America* (Chicago: Rand McNally & Co., 1963), pp. 52–56; quotation on p. 54. On McGee see *Dictionary of American Biography*, s.v. "McGee, William John"; Gifford Pinchot, *Breaking New Ground* (New York: Harcourt Brace, 1947), pp. 358–60; and Curtis M. Hinsley, Jr., *Savages and Scientists* (Washington, D.C.: Smithsonian Institution Press, 1981), pp. 231–61.

3. For elaboration on the role of the bicycle as a precursor of the automobile see David A. Hounshell, "The Bicycle and Technology in Late Nineteenth Century America"; Martha Moore Trescott, "The Bicycle"; John B. Rae, *American Automobile Manufacturers*, pp. 8–23; and Allan Nevins and Frank Ernest Hill, *Ford: The Times, the Man, the Company*, pp. 186–90.

4. The best introduction to the bicycle in the United States is Norman Leslie Dunham, "The Bicycle Era in America." Robert A. Smith, *A Social History of the Bicycle*, is also helpful but less adequately

documented. Albert A. Pope's own analysis, "The Bicycle Industry," is reliable and useful.

5. Biographical information on Pope may be found in *Dictionary of American Biography*, s.v. "Pope, Albert A."; and "Colonel Albert A. Pope."

6. Complete production figures are hard to come by. The best source is Dunham, "Bicycle Era," pp. 466–68. See also Trescott, "The Bicycle."

7. Arthur S. Dewing, *Corporate Promotions and Reorganizations*, p. 268.

8. This tradition was begun by Henry P. Woodward, "Manufactures in Hartford," p. 178.

9. "The Manufacture of Sewing Machines," p. 181; "Bicycle Manufacturing," p. 204.

10. Nathan Rosenberg, "Technological Change in the Machine Tool Industry, 1840–1910," p. 423.

11. Pope stressed the interchangeability of parts in all his promotional literature. See, for example, Pope Manufacturing Company, *Columbia Bicycle, Catalogue for January, 1881*, p. 4. The Pope company history notes that "an order was placed with the Weed Sewing Machine Company, of Hartford, Connecticut, a concern well equipped for the work and able to take it on to advantage as a supplement to their sewing machine business which was just then beginning to lag" (*An Industrial Achievement*, p. 11).

12. Guy Hubbard, "Development of Machine Tools in New England," 60 (1924): 171–73.

13. Pope Manufacturing Company, *An Industrial Achievement*, pp. 29–46.

14. Joseph W. Roe, *English and American Tool Builders*, pp. 173–76.

15. "Manufacture of Sewing Machines," p. 181, and "A Great American Manufacture." In addition, the trade catalogs of the Pope Manufacturing Company contain information on the production of Columbias. For the names and locations of these catalogs, see Lawrence B. Romaine, *Guide to American Trade Catalogs, 1744–1900* (New York: R. R. Bowker, 1960), p. 61.

16. For a brief survey of these machine tools, see L.T.C. Rolt, *A Short History of Machine Tools*, pp. 154–77.

17. "A Great American Manufacture," p. 329.

18. Robert S. Woodbury, *History of the Grinding Machine*, pp. 109–12. For a contemporary account of manufacturing ball bearings, see "Making Balls for Bearings."

19. In his report Whitworth wrote: "The complete musket is made (by putting together the separate parts) in 3 minutes" (*Special Report of Joseph Whitworth*, in Nathan Rosenberg, ed., *The American System of Manufactures*, p. 365). Compare this with "A Great American Manufacture."

20. Dunham, "Bicycle Era," p. 179. Pope wrote in 1895: "It became necessary also, at the outset to educate the people to the advantage of this invigorating sport [of bicycling], and, with this end in view, the best literature that was to be had on the subject was gratuitously distributed" ("Bicycle Industry," p. 551).

21. Philip Parker Mason, "The League of American Wheelmen and the Good-Roads Movement, 1880–1905."

22. "A Great American Manufacture," p. 326.

23. Smith, *Social History of the Bicycle*, pp. 31–33.

24. These prices appear in the annual catalogs of the Pope Manufacturing Company.

25. The clearest and most detailed account of Pope's patent strategy appears in Pope Manufacturing Company, *An Industrial Achievement*, pp. 11–14.

26. Pope, "Bicycle Industry," p. 551.

27. Dunham, "Bicycle Era," p. 196.

28. Ibid., p. 195.

29. See Overman's catalog of Victor bicycles for 1895 (Eleutherian Mills Historical Library) and Vera Shlakman, *Economic History of a Factory Town*, pp. 165–66, 199–200. Horace L. Arnold [Hugh Dolnar] wrote in his *American Machinist* series on bicycle tools that the Overman company operated strictly along the armory line of manufacture ("Bicycle Tools—XII," 19 [1896]: 52).

30. *National Cyclopaedia of American Biography*, s.v. "Johnson, Iver," 33:301.

31. For the history of resistance welding see John B. Schmitt, "The Invention of Electric Resistance Welding."

32. "Making a Bicycle"; Frederick A. C. Perrine, "Practical Aspects of Electric Welding"; and Elihu Thomson, "Electric Welding."

33. Fred H. Colvin, *60 Years with Men and Machines*, pp. 85–86. A qualification is needed here. A few bicycle manufacturers and partsmakers developed the technique of electrically welding frames. Two were the George L. Thomson Mfg. Co. and the Independent Electric Company, both of Chicago. See "Decidedly the Greatest."

34. Dunham, "Bicycle Era," p. 404. The following works also deal with the safety bicycle craze: Richard Harmond, "Progress and Flight"; Gary Allan Tobin, "The Bicycle Boom of the 1890s"; and James C. Whorton, "The Hygiene of the Wheel."

35. Dunham, "Bicycle Era," p. 468. Production figures for the bicycle vary greatly. For instance, *Literary Digest* 13 (June 13, 1896): 196 and *Scientific American* 62 (1896): 69 stated that four million bicycles would be produced in 1896. The bicycling trade journals gave even higher estimates. I have relied on Dunham's careful research on these figures and on Pope, "Bicycle Industry," p. 551.

36. U.S. Bureau of the Census, *Eleventh Census of the United States (1890), Report on Manufacturing,*

Part 1, p. 126. For the remaining years of the 1890s, Dunham, ''Bicycle Era,'' p. 468, gives annual production figures.

37. These firms have been culled out of the cycle trade literature, including the *Wheel* and the *Bicycle World;* Romaine, *Guide to American Trade Catalogs,* pp. 57–63; *American Machinist;* and *Iron Age.*

38. Pope Manufacturing Company, *An Industrial Achievement,* p. 11; Pope Manufacturing Company, *Columbia Bicycles, 1895,* p. 31; Leonard Waldo, ''The American Bicycle,'' p. 50.

39. Cleveland Moffett, ''A Visit to the Works of the Pope Tube Company,'' Part IV of *Marvels of Bicycle Making,* pamphlet in Warshaw Collection of Business Americana, Smithsonian Institution, unpaginated; reprinted from *McClure's Magazine,* 1897. See also ''The Manufacture of Bicycle Tubing.'' For information on the rubber works see Moffett, ''A Visit to the Hartford Rubber Works,'' Part III of *Marvels of Bicycle Making,* unpaginated, and Pope Manufacturing Company, *An Industrial Achievement.*

40. *National Cyclopaedia of American Biography,* s.v. ''Pope, George,'' 18:227. The secrecy behind the ownership of the Hartford company is revealed in a letter from Albert Pope to David J. Post, secretary of the Hartford company, July 24, 1880. This letter is part of a small manuscript collection of Hartford Cycle Company papers at the Connecticut State Library, Hartford, Connecticut.

41. This practice seems to have been common among bicycle manufacturers. See Arnold [Dolnar], ''Bicycle Tools—XVIII,'' 19 (1896): 231.

42. Later Pope Manufacturing Company sales literature and ''A Visit to the Works of the Hartford Cycle Company,'' one of a series of articles on the Pope Manufacturing Company in Moffett, *Marvels of Bicycle Making,* claimed independent status. Virtually none of the existing letters, however, almost all of which are from George Pope to David J. Post, fails to mention what action A. A. Pope and George Day had taken on that day (Hartford Cycle Company Papers). Cleveland Moffett described George Day as A. A. Pope's ''right-hand man'' (''Visit to the Works of the Pope Manufacturing Company'').

43. The small collecton of letters in the Hartford Cycle Company Papers suggest the agent arrangement.

44. For example, in 1895 Pope staged an impressive exhibition of posters in Washington, D.C., which included Maxfield Parrish's award-winning artwork. A copy of the exhibit catalog is in the bicycle files of the Smithsonian Institution, Washington, D.C.

45. Pope, ''Bicycle Industry,'' pp. 552–53; Pope Manufacturing Company, *An Industrial Achievement,* pp. 14–15; Mason, ''League of American Wheelmen,'' particularly chaps. 3–5. Mason suggests that Pope provided $6,000 to the LAW to begin its *Good*

Roads Magazine (p. 184). Pope wrote that ''we spent over $8,000 in the Central Park case alone'' and succeeded in getting the bicycle classed as a vehicle (''Bicycle Industry'' p. 551).

46. Arnold [Dolnar], ''Bicycle Tools—XII,'' 19 (1896): 52.

47. ''Influence of the Bicycle upon Machine Tools,'' and Colvin, *60 Years,* pp. 84–88. As these works emphasize, by 1895–96 the machine tool industry was marketing a complete line of bicycle-making machine tools. A manufacturer could go to a company such as Pratt & Whitney and purchase what was in effect a ''turnkey'' factory.

48. The company still prided itself on using ''solid steel'' ball bearing races for its hubs: ''We are not willing under any circumstances to resort to the use in Columbias of the pressed sheet steel cases now coming into common use on account of their cheapness'' (*Columbia Bicycles, 1895,* p. 25).

49. Arnold [Dolnar], ''Bicycle Tools—XX,'' 19 (1896): 276.

50. Moffett, ''A Visit to the Works of the Pope Manufacturing Company,'' in *Marvels of Bicycle Making.* Horace L. Arnold [Hugh Dolnar], ''Cycle Stampings,'' notes the movement away from dropforging on the part of armory-tradition manufacturers. Arnold covered the entire industry's techniques of hubmaking in his ''Bicycle Tools'' series, 19 (1896): 252–55, 275–76, 348–50, and 474–76. Sometime in 1895, Pope apparently abandoned its recently developed bar stock manufacturing techniques and adopted press techniques for making what was dubbed the ''barrel hub'' (see Pope Manufacturing Company, *Columbia Bicycles, 1896,* pp. 13, 35). Press techniques were also adopted for the crank hanger at the same time.

51. Arnold [Dolnar] makes this argument in his article on hubmaking in the articles cited in note 50.

52. Moffett, ''A Visit to the Works of the Pope Manufacturing Company,'' *Marvels of Bicycle Making.* This work forms the basis of subsequent discussions of Pope production technology and will not be specifically cited hereafter.

53. ''The Story of a Success.''

54. Arnold [Dolnar], ''Bicycle Tools—XXXI,'' 19 (1896): 896.

55. Arnold [Dolnar], ''Bicycle Tools—XXVIII,'' 19 (1896): 677. In an earlier article the author noted that ''the work was very easy on the man, which is a prime requirement of New England tools. The true Yankee rebels against the first quarter ounce of needless muscular effort'' (18 [1895]: 1022).

56. Arnold [Dolnar], ''Bicycle Tools—XXIX,'' 19 (1896): 738.

57. Horace L. Arnold [Hugh Dolnar], ''Bicycle Brazing,'' p. 1080. Most of my account of brazing is based upon this article.

58. According to the letters from George Pope to David J. Post, the Hartford Cycle Company often fell behind on its frame filing. See letters of April 10, 1891, February 4, 8, 1892, Papers of the Hartford Cycle Company.

59. Arnold [Dolnar], "Bicycle Tools—XXIX," 19 (1896): 739.

60. George Pope to David J. Post, April 6, 1891, Papers of the Hartford Cycle Company.

61. For example, George Pope to David J. Post, April 10, 1891, ibid.

62. The Gormully and Jeffery shops, organized along Yankee armory lines, developed automatic rim-polishing machines. They built six of these machines and, surprisingly, the entire bank of machines could polish only fifty rims per ten-hour day. This figure gives meaning to my argument that rim polishing and other processes were time-consuming and laborious. See Arnold [Dolnar], "Bicycle Tools—VIII," 18 (1895): 963–64. I do not know if Pope designed any such machinery.

63. George Pope to David J. Post, January 1, 1892, Papers of the Hartford Cycle Company.

64. Discussing the new line of bicycles shown at the 1896 Chicago Cycle Show, editors of the *Wheel* noted: "Nearly all of the innovations mentioned were introduced last year by the Pope Co. There are those who dispute Pope's leadership, but prejudice aside, it was never more apparent than now that he should have inaugurated this year's changes twelve months ago, and that his alteration of such petty detail as a name-plate should cause alterations all along the line are facts that will not down. The Columbias, and indeed, nearly all of the old wheels seem to have approached that finality of pattern and construction so long expected" (" 'I Will': Chicago True to Her Motto in the Carrying out of Her Cycle Show. . . .") A. H. Overman, Pope's chief competitor in New England, also prided himself on his company's rigorous system of "scientific testing." See "Scientific Methods Applied to Bicycle-making." For more information on Pope testing methods see "Testing the Parts of a Modern Bicycle," and Rae, *American Automobile Manufacturers,* pp. 9–10.

The development of spoke swaging was of immense importance. Until 1891, bicycle spokes were made merely by cutting off wire to spoke length and threading the ends. In their efforts to lighten the bicycle, Pope mechanics came up with the idea of cutting off all unnecessary metal from the spokes. Testing the strength of wheels assembled with various thicknesses of spokes, Pope mechanics found that small-diameter spokes would support a wheel adequately but could not be properly secured to the hub and rim. They decided that they could make a spoke with ends large enough to secure them properly and then trim off some of the steel from the middle of the spoke. Arriv-

ing at the correct shape for this lighter spoke, Pope mechanics built a special machine to manufacture it. They called it a swaging machine, and it operated through a combination of pulling wire through a set of trimming dies and hammering it with a set of rapidly striking hammers. About the same time, New England sewing machine needle manufacturers were developing swaging machines to reduce needle shanks. Formerly, they had used grinding techniques.

65. George Pope to David J. Post, January 24, 1893, Papers of the Hartford Cycle Company.

66. A. E. Harrison, "The Competitiveness of the British Cycle Industry, 1890–1914," p. 298.

67. Dunham, "Bicycle Era," p. 466.

68. The Durant-Dort Carriage Co., for instance, also made bicycles; see Lawrence R. Gustin, *Billy Durant,* p. 41; Arnold [Dolnar], "Bicycle Tools—I," 18 (1895): 781; see also, Trescott, "The Bicycle."

69. This would be an exciting aspect of the American system of manufactures to take up, particularly with the bicycle, but little solid information on precision and prices exists in the published record. The new demand for precision on the toymakers is noted by Arnold [Dolnar], "Bicycle Tools—II," 18 (1895): 801.

70. Ibid.

71. Arnold [Dolnar], "Cycle Stampings," p. 1163. The Germanic character of the Western product is suggested by the names of high-wheel bicycles made in 1887 by the Western Toy Company: the Otto and the Otto Special (*Price List for Spring 1887, The Western Toy Company*). Compare the Western Wheel Works' history with that of Crosby & Mayer Co. of Buffalo, New York. The latter claimed to be "one of the earliest and most persistent advocates of sheet steel frame connections, and produced from sheet steel metal the first crank hanger"; (*Catalogue and Price List of Sheet Steel Parts Made by Crosby & Mayer Co.,* p. 1). Unfortunately, the history of stamping or presswork is obscure. As Oberlin Smith, the creator of one of the major punch press manufacturing companies, wrote in his 1896 treatise *Press-Working of Metals:* "As far as the writer knows, the literature of this subject, outside of press-makers' catalogues, is extremely limited. It is, however, to be hoped that in the not too far off future somebody will give the world a comprehensive biography of a family of machines which are far too useful to remain much longer in the realms of literary oblivion" (p. 42). So far, no one has taken up Smith's plea. A brief and incomplete survey is Carter C. Higgins, "The Pressed Steel Industry." As early as 1851 the American Joseph Francis was stamping out metal lifeboats on a hydraulic press. See *Harper's New Monthly Magazine* 3 (June 1851): 165 and "Francis's Corrugated Metallic Boats." Other information on stamping may be obtained in Edward H. Knight, *Knight's American Mechanical Dictionary,* 3:

2302–4; Park Benjamin, *Appleton's Cyclopaedia of Applied Mechanics*, 2: 566–70; *80 Years Ferracute;* Bliss (E. W.) Company, *Catalogue and Price List of Presses, 1886;* and Bliss (E. W.) Company, *Catalogue and Price List of Presses, 1900,* esp. p. 452. As Arnold [Dolnar] pointed out in "Cycle Stampings," it appears that pressing or stamping sheet steel in particular was a technique transferred to the United States by German emigrants. Although American press manufacturers such as Ferracute and Bliss had been in business for some years, these companies had sold equipment primarily to tinware manufacturers and food canning companies. Certainly, if Arnold was correct, they had not developed presswork to the degree it had been in Germany. We desperately need a history of this important technology.

72. Western Wheel Company, *Crescent Bicycles,* pp. 18–19. A brief description of the Western Wheel Works appeared in "The Chicago Trade," *Wheel* 6 (1890): 472–74.

73. The following account is based on ibid.; Arnold [Dolnar], "Cycle Stampings," pp. 1163–67; "Press and Die Work on Bicycles," *American Machinist* 19 (1896): 1097–1100; and the following pieces in Arnold's [Dolnar] series "Bicycle Tools": 18 (1895): 781–82, 19 (1896): 50–52, 252–55, 474–76, 736–39, 871–73, and 894–97.

74. George Pope to David Post, January 12, 1891, Papers of the Hartford Cycle Company.

75. For more description of these machines, see Arnold [Dolnar], "Bicycle Tools—XVI" and "Bicycle Tools—XXXI," 19 (1896): 183, 894–97.

76. Arnold [Dolnar], "Bicycle Tools—XII," 19 (1896): 52; "Bicycle Tools—XXVII," 19 (1896): 657.

77. Pope, "Bicycle Industry," p. 553.

78. Hiram Percy Maxim, *Horseless Carriage Days,* pp. 1–2.

79. Ibid., pp. 4–5.

80. Joseph V. Woodworth, *American Tool Making and Interchangeable Manufacturing,* p. 516.

Chapter 6

1. Siegfried Giedion, *Mechanization Takes Command.*

2. Allan Nevins and Frank Ernest Hill, *Ford: The Times, the Man, the Company,* maintains that the changes made in 1913 and 1914 provided "a lever to move the world" (p. 447). On Fordism, see Charles S. Maier, "Between Taylorism and Technocracy"; Emma Rothschild, *Paradise Lost,* pp. 26–53; and Carl Rauschenbush, *Fordism.*

3. Henry Ford to the editor, *Automobile* 14 (1906): 107–19, quoted in John B. Rae, ed., *Henry Ford,* pp. 18–19.

4. See, for example, Nevins and Hill, *Ford,* pp. 323–53.

5. For more information on the design work behind the Model T, see ibid., pp. 387–93, and Charles F. Sorensen, *My Forty Years with Ford,* pp. 96–112.

6. *Nation* 88 (January 7, 1909): 7–8, quoted in Nevins and Hill, *Ford,* pp. 385–86; *Harper's Weekly,* January 1, 1910, quoted in ibid., p. 449.

7. Nevins and Hill, *Ford,* pp. 387–88.

8. Ibid., pp. 262, 265–67, 279.

9. Ibid., p. 281; Sorensen, *My Forty Years with Ford,* p. 283. See Flanders's obituary in the *New York Times,* June 25, 1923, and *Automobile Topics* 70 (1923): 517, 525.

10. Nevins and Hill, *Ford,* p. 281. On the importance of the Landis grinding methods, see Robert S. Woodbury, *History of the Grinding Machine,* p. 123.

11. Reminiscences of Max F. Wollering, Ford Archives, Edison Institute, Dearborn, Michigan. The importance of considering interchangeability—even at this late date—cannot be overemphasized. As late as 1915 the general manager of a major machine tool company raised the question, "How many are there engaged in manufacturing enterprises today who can truthfully apply the word interchangeable to their product?" (J. P. Brophy, "Interchangeability").

12. *Cycle and Automobile Trade Journal* 10 (January 1, 1906), quoted in Nevins and Hill, *Ford,* p. 282.

13. Nevins and Hill, *Ford,* pp. 282, 324.

14. Apparently, many of the shops in the Detroit area had arranged their machine tools according to class. But for a long time the best Yankee shops had placed their equipment according to "jobs," such as the crank job or the hub job at the Pope bicycle works. Ford employees found Flanders's arrangement of tools novel. Seen against a proper background, however, his work simply followed the best New England practice. Writing in an autobiographical typescript, Fred Colvin said of a visit to the Ford Piquette Avenue plant: "Here I saw the beginnings of a new idea in automobile construction, for the shop was arranged for continuous line production, the first I had ever seen. Instead of being departmentalized by operations the departments were laid out to handle the operations of every part in sequence to produce them with a minimum of handling ("Automobiles," Box 3, Fred H. Colvin Papers, Freiberger Library, Case Western Reserve University, Cleveland, Ohio). For a more fully developed version of Flanders's ideas on sequential arrangement of tools, see the paper by Ford's chief tool designer, Oscar C. Bornholdt, "Continuous Manufacturing by Placing Machines in Accordance with Sequence of Operations."

15. Nevins and Hill, *Ford,* pp. 325, 335–36.

16. Sorensen, *My Forty Years with Ford,* pp. 93, 45.

17. Nevins and Hill, *Ford,* p. 364. There is great confusion in Ford history (written primarily from reminiscences of old employees who gave their thoughts to interviewers in the 1950s) about the use of gravity slides and conveyor systems at the Ford factories. Nevins and Hill vacillate on this matter but are firm that gravity slides were in place at Piquette before Flanders left in 1908 (ibid., pp. 370, 469). The visual evidence suggests otherwise, however.

18. Ibid., p. 365.

19. An expression used by Sorensen, *My Forty Years with Ford,* p. 45.

20. Information for this paragraph has been gleaned from Nevins and Hill, *Ford,* and from Sorensen, *My Forty Years with Ford.* I have tried to find published material by or about these men in the technical literature, but to the technical community, these important figures at Ford remained virtually anonymous.

21. Nevins and Hill, *Ford,* pp. 336, 338, 397.

22. Ibid., p. 396. For a glimpse of how these operations sheets worked and how they were revised, see Accession 166 (seven boxes), Ford Archives.

23. Nevins and Hill, *Ford,* p. 396.

24. Sorensen, *My Forty Years with Ford,* pp. 46–47, 106–7; Nevins and Hill, *Ford,* pp. 458–61; and Reminiscences of E. A. Walters, Ford Archives. Walters's Reminiscences provide an excellent history of the Keim plant and its personnel from 1899 through its absorption by the Ford Motor Company. For a contemporary article on the Keim company, see "Pressed Steel Automobile Parts."

25. Walter Flanders, "Large Capital Now Needed to Embark in Automobile Business," *Detroit Saturday Night,* January 22, 1910, quoted in James J. Flink, *America Adopts the Automobile, 1895–1910,* p. 300.

26. Nevins and Hill, *Ford,* pp. 340, 452.

27. Ibid., pp. 451–56. For an additional description of the Highland Park factory, see Horace Lucien Arnold and Fay Leone Faurote, *Ford Methods and the Ford Shops,* pp. 22–30.

28. Nevins and Hill, *Ford,* p. 452. Writing in 1912, O. J. Abell explained the reasons for Ford's decision: "The restriction of the company's product to a single model chassis was a matter of development to which at least three influences contributed: First, the study of alloy steels together with their proper heat treatment from which the design adopted has been made possible. Second, a remarkably far-sighted recognition of the magnitude of the market for a low-priced car. Third, the advantage and the necessity, from a manufacturing standpoint, of adhering to a single product if the quantity demanded by the market was to be produced at a minimum selling price with a satisfactory profit" ("Making the Ford Motor Car," 89 [1912]: 1383).

29. Sorensen, *My Forty Years with Ford,* p. 125.

Sorensen claims—probably incorrectly—that these layouts determined the floor plans of the Highland Park factory, that is, they provided a guide for Kahn. Most likely the layouts were made after the Highland Park factory had been built. The practice of tagging machine tools was carried over into machine tool acquisition practices, on which see the Reminiscences of A. M. Wibel, Ford Archives.

30. This work originally appeared as a series of articles in *Engineering Magazine* in 1914 and 1915, stimulated by the five-dollar day. *Engineering Magazine*'s thought and concept for a treatment of the Ford Motor Company appear in Charles B. Going, editor, *Engineering Magazine,* to Horace L. Arnold, which wound up in Henry Ford's hands (Acc. 2, Henry Ford Office, Misc. Correspondence, Box 38, Ford Archives).

31. Sixteen articles, May 8–November 27, 1913, vols. 38 and 39.

32. An autobiographical account of Colvin's visit to the Highland Park factory appears in a two-page typescript, "Automobiles," in Box 3 of the Colvin Papers. Colvin notes here that he visited Highland Park "in the spring of 1913." In his published autobiography, however, Colvin said that he was invited to the factory in January of 1913 and "lost no time" in getting there (*60 Years with Men and Machines,* p. 130). For additional material on Colvin, see *National Cyclopaedia of American Biography,* s.v. "Colvin, Fred Herbert." *Iron Age* ran a four-part series written by Oliver J. Abell, "Making the Ford Motor Car."

33. "Building an Automobile Every 40 Seconds," pp. 757, 759.

34. Ibid., p. 761.

35. Ibid.

36. Ibid. Colvin and later journalists emphasized the close grouping of machine tools because Henry Ford and his production engineers did so. See the Reminiscences of William Pioch, Logan Miller, and A. M. Wibel, Ford Archives. These men later claimed that close grouping of machines proved to be a bad idea because of expansion and model changes.

37. Quoted in Nevins and Hill, *Ford,* p. 461.

38. Colvin, "Building an Automobile," p. 759.

39. Bornholdt, "Continuous Manufacturing," p. 1672.

40. Colvin, "Building an Automobile," p. 761. The planning, layout, and scheduling of Ford production during this era is marvelously detailed in the Reminiscences of A. M. Wibel, who joined the company in May 1912 as assistant to Oscar Bornholdt. Wibel maintained that the development of these procedures "was the beginning of mass car" production.

41. "Machining the Ford Cylinders—I." Colvin's statement that the Ford engines required no scraping appears on p. 843.

42. Sorensen, *My Forty Years with Ford,* p. 54.

43. Wibel, Reminiscences. Wibel worked in what was later called the engineering procurement department, of which he eventually served as head for many years. When he started at Ford in 1912, he worked as assistant to Bornholdt and then to Bornholdt's successor, Charles Morgana. Allan Nevins and Frank Ernest Hill cite other Ford employees' reminiscences that deal with this matter. Carl Emde apparently designed the notable drilling machine that put forty-five holes in the engine block in one operation, but it was Charles Morgana's responsibility to convince a toolbuilder to construct such a machine. One Ford pioneer remembered that Morgana went from machine tool company to machine tool company in search of ideas for his new processes. Morgana then told Emde about these ideas, and Emde used them to design Ford production machines (Nevins and Hall, *Ford,* p. 465). In addition, the Reminiscences of tool designers William Pioch and Logan Miller are perhaps the most reliable on the question of Ford versus outside design of machine tools.

44. An important starting point in this history is the American Society of Mechanical Engineers' Report of the Sub-Committee on Machine Shop Practice, "Developments in Machine Shop Practice during the Last Decade."

45. As the ASME Sub-Committee on Machine Shop Practice stressed in its 1912 report on the previous ten years of development in machine shop practice: "The development of jigs and fixtures for interchangeable manufacturing has been remarkable. The expansion of automobile manufacture has been enormous and most of the leading concerns employ jigs and fixtures exclusively, thus insuring interchangeability, low production cost and systematic production. Many improvements have been made in the way of clamping devices, standardization of bushings, handles, levers, frames, etc. too numerous to mention specifically. Toolmaking has been developed on manufacturing lines" (ibid., p. 858). This report also singled out the importance of greater weight and rigidity in machine tools, resulting "not only in increased production but in accuracy of product" (ibid., p. 848). As late as 1909, however, Henry Leland's Cadillac plant was still hand-scraping crankshaft bearings (Nevins and Hill, *Ford,* p. 468).

46. Nevins and Hill, *Ford,* p. 464.

47. Colvin, "Machining the Ford Cylinders—I," p. 845. For a general discussion of Ford's special-purpose machine tools see the Reminiscences of William Pioch, A. M. Wibel, William Klann, and Logan Miller, Ford Archives.

48. Fred H. Colvin, "Making Rear Axles for the Ford Auto."

49. Fred H. Colvin, "Special Machines for Making Pistons," p. 352. For a reasonably complete view of the Keim operations and men at Highland Park, see the Reminiscences of E. A. Walters.

50. Colvin, "Building an Automobile," p. 762.

51. Fred H. Colvin, "Methods Employed in Making the Ford Magneto."

52. Fred H. Colvin, "Machining the Ford Cylinders—II." In his Reminiscences, William Klann, the superintendent of the engine department, described these stands: "The assembly benches were made two feet wide by five feet long and we had twenty of these benches with two men at each, or forty men, fitting the crankshaft to the cylinder block. At this time the tops of the cylinder bearings were cast iron, and the bottom crankshaft bearing had one-fourth inch of babbitt in the bearing caps. We had twenty benches with forty men to fit crankshaft bearings. They would get the cylinder block and bore them out on lathes in this Machine Shop and we'd get the bearing cap separate and fit the cap on and scrape them in. Each man had a set of wrenches for the bearings and each man had a rod for the crankshafts and each man had a bench to put that motor on. He had a separate bench to scrape his bearings in. Each man had two vises to look after all the time."

53. Fred H. Colvin, "Ford Crank Cases and Transmission Covers," p. 53.

54. Colvin, "Making Rear Axles," p. 148.

55. Fred H. Colvin, "Ford Radiators and Gasoline Tanks."

56. Fred H. Colvin, "Special Machines for Auto Small Parts," p. 442; italics added. Contrast Ford's final assembly methods with those described in Harold W. Slauson, "Efficient System for the Rapid Assembly of Motor Cars."

57. See Colvin's section, "Assembling 800 Cars Per Day," in "Building an Automobile," pp. 761–62, for a description of static assembly with moving gangs of assemblers.

58. See the chapter, "A Lever to Move the World," in *Ford,* pp. 447–80; quotation on p. 469. Although Nevins and Hill relied on the oral history accounts of many former Ford employees, it appears that they accepted in broad outlines the account by William Klann. Yet there are some discrepancies in Klann's account, which in general is self-serving. The uncertainty about what actually happened in 1913 is best reflected in the Reminiscences in the Ford Archives of James O'Connor, who worked in Ford's assembly department from 1909 until his retirement in 1952. O'Connor was a foreman of assembly in 1913.

59. On the question of exactly when Colvin visited the Ford factory, see note 32. From this evidence, it may be concluded that Colvin toured Highland Park no more than three months and possibly only days before the first experiments were conducted on a magneto assembly line.

60. Colvin, "Building an Automobile," p. 759. Prints of many of the photographs published in Colvin's *American Machinist* series survive in the Ford Archives. They were taken by the commercial pho-

tographers Spooner & Wells. It is unclear whether they were commissioned by Colvin or Henry Ford. On the uniqueness of these slides and conveyor systems, see Arnold and Faurote, *Ford Methods,* pp. 162–63. This work supports the contention that work slides and conveyors were added after the assembly line innovation. Well after magneto, motor, transmission, and chassis assembly lines had been installed, Arnold wrote, "Besides these almost unbelievable reductions in assembly time, the Ford shops are now making equally surprising gains by the installation of component-carrying slides, or ways, on which components in process of finishing slide by gravity from the hand of one operation-performing workman to the hand of the next operator" (p. 103). In October of 1914, Arnold noted that "it was not until the beginning of 1914 that it was found that, in some special instances, the convenience of the workmen could be served by the installation of gravity work-slides" (p. 272). See also other specific dates on materials handling installations in ibid., pp. 271–86.

61. Sorensen, *My Forty Years with Ford,* pp. 117–19. On Sorensen's account, see the Reminiscences of Richard Kroll and William Pioch, Ford Archives. Pioch remarked that Sorensen's account was "quite a picturesque story, but I never heard about it until now [early 1950s]."

62. This view differs from Klann's, as expressed in his Reminiscences.

63. For details of the Ford foundry, see Fred H. Colvin, "Continuous Pouring in the Ford Foundry," and Arnold and Faurote, *Ford Methods,* pp. 327–59. I have relied upon Arnold and Faurote for precise dating of installations. It is noteworthy that Arnold actually makes the connection between the Westinghouse foundry and Ford's assembly line (see pp. 130–31). On the Westinghouse foundry, see "The Westinghouse Foundry, Near Pittsburg, Pa." The idea of continuous-process parts production and assembly, as suggested by the Ford foundry, is discussed in the Reminiscences of William Klann: "We saw these conveyors in the Foundry and we thought, 'Well, why can't it work on our job?' That is where we got the idea from, from the conveyor in the Foundry."

64. Henry Ford, in collaboration with Samuel Crowther, *My Life and Work,* p. 81.

65. Klann, Reminiscences.

66. Ibid. I have examined a number of catalogs of foundry milling and brewing equipment manufacturers in the Eleutherian Mills Historical Library, and almost all dating from the period 1900–1910 include conveyor systems of one kind or another.

67. On Oliver Evans, see Eugene S. Ferguson, *Oliver Evans.*

68. John Storck and Walter Dorwin Teague, *Flour for Man's Bread,* pp. 196–222.

69. Klann, Reminiscences.

70. "Norton's Automatic Can Making Machinery." This can line was the result of efforts by Edwin Norton, who was instrumental in the formation in 1901 of the American Can Company, a consolidation of ninety-five canmaking firms. In 1904 Norton organized the Continental Can Company. See Earl Chapin May, *The Canning Clan* (New York: Macmillan, 1937).

71. Sorensen, *My Forty Years with Ford,* p. 128, and Nevins and Hill, *Ford,* p. 470. The Reminiscences of Richard Kroll, James O'Connor, William Pioch, and A. M. Wibel support Sorensen's claims.

72. Colvin, "Ford Radiators and Gasoline Tanks."

73. Nevins and Hill, *Ford,* p. 475.

74. Ibid., pp. 471–72. The photograph appears opposite p. 544. Other accounts of the development of the assembly line which rely on Arnold and Faurote and on Nevins and Hill include Jack Russell, "The Coming of the Line," and David Gartman, "Origins of the Assembly Line and Capitalist Control of Work at Ford."

75. Arnold and Faurote, *Ford Methods,* p. 109. Arnold's analysis of Ford's assembly processes was first published in July 1914 as an article, "Ford Methods and the Ford Shops—IV," *Engineering Magazine* 47 (1914): 507–32. See also Fred Colvin's article, "Assembling Magnetos, Motors and Transmissions."

76. Reminiscences of William Klann.

77. Horace L. Arnold, "Ford Methods and the Ford Shops," *Engineering Magazine* 47 (1914): 1–26.

78. Arnold and Faurote, *Ford Methods,* pp. 112–15. But as A. M. Wibel noted in his Reminiscences, "We have no real way of pinning down what came first."

79. As Nevins pointed out, there is no contemporary documentation of the development of the assembly line in the Ford Archives. This fact is underlined by close examination of Henry Ford's *My Life and Work.* In his discussion of the assembly line, Ford's ghostwriter, Samuel Crowther, drew exclusively from Horace Arnold's account in *Ford Methods and the Ford Shops.* See Ford, *My Life and Work,* pp. 81–83.

80. As Arnold noted, "The desirability of general application of the moving assembly line . . . was not at once fully conceded by all the Ford engineers" (Arnold and Faurote, *Ford Methods,* p. 112); see also ibid., pp. 114–15.

81. Reminiscences of William Klann; Arnold and Faurote, *Ford Methods,* pp. 115–16.

82. Colvin's article, cited in note 75, describes the operations illustrated in a photograph: "The planetary transmissions are being assembled as the flywheel and transmission units are moved along a track made of channel iron and supported by piping. . . . The flywheels are then turned so that the

transmission portions go down between the rails. . . . They are moved along this track by the continuously moving chain shown in the framework beneath it. While the parts are moving slowly, the magnets are assembled, fastened into place and tested—all without stopping the procession for one moment" (p. 557).

83. Reminiscences of William Klann. Data on assembly times are from Arnold and Faurote, *Ford Methods,* p. 115.

84. Reminiscences of William Klann.

85. Arnold and Faurote, *Ford Methods,* p. 116.

86. Ibid., p. 135; Sorensen, *My Forty Years with Ford,* p. 130; Nevins and Hill, *Ford,* pp. 474–75. Also see Avery's dinner address for thirty-five-year employees of the Ford Motor Company, December 19, 1944, Acc. 1, Fair Lane Papers, IV, Ford Motor Company, Box 117, Ford Archives. The Avery quote appears in "How Mass Production Came into Being," which was taken from his introductory remarks in a paper on safety glass manufacture. Avery spoke at an ASME meeting in Detroit on the application of motor car manufacturing methods in other industries.

87. Such instructions were common. Logan Miller, a thirty-eight-year employee of Ford Motor Company, recalled that while he was working in the toolroom in 1914, "Mr. Martin came up to the tool room one day and asked me what a particular die was for and where the part went on the car, and I didn't know. He said, 'Well, before you can do the job, I think you had better spend some time going through the various departments and finding out where these various parts are made, and maybe you can improve on them.'" Miller took two weeks to tour the factory and "filled notebooks" with ideas for improving production (Miller, Reminiscences).

88. Colvin, "Building an Automobile," pp. 761–62. Stationary chassis assembly in 1913 required the routing of six hundred men—five hundred assemblers and one hundred parts carriers.

89. Stephen Meyer III, *The Five Dollar Day,* pp. 11, 20; Nevins and Hill, *Ford,* pp. 468, 474. Daniel Nelson discusses developments at Ford in the context of scientific and systematic management in *Managers and Workers.* Harry Braverman, on the other hand, does not explicitly link Taylorism and Fordism in terms of their development but only as manifestations of the "capitalist mode of production" (*Labor and Monopoly Capital,* pp. 139–51).

90. Reminiscences of Anthony Harff and A. M. Wibel, Ford Archives.

91. This was the fourth of Frederick W. Taylor's four principles of scientific management as elaborated in *The Principles of Scientific Management,* p. 36.

92. Arnold and Faurote, *Ford Methods,* p. 20. Ford had argued the same thing on January 11, 1914, in an interview with a reporter from the *Philadelphia Ledger.* See "Henry Ford to give Millions to Employes [sic] Because He Watched Gasoline Engine at

Work," *Philadelphia Ledger,* January 11, 1914, sec. 5, p. 4. I thank John Rumm for this reference. Perhaps Ford's views were expressed earlier when Fred H. Colvin visited the Ford factory in early 1913. Certainly some discussion with Ford or his employees is reflected in Colvin's statement, quoted above, that rear axle assembly at Highland Park "show[s] that motion study has been carefully looked into, whether it is called by that name or not." Henry Ford's statement was supported by James O'Connor, a foreman of assembly at Ford from 1913 to 1951. He said in the early 1950s that "there was no stop-watching then [1913]. We didn't see much of that until around '26" (Reminiscences of James O'Connor). Ford tool expert William Pioch also argued that there was no time study at Ford during this period (Pioch, Reminiscences). On the other hand, Max Wollering claimed in the early 1950s that he used a stopwatch at Ford's Piquette Avenue plant as early as 1907 (Wollering, Reminiscences). For a brief survey of Taylorism and the efficiency craze of the period 1910–20, see Samuel Haber, *Efficiency and Uplift.*

93. Frank B. Copley, *Frederick W. Taylor,* 2: 445; Nevins and Hill, *Ford,* p. 468.

94. Henry Ford, "Mass Production," p. 821. According to this article, Taylor and his disciples "did not see that another and better method might be devised which would make it unnecessary for a working man to carry 106,400 lb. of pig-iron to earn $1.85." For details on Schmidt, see Taylor, *Principles of Scientific Management,* pp. 57–62.

95. The seedbed of Taylor's system of management was the large industrial machine shop, which was heavily dependent upon skilled machinists. Taylor's work and the early work of his disciples took place in manufacturing establishments that differed significantly from the light-metal, consumer durable goods factories in which the American system of manufactures arose. As has been stressed throughout this study, emphasis in American system shops was placed on the development of special-purpose machine tools and fixtures, jigs, and gauges that removed the need for the worker to exercise judgment on his workpiece. Taylor revised the motions and procedures of workers to improve efficiency; those steeped in the American system developed new hardware to increase productivity or precision.

96. Sorensen, *My Forty Years with Ford,* p. 131. Speaking to a group of mechanical engineers in Detroit in 1929, Clarence Avery explained why the chain system was resorted to in moving chassis assembly operations: "The first continuous assembly line had no mechanical means of movement. The wheels of the car were assembled at a very early stage and channel iron tracks provided for them. At intervals, giving sufficient time for the operations to be performed, the foreman blew a whistle and all hands pushed the cars forward to the next position, and then returned to their original locations to perform their next opera-

tions. . . . In the next stage we provided rigid spacers between the cars, and introduced a pusher chain about three cars long at the beginning of the line. This worked well for a few weeks. The cumulative resistance, however, was too close to the safety factor. One day the complete line buckled and pushed a section from the side wall of the building. It was then that the continuous chain was introduced'' (''How Mass Production Came into Being'').

97. This account of chassis assembly lines is based primarily on Arnold and Faurote, *Ford Methods,* pp. 135–42; quotation on p. 139.

98. Figures from ibid., p. 193; see also pp. 114–15, 136. Other subassemblies included body assembly, upholstery work, piston assembly, and other components.

99. Horace L. Arnold, ''Ford Methods and the Ford Shops—IV,'' p. 513. Details of the Ford conveyors, work slides, and rollways are in Arnold and Faurote, *Ford Methods,* pp. 271–86. Henry Ford (or Samuel Crowther) wrote in 1922: ''Every piece of work in the shops moves; it may move on hooks on overhead chains going to assembly in the exact order in which the parts are required; it may travel on a moving platform, or it may go by gravity, but the point is that there is no lifting or trucking of anything'' (*My Life and Work,* p. 83).

100. Keith Sward, *The Legend of Henry Ford,* p. 49. Sward based these figures on information provided in Ford, *My Life and Work,* pp. 129–30.

101. These paragraphs on the five-dollar day are based on Nevins and Hill, *Ford,* pp. 512–41; Sward, *Legend,* pp. 49–63; David L. Lewis, *The Public Image of Henry Ford,* pp. 69–77; and Meyer, *Five Dollar Day.* Meyer argues that the five-dollar day was not simply a wage but a profit-sharing plan.

102. Acc. 1, Fair Lane Papers, IV, Ford Motor Company, ''Personnel Complaints,'' Box 120, Ford Archives.

103. Although the work of the Ford sociological department is treated in Nevins and Hill, *Ford,* and Sward, *Legend,* the best recent analysis is Stephen Meyer III, ''Mass Production and Human Efficiency,'' and his resulting book, *Five Dollar Day.* Meyer's work also provides an excellent discussion of the attempts to lower worker turnover before the announcement of the five-dollar day.

104. In an autobiographical note, Fred H. Colvin wrote that he had followed Ford developments at the Piquette Avenue plant and during the early days of the Highland Park plant. Since he was a journalist, he wanted to write about Ford Motor Company operations, ''but I was not permitted to write a line about the new shop until Ford was ready for it to be described in detail.'' Ford's permission did not come until early in 1913 (Colvin, ''Automobiles,'' Colvin Papers).

105. Arnold and Faurote, *Ford Methods,* p. 8.

106. See the *Iron Age* series in 89 (1912): 1383–91, 1454–60; 92 (1913): 1–7; and 93 (1914): 902–4. See also Harold Whiting Slauson, ''A Ten-Million Dollar Efficiency Plan,'' and two additional articles by Fred H. Colvin, ''Assembling Magnetos, Motors, and Transmissions'' and ''Continuous Assembling in Modern Automobile Shops.''

107. By August 26, 1915, Fred Colvin was able to report that the Studebaker Company had adopted continuous line assembly techniques for its four-cylinder automobiles. In his article ''Continuous Assembling in Modern Automobile Shops,'' Colvin suggested that any automobile manufacturer that had not adopted the assembly line had failed ''to grasp the modern tendency to its fullest extent'' (p. 370). The Hudson Company also adopted the assembly line, as did Packard. See ''Continuous Assembling Frame,'' ''Assembling Motor Cars in the Packard Plant,'' and Allan Nevins and Frank Ernest Hill, *Ford: Expansion and Challenge, 1915–1933,* pp. 391–92. Horace L. Arnold had suggested as early as July 1914 that Ford's assembly and conveyor methods could be ''applied to any and all small-machine manufacturing, with very large reductions of labor-cost'' (''Ford Methods and the Ford Shops—IV,'' p. 513).

108. As Warren Ordway suggested, ''This new system is applicable to practically any assembly operation as well as to filling containers with liquids, inspecting, packing and boxing, and a hundred other uses. One advantage of the system . . . is that it can be installed in an old plant as well as in a new one'' (''Assembling by Conveyor,'' p. 103). This article illustrates automobile component assembly lines as well as lines for radio component assembly. The Hoover Suction Sweeper Company used conveyors for inspection as early as 1923. See Fred Colvin, ''Methods Used in Assembling Sweepers.'' The German electrical manufacturing company, AEG, assembled its Vampyr vacuum sweeper on a moving line (Fließband) as early as 1926. See Ulrich Troitsch and Wolfhard Weber, eds., *Die Technik,* p. 405; also the recommendations of Albert A. Dowd and Frank W. Curtis, ''Saving Time in Assembling.''

109. Reginald McIntosh Cleveland, ''How Many Automobiles Can America Buy?,'' pp. 679–80, 682; Edward A. Rumely, ''The Manufacturer of Tomorrow''; and Harry Franklin Porter, ''Four Big Lessons from Ford's Factory.''

110. Charles F. Kettering and Allen Orth, *The New Necessity.*

Chapter 7

1. Daniel J. Boorstin, *The Americans,* pp. 546–55.

2. General Motors' leadership in the innovation of the annual model change is generally recognized in both the contemporary literature and current scholar-

ship. A penetrating analysis of GM's "lead in the habit-forming campaign that has made the annual model a practical necessity" appeared in "General Motors II: Chevrolet," pp. 39–40, 46, 103–4, which was partially reprinted in Alfred D. Chandler, ed., *Giant Enterprise*, pp. 153–70. Boorstin's *The Americans* provides an excellent social analysis of the annual model change and Ford's paradoxical contribution to it. Robert Paul Thomas, "Style Change and the Automobile Industry during the Roaring Twenties," attributes the coming of the annual model change to the development of the closed body automobile, an innovation of the Hudson Brothers.

3. Contemporaries adopted the expression *Fordism* to describe these simultaneous developments before *mass production* gained currency. See, for instance, Paul M. Mazur, "The Doctrine of Mass Production Faces a Challenge," *New York Times*, November 29, 1931, sec. 9, p. 3, and Anne O'Hare McCormick, "The Future of the Ford Idea." More recently, Emma Rothschild has contrasted Fordism with "Sloanism" in *Paradise Lost*, pp. 26–53.

4. Peter F. Drucker, *Concept of the Corporation*, pp. 219–20.

5. An excellent study of this decline in market share is James Dalton, "What Will Ford Do Next?," *Motor* 45 (May 1926), pp. 30–31, 84, 102ff., reprinted in Chandler, ed., *Giant Enterprise*, pp. 104–11. The standard, authoritative source is Allan Nevins and Frank Ernest Hill, *Ford: Expansion and Challenge, 1915–1932*, pp. 379–436.

6. Sloan quoted in Boorstin, *The Americans*, p. 554. General Motors' explicitly stated marketing policy was "of building a car for every purse and purpose," which allowed the American to "climb the ladder of consumption" (*Annual Report of General Motors Corporation for 1925*, p. 7, reprinted in Chandler, ed., *Giant Enterprise*, p. 151). Emma Rothschild called this phenomenon "Sloanism" in *Paradise Lost*, pp. 26–53. Planning for change is a theme that runs throughout the essays reprinted in Alfred D. Chandler, ed., *Managerial Innovation at General Motors*. The overriding objective at GM, unlike Ford (it seems), was that "business is operated only to make a profit," as elaborated in Thomas B. Fordham and Edward H. Tingly, "Control through Organization and Budgets," reprinted in ibid.

7. "General Motors II," p. 103. Production figures are dispersed in Nevins and Hill, *Ford: Expansion*, pp. 379–596.

8. For Knudsen's contributions to both Ford and General Motors, see Nevins and Hill, *Ford: Expansion*, and Norman Beasley, *Knudsen*.

9. Alfred P. Sloan, Jr., *My Years with General Motors*, p. 83; Alfred D. Chandler, Jr., and Stephen Salsbury, *Pierre S. duPont and the Making of the Modern Corporation*, p. 529. The history of GM's copper-cooled engine fiasco is detailed in ibid., pp.

511–59, and Stuart W. Leslie, "Charles F. Kettering and the Copper-cooled Engine."

10. "General Motors II," p. 39, and Beasley, *Knudsen*, p. 149.

11. William S. Knudsen, "'For Economical Transportation.'"

12. Ibid., pp. 65–66; Beasley, *Knudsen*, p. 124.

13. Knudsen, "'For Economical Transportation,'" p. 66.

14. Ibid., pp. 67–68.

15. Beasley, *Knudsen*, pp. 139–40; *New York Times*, November 17, 1928, February 3, August 8, 1929; "Chevrolet Begins Production of New Six"; "Chevrolet Production Change Over to 6-Cylinder Car Held Industrial Feat."

16. Sloan quoted in Boorstin, *The Americans*, p. 552; Charles F. Kettering, "Keep the Consumer Dissatisfied."

17. *Detroit Journal*, June 19, 1915, quoted in Nevins and Hill, *Ford: Expansion*, p. 201. On the "industrial colossus," see ibid., pp. 279–99, and pp. 200–216; these chapters provide the best history of the development of the River Rouge complex.

18. This interpretation is that of Nevins and Hill and of Sorensen himself in *My Forty Years with Ford*, pp. 150–79.

19. The "Topsy" expression was used by R. T. Walker in his Reminiscences and quoted by Nevins and Hill, *Ford: Expansion*, p. 207; on the foundry, see p. 212.

20. The Reminiscences of Logan Miller in the Ford Archives indicate that direct pouring created immense problems of quality control and that eventually the company went back to charging cupolas with pig iron that had been sorted according to analysis characteristics.

21. Sorensen claims that he thought of having founding and machining operations on one floor under the same roof (*My Forty Years with Ford*, pp. 163–64); Nevins and Hill, *Ford: Expansion*, p. 216.

22. Production figures vary significantly from source to source, even within the company's own records, usually because some figures cover fiscal years, others, calendar years. Those used in this chapter are derived from Appendix I of Allan Nevins and Frank Ernest Hill, *Ford: Decline and Rebirth, 1933–1962*, p. 478.

23. The following information is derived chiefly from correspondence, notes, reports, and memorandums in the records maintained by A. M. Wibel, who headed the engineering department during this era (Accession 390, Ford Archives; see esp. Boxes 9, 45, 52, 66, and 75). The unedited Reminiscences of Wibel, also in the Ford Archives, nicely complement these manuscripts and are drawn upon here. They also reveal Wibel's penchant for details and excellent memory.

24. These operations sheets were prepared by an

informal committee consisting of the superintendents (Sorensen, Martin, and Kanzler), the heads of the tool department, the engineering department, and the foremen of the various parts departments (each part was manufactured in a department, e.g., T-400, the engine block department). Samples of these sheets and the engineering department's use of them may be seen in Sorensen's papers, Acc. 38, Box 50, and in seven boxes (for the Model T) of Acc. 166, Ford Archives. Assembly and subassembly operations sheets are also in Acc. 166.

25. On ordering machine tools, see, for example, "Ford Motor Company, Rouge Plant, T 400—6000 per 16 hrs.," [1923?], Acc. 38, Box 50, and the large number of manuscripts cited in note 23. An excellent example of the engineering department's control over the machine tool inventory and activities in the plant is seen in a departmental communication of May 21, 1925, listing machine tools for roller bearing production freed up by a change in the rear axle design (Acc. 390, Box 45, Ford Archives). See also "List of Centerless Grinders at Rouge and Highland Park Plants," May 7, 1925, Acc. 390, Box 45; cf. Wibel's Reminiscences. For evidence of pressure on machine tool makers, see the Reminiscences of Wibel and Miller and Sorensen, *My Forty Years with Ford*, pp. 177–78.

26. Wibel, Reminiscences. Since the Ford Motor Company had no rigidly established departments and certainly no men who maintained titles, the procedures for preparing for new tools are open to some question, although they are suggested in the manuscripts in Acc. 390—"Wibel desk files"—and Wibel's Reminiscences. Logan Miller's Reminiscences suggest slightly different arrangements, as do the manuscripts in Acc. 680, "Plant Engineering," Box 5, Ford Archives, which suggest that horsepower and space requirements were calculated by the tool department and passed on to layout, which was under the direction of the tool department head.

27. Sorensen, who directed all of these departments and who probably coordinated more information than Wibel's engineering department, put all of this under the general rubric of "planning" (*My Forty Years with Ford*, p. 178). An excellent example of Wibel's recordkeeping appears in a memorandum he wrote to Martin and Sorensen, October 29, 1924, on how many machine tools had been moved in the Ford enterprise since January 1, 1924, including new machine tools, scrapped ones, and ones just moved within the Highland Park plant (Acc. 572, Box 23, File 11.22.2.2, Ford Archives).

28. Logan Miller, one of Sorensen's deputies, acknowledged this point in his Reminiscences. This system of planning production increases and estimating costs worked exceptionally well when the change in the product was not radical.

29. Years of study of Ford records, which are not fully complete, would be required to determine the precise reasons for the general trend of cost reductions of the Model T. Records suggest this decline, although increases took place in 1923 and 1925, when style changes were made. General price trends in the American economy may have operated. Within the company, materials handling improvements, the Rouge developments, machine tool changes, and stretching out contributed to the price trends. The Reminiscences of Pioch, Miller, and Wibel and Sorensen's *My Forty Years with Ford* suggest that tooling innovations were made.

30. Documentation on cost accounting at the Ford Motor Company is found in the following collections in the Ford Archives: Acc. 1, Fair Lane Papers, Box 122; Acc. 38, Charles Sorensen files, Box 63; Acc. 125, Ford Motor Company finances, 1913–24, 45 boxes; Acc. 157, Martindale papers, Box 157; Acc. 390, Wibel files, Box 52; Acc. 488, Frank Hodas papers, Box 1; Acc. 542, Cost accounting–general files, Boxes 1–34, 128–33; Acc. 572, Nevins project, Boxes 22–23; Acc. 680, Plant Engineering, Box 5; Acc. 735, Cost estimates, 1927–36, eight transfer cases; Acc. 736, Car costs, four transfer cases. See also the Reminiscences of Anthony Harff, Logan Miller, and especially Herman L. Moekle, who describes the work of the manufacturing cost department.

31. Acc. 157, Martindale papers, Box 273, Ford Archives.

32. Moekle says, "Costs were computed and those costs were made available to Mr. Edsel Ford and Mr. Sorensen and perhaps Mr. Knudsen [branch manager] and maybe even Mr. Mayo [plant engineering]. Mr. Henry Ford, so far as I knew, was made aware of what the book costs showed" (Reminiscences).

33. Wibel, Reminiscences. The accounting department, at the hands of W. E. Carnegie, also issued comparative cost studies. See, for example, "Comparative Costs of Parts Made Here and Bought Outside," March 1930, Acc. 38, Box 63, Ford Archives. Examples of graphs are in Acc. 390, Wibel files, Box 52, ibid.

34. Documentation on the use of mechanical drawings is spread thinly throughout much of the manuscript material in the Ford Archives. See especially the specifications records in Acc. 166, Box 1, which continually note, "as per drawing"; also the "factory letters" in Acc. 575, Boxes 1–15. The Reminiscences of Pioch, Richard Kroll, John F. Wandersee, William Klann, and Laurence S. Sheldrick provide a much clearer view. Almost no drawings survive in the Ford Archives because of a fire in the Rotunda, which once housed the drawings. A couple of blueprints of gauges are in Sorensen's papers, Acc. 38, Box 49. Ford's open-door policy to technical journalists resulted in the publication of many of those drawings. For the best examples see the Fay Leone Faurote series on Ford Model A production published

in the *American Machinist* in 1928. Good examples appear in 68 (May 10, 1928): 761 and 69 (July 12, 1928): 64. Faurote also described the company's use of drawings in his 1928 series in *Factory and Industrial Management*. See especially, "Planning Production through Obstacles, Not around Them," p. 302; "Make Time and Space Earn Their Keep," p. 544; and "Planning and Mass Production Coordinated," p. 985. Pioch's Reminiscences, pp. 107–10, contain a lengthy account of the Ford system of numbering its drawings, which suggests the nature and use of the drawings. The policy of no changes was underlined by Richard Kroll, the chief inspector at the Rouge Plant, in his Reminiscences.

35. Almost all of the factory letters issued between 1908 and 1921 survive in Acc. 575 of the Ford Archives. Some were pulled by Nevins's research team and are now in Acc. 572 (Box 14). Letters for the last six years of T production are in Acc. 572, Box 14, but I know of no other run of letters such as those in Acc. 575. See, for example, "Instructions & Assembly Letter #75," October 18, 1926. Some letters are in the Martindale papers, Acc. 157 (e.g., Box 260). General letters are in Ford Motor Company—General letters file, 1915–46, Acc. 78, 81 boxes.

36. There were exceptions, of course, such as was recounted in Harry Franklin Porter, "Four Big Lessons from Ford's Factory," p. 640.

37. Leslie R. Henry, "The Ubiquitous Model T," Philip Van Doren Stern, *Tin Lizzie*, pp. 166–67.

38. This section is based primarily on Henry, "Ubiquitous Model T," and Stern, *Tin Lizzie*. The *New York Times* picked up many of these changes.

39. "All metal" is a slight misnomer. William J. Abernathy's statement in his case study of the closed steel body that "Ford introduces all-steel closed bodies for the Model T" in 1925 is not fully accurate (*The Productivity Dilemma*, p. 184). Some wood was used for flooring and the top, but pressed sheet steel predominated. For an illustration of this body, see the lower photograph in Floyd Clymer, *Henry's Wonderful Model T, 1908–1927*, p. 95; see also the Reminiscences of Logan Miller and Ernest A. Walters.

40. Joseph Galamb, Reminiscences, Ford Archives; Thomas, "Style Change," notes that initially the enclosed all-metal body drove prices upward, but by mid-decade companies that had had a few years to work on the problems reduced costs significantly and brought prices down close to those for open bodies (pp. 118–38). For Ford cost data between 1920 and 1926, presented graphically, see Acc. 157, Martindale Papers, Box 273, Ford Archives.

41. Two substantial contemporary views are James C. Young, "Ford to Fight It out with His Old Car," *New York Times*, December 26, 1926, sec. 8, p. 1, and Dalton, "What Will Ford Do Next?"

42. Thomas, "Style Change," p. 130, points out

that Dodge Brothers pursued the same marketing approach after splitting with Ford but was forced to make changes in 1924.

43. Reminiscences of Pioch and Klann; Nevins and Hill, *Ford: Expansion*, p. 406.

44. This observation is based on a close reading of the *Times*, 1920–32.

45. See Thomas, "Style Change," and Nevins and Hill, *Ford: Expansion*, pp. 379–408.

46. Young, "Ford to Fight It out with His Old Car." On the extent of buying on credit in 1926, see "Installment Selling to the Front."

47. Young, "Ford to Fight It out with His Old Car."

48. Dalton, "What Will Ford Do Next?" p. 111; Young, "Ford to Fight It out with His Old Car."

49. Ford's profit margin in 1926 was believed to be as low as $29, down from $40 the previous year ("Ford Made $29 on Each Car," *New York Times*, April 28, 1926). By the time Dalton raised his question, Ford had already made two price cuts in 1926 (Nevins and Hill, *Ford: Expansion*, pp. 414–15). Dalton, "What Will Ford Do Next?" cogently discusses the effect of Ford's price cuts on his market share and the unlikelihood of Ford's ability to offer substantial additional cuts.

50. [Ernest C. Kanzler] to [Henry Ford], January 26, 1926, Acc. 1, Fair Lane Papers, Box 116, Ford Archives.

51. Nevins and Hill, *Ford: Expansion*, pp. 409–11.

52. *Annual Report of General Motors Corporation for 1925* (February 24, 1926), p. 7.

53. Abernathy, *Productivity Dilemma*, p. 43. These changes are treated excellently in Chandler, ed., *Giant Enterprise*.

54. [Kanzler] to [Ford], January 26, 1926.

55. The first quote in the paragraph is from Eugene J. Farkas, Reminiscences, Ford Archives; see also Nevins and Hill, *Ford: Expansion*, p. 442. Charles Sorensen's account of the pressures and decision is in *My Forty Years with Ford*, pp. 217–31. Nevins and Hill discuss the intense pressure on Henry Ford to change models in *Ford: Expansion*, pp. 409–36. Wibel, Reminiscences, gives the view of the head of procurement for the production engineering department.

56. *New York Times*, May 26, 1927, p. 4.

57. The day's events are recounted in Nevins and Hill, *Ford: Expansion*, p. 431.

58. Details about the design of Model A appear in Nevins and Hill, *Ford: Expansion*, pp. 437–58.

59. Dalton, "What Will Ford Do Next?," p. 110.

60. Apparently Ford did not use any of the design work completed for the X-car.

61. Pioch, Reminiscences.

62. Ibid. See also the Reminiscences of Galamb, Sheldrick, J. L. McCloud, and Wibel.

63. Sheldrick, Reminiscences.

64. Reminiscences of Wibel, Sheldrick, and Klann.

65. *New York Times,* June 12, 1927, sec. 8, p. 17, and ibid., June 22, 1927, p. 12.

66. James C. Young, "Ford's New Car Keeps Motor World Guessing," ibid., June 26, 1927, sec. 9, p. 2; ibid., July 25, 1927, p. 1; July 31, 1927, p. 1; August 5, 1927, p. 36; August 11, 1927, p. 23.

67. Ibid., October 15, 1927, p. 6; October 23, 1927, p. 1; December 2, 1927, p. 3; October 25, 1927, p. 6.

68. Ibid., October 23, 1927, p. 1; "Fordson Plant Undergoes Huge Expansion Program," *Ford News* 7 (October 15, 1927): 1, reprinted in the *New York Times,* November 1, 1927, p. 2.

69. *New York Times,* December 2, 1927, p. 3; February 19, 1928, p. 1 (production on this date averaged eight hundred cars per day); March 27, 1928, p. 30; August 29, 1928, p. 29; October 9, 1928, p. 37; February 12, 1928, sec. 2, p. 5.

70. On Ford's gas tank design and its production problem, see the Reminiscences of Galamb, Klann, Miller, Pioch, Sheldrick, Wibel, and Walters.

71. Fay Leone Faurote, "A Gasoline Tank That Serves Also as Cowl and Dashboard," *American Machinist* 68 (1928): 807. This article is part of a large series on production of the Model A written by Faurote and published in the *American Machinist* between April 19 and August 16, 1928. Faurote also published a different series of articles in *Factory and Industrial Management* between October 27 and August 1928, which I have drawn from.

72. Reminiscences, of Wibel and Walters.

73. Faurote, "Gasoline Tank," p. 807. See also Faurote, "Planning and Mass Production Coordinated," p. 986. For a detailed account of Ford welding problems, see Miller, Reminiscences.

74. Walters, Reminiscences. Richard Kroll recalled that sometimes machining was begun before proper gauges for inspection had been completed (Reminiscences).

75. Wibel, Reminiscences.

76. Miller, Reminiscences; Henry Ford with Samuel Crowther, *Moving Forward,* pp. 189–90.

77. Fay Leone Faurote, "Machining and Welding Operations on the Rear Axle Housings."

78. Ibid.

79. Sheldrick, Reminiscences; Fay Leone Faurote, "Single-Purpose Manufacturing," p. 771. See also the departmental communication, March 11, 1930, Acc. 38, Sorensen Papers, Box 62, Ford Archives.

80. Fay Leone Faurote, "Preparing for Ford Production," p. 637; Faurote, "Producing the New Ford Model," pp. 64–65; and *Ford News* 11 (March 15, 1929): 69. See also *New York Times,* December 2, 1927, p. 2; December 4, 1927, sec. 11, p. 13; and February 26, 1928, sec. 9, p. 14. Precision is discussed more rigorously by Fay Leone Faurote, "Splitting an Inch a Million Ways." Finally, a penetrating article, "Mr. Ford Doesn't Care," in *Fortune* points out that parts sales for the Model A decreased significantly as compared to the Model T because of increased precision of manufacture (p. 67).

81. In addition to sources cited in note 80, see the following manuscript material relating to Ford and Johansson: Acc. 1, Fair Lane Papers, Box 82; Acc. 23, Henry Ford Office Files, Box 5; Acc. 38, Sorensen Papers, Box 49; Acc. 44, W. J. Cameron Files, Box 16; Acc. 157, Martindale Papers, Box 188; Acc. 285, Henry Ford Office Correspondence, Boxes 154 and 238, Ford Archives. For more information on Johansson and his gauges, see Torsten K. W. Althin, *C. E. Johansson, 1864–1943.* See also the Reminiscences of Richard Kroll, chief of inspection at Ford.

82. Pioch, Reminiscences.

83. Ibid.; Wibel, Reminiscences. The expense of the frequent machinery moves is clear in a memorandum from A. M. Wibel to P. E. Martin and Charles Sorensen, October 24, 1924. In ten months of 1924, some three thousand machine tools were moved in the Highland Park factory alone, costing an average of $50 for each move—a total of $150,000 (Acc. 572, Nevins and Hill File, Box 23, Ford Archives).

84. *New York Times,* July 25, 1927, p. 1; Fay Leone Faurote, "Henry Ford Still on the Job with Renewed Vigor," pp. 193–94; Pioch, Reminiscences.

85. Faurote, "Preparing for Ford Production," p. 636; Wibel, Reminiscences. The cost figure could be anywhere between $15 million, as reported in *Ford News* 7 (September 1, 1927): 1, and $25 million, as reported by Faurote, "Preparing for Ford Production," p. 635. The final cost of the changeover, including design and tooling costs and lost profits, totaled about $250 million (Nevins and Hill, *Ford: Expansion,* p. 458).

86. Philip E. Haglund, Reminiscences, Ford Archives.

87. For a more extensive treatment, see Faurote, "Single-Purpose Manufacturing."

88. See illustrations and discussions of these machines in the following articles by Fay Leone Faurote: "Cylinder Block and Head Operations," *American Machinist* 68 (1928): 679–84; "Operations on the Transmission Case . . . ," *American Machinist* 68 (1928): 874–79; "Machining Operations on the Flywheel," *American Machinist* 68 (1928): 917–21; and "Machining and Welding Operations on the Rear Axle Housings." See also A. M. Wibel's discussion of these machines in his Reminiscences.

89. James O'Connor, Reminiscences, Ford Archives.

90. "Final Assembly Line Ready . . . ," p. 8; Fay Leone Faurote, "Final Assembly," p. 273.

91. Haglund, Reminiscences.

92. Klann, Reminiscences; see also O'Connor, Reminiscences.

93. Klann, Reminiscences. See Avery's address at the Ford Motor Company dinner for thirty-five year employees, December 19, 1944, Acc. 1, Fair Lane Papers, Box 117, Ford Archives. Contemporary evidence includes memorandums, Pioch to Avery, December 6, 1922, and Crittenden to Avery, April 27, 1923, Acc. 680, Plant Engineering, Box 5, ibid.

94. O'Connor, Reminiscences.

95. Ibid.

96. Linn Bryson to Charles Sorensen, January 10, 11, 1928, Acc. 572, Nevins and Hill papers, Box 22, Ford Archives.

97. See Sorensen's account of the Rouge in My Forty Years with Ford, pp. 170–79.

98. P. E. Martin notebook entry, October 21, 1927, Acc. 823, P. E. Martin Papers, Ford Archives. Cf. Henry Ford's public prediction of fifteen million in New York Times, July 31, 1927, p. 1.

99. Jas. Guinon to Charles Sorensen, March 8, 1925, Acc. 572, Nevins and Hill Papers, Box 20, Ford Archives. For other aspects of Ford production delays, see L. C. Dibble, "Slow Progress in Plant Changes Delays New Ford Model." For a more triumphant view, see Waldemar Kaempffert, "The Dramatic Story Behind Ford's New Car," New York Times, December 18, 1927, sec. 10, p. 5.

100. New York Times, December 21, 1927, p. 30, February 19, 1928, p. 1.

101. On Universal Credit Corporation, see Nevins and Hill, Ford: Expansion, pp. 465–66 and New York Times, September 19, 1928, p. 3.

102. Bennett's rise and Sorensen's slippage are admirably treated in Allan Nevins and Frank Ernest Hill, Ford: Decline.

103. Pioch, Reminiscences.

104. Ibid.

105. Ibid. and Walters, Reminiscences.

106. Wibel, Reminiscences; Henry Ford with Samuel Crowther, Moving Forward, pp. 170–86. Ford's machine tool acquisitions followed a general trend in the American automobile industry in the period 1927–30, which is indicative of the increasing importance of the annual model change and the production problems it engendered. See Frederick B. Heitkamp, "Using More Semi-Special Tools"; and the anonymous articles "Flexibility in Making Automobile Frames"; "Many Highly Specialized Jobs Now Being Performed with Standard Machine Tools"; and "Short Cuts on Standard Machines for Common Automotive Jobs Are Being Developed to Reduce Use of Special Tools."

107. Sheldrick, Reminiscences.

108. For other diagnoses of the problems with the Model T to Model A changeover, see the Reminiscences of Wibel, Pioch, Walters, Kroll, Sheldrick, Klann, and Miller.

109. New York Times, June 30, 1929, sec. 2, p. 7; August 18, 1929, sec. 9, p. 14; September 19, 1929, p. 40. For rumors of a shutdown, see New York Times, October 3, 1929, p. 43.

110. These figures come from Nevins and Hill, Ford: Expansion, p. 571. Cf. Fortune's in "Mr. Ford Doesn't Care," p. 68.

111. New York Times, December 29, 1929, p. 3. On price cutbacks in November, see ibid., November 19, 1929, p. 41.

112. Ibid., January 5, 1930, sec. 10, p. 2; February 9, 1930, sec. 9, p. 14; March 30, 1930, sec. 9, p. 11; April 27, 1930, sec. 10, p. 10; August 31, 1930, sec. 9, p. 6; December 14, 1930, sec. 10, p. 10; and Nevins and Hill, Ford: Expansion, pp. 575, 583.

113. Production figures are in Appendix I, Nevins and Hill, Ford: Decline, p. 478. Sales and profit figures are in Nevins and Hill, Ford: Expansion, pp. 571, 577, and "Mr. Ford Doesn't Care," p. 68.

114. Nevins and Hill, Ford: Expansion, p. 577.

115. New York Times, July 30, 1931, p. 2; August 30, 1931, p. 1.

116. See, for example, Nevins and Hill, Ford: Expansion, pp. 578–96.

117. New York Times, July 30, 1931, p. 2, and August 30, 1931, p. 1.

118. Nevins and Hill, Ford: Expansion, pp. 594–95; New York Times, February 12, 1932, p. 15.

119. Nevins and Hill, Ford: Expansion, p. 595. Nevins and Hill's account of the V-8 drew heavily from James Sweinhart's "authoritative" story in the Detroit News, February 11, 1932, reprinted in Ford News 12 (March 1932): 3, 17.

120. New York Times, March 31, 1932, p. 44.

121. Pioch, Reminiscences.

122. Ibid.; Wibel, Reminiscences.

123. From July 1931 to April 1932 Rouge assembly operations were virtually nonexistent, and branch assembly plants wound down their production from three to four thousand cars per week to zero. See the weekly assembly records for 1931 and 1932 in Acc. 622, Ford Archives.

124. New York Times, March 13, 1932, p. 6. Philip E. Haglund's diagnosis of the T to A changeover may obtain for the changeover to the V-8: "I assume it was the lack of engineering ability that resulted in such a long period of time to make the changeover in 1927. . . . When we employees compared our Engineering Department with other auto-

mobile plants, we came to the belief we didn't have an Engineering Department in the same sense as other plants. We had no depth. Our engineering was done by a few who had experience. It was apparently inadequate'' (Reminiscences).

125. "1932 Production," Acc. 622, Ford Archives.

126. *New York Times*, March 11, 1932, p. 27.

127. "Mr. Ford Doesn't Care," p. 66.

128. Nevins and Hill, *Ford: Expansion*, pp. 596, 458.

129. Wibel, Reminiscences; see also *New York Times*, March 24, 1930, p. 2.

130. Sorensen to Edsel Ford, August 10, 1928, quoted in Nevins and Hill, *Ford: Expansion*, p. 466.

131. "Mr. Ford Doesn't Care," p. 67.

132. See, for example, Kaempffert, "The Dramatic Story Behind Ford's New Car''; and a *New York Times* editorial comment, "Ford the Extraordinary," December 18, 1927, sec, 3, p. 4.

133. "Where the Ford Method Halted," *New York Times*, May 30, 1932, p. 12. Earlier, Reinhold Niebuhr, who was then pastor of Detroit's Bethel Evangelical Church, had sharply criticized Ford for failing to plan adequately for change, thereby laying off tens of thousands of employees during the changeover with no unemployment compensation (*Leaves from the Notebook of a Tamed Cynic*, pp. 154–55). I am indebted to Alan Neely for this reference.

Chapter 8

1. Henry Ford, "Mass Production''; *New York Times*, September 19, 1926, sec. 10, p. 1. On the circumstances leading to Ford's article, see the Introduction, above.

2. See, for example, "Love of Nice Things," *New York Times*, December 3, 1927, p. 14; "Says Ford Needed Beauty," ibid., February 23, 1928, p. 12; and "Says Beauty Aids Trade," ibid., April 17, 1928, p. 48.

3. Jeffrey L. Meikle, *Twentieth Century Limited*, pp. 26, 27. Meikle called the changeover to Model A "the most expensive art lesson in history'' (p. 10; see also pp. 10–14).

4. Paul M. Mazur, *American Prosperity*, p. 125.

5. Henry Ford with Samuel Crowther, *My Life and Work*, opposite p. 130.

6. Edward A. Filene, *The Way Out*, pp. 93, 96.

7. Ibid., pp. 99, 180.

8. Ibid., pp. 176–77, 179. An extensive biographical sketch of Filene appears in *National Cyclopaedia of American Biography*, 45:17–19.

9. "At the Wheel: Mass Production and Indi-

viduality," *New York Times*, March 31, 1929, sec. 9, p. 16.

10. *New York Times*, September 19, 1926, sec. 10, p. 1; Henry Ford, "Mass Production."

11. On Filene, see note 6. Hoover quoted in Stuart Ewen, *Captains of Consciousness*, p. 28.

12. Edward A. Filene, "Mass Production Makes a Better World," p. 629. Not everyone agreed that, with the Model A, Ford had been able to preserve mass production and to respond to changing consumer tastes. In 1928 Paul M. Mazur went as far as to call mass production—and Ford—dead: "The King is dead! Long live the King! Mass production as an autocratic ruler of the destinies of American industry has passed beyond'' (*American Prosperity*, p. 130). Mazur saw the future not in mass production but in "mechanized production" (ibid., pp. 125–60). In a similar vein, see Anne O'Hare McCormick, "The Future of the Ford Idea."

13. Edward A. Filene with Charles W. Wood, *Successful Living in This Machine Age*, p. 1.

14. Filene, "Mass Production," p. 631; Filene, *Successful Living*, p. 144.

15. Harvey N. Davis, "Spirit and Culture under the Machine," pp. 283, 286, 288–90. Davis was a highly respected engineer, educator, and president of Stevens Institute of Technology. See his biographical sketch in the *National Cyclopaedia of American Biography*, 40:132–33.

16. Filene, *Successful Living*, p. 98; Tennessee Valley Authority, *Annual Report, 1936* (Washington, D.C.: U.S. Government Printing Office, 1937), p. 26.

17. Meikle, *Twentieth Century Limited*, pp. 21, 24, 26. The words quoted at the beginning of the paragraph came from a contemporary review.

18. Ibid., pp. 25–27; Kiesler quoted in ibid., p. 38. Meikle's *Twentieth Century Limited* is primarily a treatment of these industrial designers. See especially his chapter, "The New Industrial Designers," pp. 39–67.

19. "Ford Urges for Farms New Mass Production," *New York Times*, March 7, 1930, p. 48.

20. Allan Nevins and Frank Ernest Hill, *Ford: Expansion and Challenge, 1915–1933*, pp. 490–91.

21. "Ford Urges for Farms New Mass Production." While in Ft. Myers, Florida, a year later, Ford reiterated these notions, as reported in "Ford Says Machine Can Never Oust Man," ibid., March 17, 1931, p. 31.

22. "'Mass Production' in Agriculture," ibid., June 2, 1930, p. 22.

23. Trip Report of William Blitzer, March 28, 1947, Papers of the Albert Farwell Bemis Foundation, MIT Institute Archives and Special Collections, Cambridge, Massachusetts.

24. The bulk of this section on Gunnison is based on my article in *Dictionary of American Biography,* 7th Suppl., s.v. "Gunnison, Foster."

25. *Architectural Forum* quoted in William Blitzer, "Case Study of an Entrepreneur: Foster Gunnison," p. 3, Bemis Foundation Papers, MIT.

26. Ibid., pp. 7–8.

27. Alfred Bruce and Harold Sandbank, *A History of Prefabrication,* p. 10.

28. Blitzer, "Case Study of an Entrepreneur," p. 15; quotation from *Architectural Forum* appears in Bruce and Sandbank, *History of Prefabrication,* p. 64.

29. Blitzer, "Case Study of an Entrepreneur," pp. 20, 34. Blitzer also points out that when in 1939 Gunnison introduced the "Miracle Home," a house comparable to the Model T in its price, he found that it failed: "Through the years Gunnison . . . seems to have become more and more convinced that cutting corners to sell a 'minimum' house did not pay" (p. 19).

30. See Gilbert Herbert, *Pioneers of Prefabrication,* and Charles E. Peterson, "Early American Prefabrication." Lewis Mumford later said the ethos of mass production had created "enthusiastic and often fanatical minds that regarded prefabrication as a cure-all" (*City Development,* p. 61).

31. Bruce and Sandbank, *History of Prefabrication,* p. 7. See also some of the newspaper clippings offering "typical examples of the widespread publicity and enthusiasm which greeted the idea of prefabrication in the market-hungry thirties," in ibid.

32. Mumford, *City Development,* pp. 63–83.

33. Ibid., p. 72.

34. Ibid., p. 73.

35. Edward A. Filene, "Foresees Ford Prosperity for Furniture Industry." Filene stated: "If the furniture business will adopt the Ford and Chevrolet methods of production and similarly improved methods of distribution—that is, if it will adopt scientific mass production and mass distribution—it can and will attain a prosperity far beyond the rational hopes of the present." For the most part, furniture manufacturers and dealers reacted negatively to Filene's rhetoric of mass production. In two successive issues, *Furniture Record and Journal* published the furniture industry's response: "Furniture Men Flay Filene's Proposal of Mass Production" and "Dealer-Opinion on Filene's Mass-Production Forecast." Almost all of the furniture industry's representatives argued that Filene overlooked stylistic and taste considerations and overemphasized mass production methods. I am indebted to Michael Ettema for these references.

36. Upton Sinclair, *The Flivver King,* p. 174.

37. For instance, the *New York Times* reported that five hundred thousand people annually visited Ford's River Rouge factory "to see the wonders of modern manufacturing" (October 20, 1929, p. 14). The final quote is from *Tri-City Labor Review,* Rock Island, Illinois, 1932, as quoted in Ewen, *Captains of Consciousness,* p. 12.

38. Aldous Huxley, *Brave New World;* the quotations are from pp. 29, 4. For more information on the background to Huxley's satire of Ford and Fordism, see Jerome Meckier, "Debunking Our Ford," an article I learned about only after completing this chapter.

39. Other authors also used fiction to comment critically on the ethos of mass production. See especially John Dos Passos, *The Big Money* (1936), and Sherwood Anderson's *Marching Men* (1917), *Winesburg, Ohio* (1918), *Poor White* (1920), and *Perhaps Women* (1930).

40. George Basalla, "Science, Technology, and Popular Culture," pp. 49–50. The full script appears in René Clair, *A nous la liberté and entra'acts.*

41. Clair, *A nous la liberté and entra'acts,* pp. 96–97.

42. Charles Chaplin, *My Autobiography,* p. 383. On the film's original title, see "Chaplin Tells Theme of His New Picture," *Los Angeles Times,* March 17, 1935.

43. John McCabe, *Charlie Chaplin,* p. 182.

44. *Detroit News,* October 16, 1923, and *Detroit Free Press,* October 16, 1923. I am indebted to David R. Crippen, reference archivist of the Ford Archives, for help on Chaplin's visit to Highland Park. Between the time he toured the Ford plant and his production of *Modern Times,* Chaplin's views on mass production may have been influenced by his meeting in 1931 of Mahatma Gandhi. See John H. Van Deventer, " 'Dat Ole Debil,' Machinery." I thank John K. Smith for this reference.

45. Extensive citations for reviews of *Modern Times* appear in Timothy James Lyons, *Charles Chaplin,* pp. 121–27.

46. James Agee, *Agee on Film,* 1:288; George Basalla, "Keaton and Chaplin," p. 199.

47. Richard Müller-Freienfels, *The Mysteries of the Soul,* pp. 235–92; Julius Klein, "Business," p. 100.

48. Stuart Chase, "Play," pp. 347, 353. A prolific writer, Chase vacillated on the ethos of mass production, sometimes prophesying its benefits, sometimes critical. Compare his article "Play" with his review of Henry Ford's *Today and Tomorrow:* "Henry Ford's Utopia." See also his *Men and Machines* (1929) and *The Nemesis of American Business* (1931). On Chase as a writer and thinker see Robert B. Westbrook, "Tribune of the Technostructure."

49. Quoted in Ewen, *Captains of Consciousness,* pp. 11–12.

50. McCormick, "The Future of the Ford Idea," p. 1.

51. Paul Mazur, "The Doctrine of Mass Produc-

tion Faces a Challenge,'' *New York Times,* November 29, 1931, sec. 9, p. 3. Mazur had earlier published an article, ''Mass Production, Has It Committed Suicide?'' much of which appeared in his *American Prosperity.*

52. Harold Callender, ''Gandhi Dissects the Ford Idea.''

53. Mazur, ''Doctrine of Mass Production.''

54. E. E. Calkins, ''The New Consumption Engineer and the Artist.'' See also Roy Sheldon and Egmont Arens, *Consumer Engineering.*

55. Some of this literature includes Harry Tippen, *The New Challenge of Distribution: The Paramount Industrial Problem* (New York: Harper & Brothers, 1932); John B. Cheadle et al., *No More Unemployed* (Norman, Okla.: University of Oklahoma Press, 1934); Lewis Corey, *The Decline of American Capitalism* (New York: Covici, 1934); Maurice Leven et al., *America's Capacity to Consume* (Washington, D.C.: Brookings Institution, 1934); William H. Lough, *High Level Consumption: Its Behavior; Its Consequences* (New York: McGraw-Hill, 1935); Frederick Purdy, *Mass Consumption* (New York: Talisman Press, 1936); Carle C. Zimmerman, *Consumption and Standards of Living* (New York: D. Van Nostrand Co., 1936); Charles S. Wyand, *The Economics of Consumption* (New York: Macmillan, 1937); Elizabeth Ellis Hoyt, *Consumption in Our Society* (New York: H. Holt, 1938); Roland Snow Vaile, *Income and Consumption* (New York: H. Holt, 1938); and Alfred P. Sloan, Jr., *The Creation of Abundance,* pamphlet, March 11, 1939, Eleutherian Mills Historical Library.

56. Ewen, *Captains of Consciousness,* p. 57, quoting Leverett S. Lyon, ''Advertising,'' *The Encyclopedia of the Social Sciences,* 1 (1922): 475.

57. Helen Woodward, *Through Many Windows* (New York: Harper & Brothers, 1926), as quoted by Ewen, *Captains of Consciousness,* p. 80; David M. Potter, *People of Plenty,* pp. 166–88.

58. Ewen, *Captains of Consciousness,* offers a profound study of advertising in the era of the ethos of mass production.

59. This section on Rivera is based on Linda Downs, ''The Rouge in 1932,'' pp. 47–91.

60. Quoted in ibid., p. 48.

61. The Rouge was attracting more than five hundred thousand visitors annually (*New York Times,* October 20, 1929, p. 14).

62. Compare Rivera's treatment of the fender press in his initial sketch of the south wall panel with his later sketch of the same panel and the actual panel. See Downs, ''The Rouge in 1932,'' pp. 80, 83.

63. ''Will Detroit, Like Mohammed II, Whitewash Its Rivera Murals?'' *Art Digest* 7, no. 5 (1933): 6, quoted in ibid., p. 52.

64. *Special Report of Joseph Whitworth* in

Nathan Rosenberg, ed., *The American System of Manufactures,* p. 307.

65. On the industrial designers, see Meikle, *Twentieth Century Limited.* Both the proponents and critics of mass production worked in virtually every medium of communication. There is a clear need for a major study of technology and its critics in the Great Depression which would pay attention to journalistic, literary, radio, film, and exhibition material.

66. Lewis Mumford, *Technics and Civilization,* pp. 93–94.

Appendix 1

1. Great Exhibition of the Works of Industry of All Nations, 1851, *Official Description and Illustrated Catalogue,* 3: 1455.

2. Samuel Colt, ''On the Application of Machinery to the Manufacture of Rotating Chambered-Breech Fire-Arms, and the Peculiarities of Those Arms,'' p. 61.

3. ''Institution of Civil Engineers,'' p. 611.

4. Charles Tomlinson, ed., *Rudimentary Treatise on the Construction of Locks,* pp. 154–63.

5. *Special Report of Joseph Whitworth,* in Nathan Rosenberg, ed., *The American System of Manufactures,* p. 387. The *Special Report of George Wallis* also appears in this volume.

6. See Chapter 1.

7. The full report is reprinted in Rosenberg, ed., *The American System of Manufactures,* pp. 87–197.

8. Ibid., p. 143.

9. ''Revolvers,'' *Household Words* 9 (1855), reprinted in Charles T. Haven and Frank A. Belden, *A History of the Colt Revolver,* pp. 345–49.

10. ''Repeating Pistols,'' reprinted in ibid., pp. 350–67.

11. John Anderson, ''On the Application of Machinery in the War Department.''

12. John Anderson, ''On the Application of the Copying or Transfer Principle in the Production of Wooden Articles.''

13. ''Hobbs' Lock Manufactory.''

14. ''The Royal Small-Arm Manufactory, Enfield,'' p. 204.

15. John Anderson, ''On the Application of the Copying Principle in the Manufacture and Rifling of Guns.''

16. Charles Hutton Gregory, ''Address of the President,'' p. 188.

17. Ibid., p. 186.

18. Thomas Brassey, *Lectures on the Labour Question,* pp. 46–47.

19. Thomas A. Edison to George S. Nottage, March 23, 1878, Letterbook 750806, pp. 463, 465,

Papers of Thomas A. Edison, Edison National Historic Site, West Orange, New Jersey.

20. Charles H. Fitch, "Report on the Manufactures of Interchangeable Mechanism," pp. 618–20.

21. Sir Edmund Beckett, "Lock," *Encyclopaedia Britannica* (Philadelphia: J. M. Stoddart & Co., 1882) 14: 758. Beckett, first baron Grimthorpe, was an expert on clock- and watchmaking, though he had invented locks and published some on them as well. Beckett also contributed the article on clocks in the ninth edition of the *Britannica*, but he did not discuss the manufacture of American clocks. It is instructive to compare this edition's article on the lock with those of the earlier eighth edition and the later eleventh edition.

22. *James Nasmyth, Engineer: An Autobiography*, ed. Samuel Smiles (New York: Harper & Brothers, 1883), p. 366.

23. "Cutlery at Sheffield," p. 20.

24. "Extracts from Chordal's Letters," 6 (May 5, 1883): 4.

25. "Extracts from Chordal's Letters," 6 (June 16, 1883): 2–3. See also Chordal's letters in *American Machinist* 6 (June 23, 1883): 1–3.

26. S. W. Goodyear, "Working to Standards in Large and Small Shops—Early Sewing-Machine Economies"; Goodyear, "Personal Recollections," 6 (October 13, 1883): 4–5; and 6 (December 1, 1883): 4–5.

27. Charles H. Fitch, "The Rise of a Mechanical Ideal," p. 517.

28. "Obituary of John Anderson," p. 350.

29. B. F. Spalding, "The 'American System' of Manufacture," pp. 2–4; "The 'American' System of Manufacture," pp. 2–3, 11–12.

30. W. F. Durfee, "The History and Modern Development of the Art of Interchangeable Construction in Mechanism," esp. pp. 1255–56.

31. John Rigby, "The Manufacture of Small Arms."

32. Horace L. Arnold [Henry Roland], "The Revolution in Machine-Shop Practice."

33. Frederick A. McKenzie, *The American Invaders*.

34. Joseph V. Woodworth, *American Tool Making and Interchangeable Manufacturing*, esp. pp. 20–27.

35. Joseph W. Roe, "Development of Interchangeable Manufacture."

36. Joseph W. Roe, *English and American Machine Tool Builders*, pp. 140–41.

37. Guy Hubbard, "Development of Machine Tools in New England," pp. 1–4, 463–67.

38. L. P. Alford, "Duplicate and Interchangeable Manufacture."

39. Joseph W. Roe, "Interchangeable Manufacture." Roe's address later appeared in the Newcomen Society of North America's published series of addresses as "Interchangeable Manufacture in American Industry" (1939).

Bibliography

Manuscripts

Cambridge, Massachusetts. MIT Institute Archives and Special Collections. Papers of the Albert Farwell Bemis Foundation.

Cleveland, Ohio, Freiberger Library, Case Western Reserve University. Fred H. Colvin Papers.

Dearborn, Michigan. Edison Institute. Ford Archives.

Hartford, Connecticut. Connecticut State Library. Papers of the Hartford Cycle Company.

Madison, Wisconsin. State Historical Society of Wisconsin. McCormick Collection.

———. State Historical Society of Wisconsin. Papers of the Singer Manufacturing Company.

North Kingstown, Rhode Island. Patent Library, Brown & Sharpe Manufacturing Company. Historical Data Files.

Paris, France. Archives Nationales. Report of William John Macquorn Rankine on the Singer Manufacturing Company's Glasgow Factory. Report F12-6907.

South Bend, Indiana. Northern Indiana Historical Society. "Cabinet Factory and Western Foundry—South Bend, Ind.," 1920.

South Bend, Indiana. South Bend Public Library. Vertical File on Leighton Pine and Singer Manufacturing Company, South Bend Works.

West Orange, New Jersey. Edison National Historic Site. Papers of Thomas A. Edison. Thomas A. Edison to George S. Nottage, March 23, 1878. Letterbook 750806, pp. 463, 465.

Public Documents

Fitch, Charles H. "Report on the Manufactures of Interchangeable Mechanism." In U.S. Department of the Interior, Census Office. *Tenth Census of the United States (1880).* Vol. 2: *Report on the Manufactures of the United States,* pp. 611–704. Washington, D.C.: U.S. Government Printing Office, 1883.

Great Britain. Parliament. House of Commons. Reports from committees, vol. 18. Select Committee on Small Arms. *Report from the Select Committee on Small Arms.* London, 1854.

U.S. Bureau of the Census. *Manufactures 1905, Part 1, United States by Industries.* Washington, D.C.: U.S. Government Printing Office, 1907.

U.S. Department of the Interior, Census Office. *Eleventh Census of the United States (1890).* Vol. 11: *Report on Manufacturing Industries of the United States. Part 1. Totals for States and Industries.* Washington, D.C.: U.S. Government Printing Office, 1895.

———. *Twelfth Census of the United States (1900).* Vol. 7: *Manufactures. Part 1. United States by Industries.* Washington, D.C.: U.S. Government Printing Office, 1902.

U.S. Tennessee Valley Authority. *Annual Report,* 1936. Washington, D.C.: U.S. Government Printing Office, 1937.

Books

Abernathy, William J. *The Productivity Dilemma: Roadblock to Innovation in the Automobile Industry.* Baltimore: Johns Hopkins University Press, 1978.

Agee, James. *Agee on Film.* 2 vols. New York: Grosset and Dunlap, Universal Library, 1969.

Althin, Torsten, K. W. *C. E. Johansson, 1864–1943.* Translated by Cyril Marshall. Stockholm: Privately printed, 1948.

Anderson, Sherwood. *Marching Men*. New York: John Lane Co., 1917.

———. *Perhaps Women*. New York: H. Liveright, 1930.

———. *Poor White*. 1920. New York: Viking, Compass Books. 1966.

———. *Winesburg, Ohio*. 1919. Rev. ed. New York: Viking, 1960.

Ardrey, Robert L. *American Agricultural Implements*. 1894. Reprint. New York: Arno Press, 1972.

Arnold, Horace Lucien, and Faurote, Fay Leone. *Ford Methods and the Ford Shops*. New York: Engineering Magazine, 1915.

Asher & Adams' Pictorial Album of American Industry. 1876. Reprint. New York: Rutledge Books, 1976.

Babbage, Charles. *On the Economy of Machinery and Manufactures*. 4th ed. London: Charles Knight, 1835. Reprint. New York: Augustus M. Kelly, 1963.

Battison, Edwin A. *From Muskets to Mass Production: The Men and the Times That Shaped American Manufacturing*. Windsor, Vt.: American Precision Museum, 1976.

Beasley, Norman. *Knudsen: A Biography*. New York: McGraw-Hill, 1947.

Benjamin, Park. *Appleton's Cyclopaedia of Applies Mechanics*. 2 vols. New York: D. Appleton, 1881.

Bigelow, David. *History of Prominent Merchants and Manufacturing Firms in the United States. . . .* Vol. 6. Boston: D. Bigelow, 1857.

Bishop, Robert, and Coblentz, Patricia. *The World of Antiques, Art, and Architecture in Victorian America*. New York: E. P. Dutton, 1979.

Bliss (E. W.) Company. *Catalogue and Price List of Presses, Dies and Special Machinery, for Working Sheet-Metal. . . . New York: Lockwood Press*, 1886.

———. *Catalogue and Price List of Presses, Drop Hammers, Shears, Dies and Special Machinery. . . .* 12th ed. New York: N.p., 1900.

Boorstin, Daniel J. *The Americans: The Democratic Experience*. New York: Random House, 1972.

Brassey, Thomas. *Lectures on the Labour Question*. 3d ed. rev. London: Longmans, Green, 1878.

Braverman, Harry. *Labor and Monopoly Capital: The Degradation of Work in the Twentieth Century*. New York: Monthly Review Press, 1974.

Brown & Sharpe Manufacturing Company. *Catalogue and Price List*. Providence, R. I.: Brown & Sharpe, 1885.

Bruce, Alfred, and Sandbank, Harold. *A History of Prefabrication*. 1944. Reprint. New York: Arno Press, 1972.

Burlingame, Roger. *Engines of Democracy*. New York: C. Scribner's Sons, 1940.

Camp, Hiram. *Sketch of the Clock Making Business, 1792–1892*. New Haven: Privately published, 1893.

Catalogue and Price List of Sheet Steel Bicycle Parts Made by Crosby & Mayer Co. Buffalo, N.Y.: Crosby & Mayer Co., 1898.

Chandler, Alfred D., Jr. *The Visible Hand: The Managerial Revolution in American Business*. Cambridge, Mass.: Harvard University Press, 1977.

———, ed. *Giant Enterprise: Ford, General Motors, and the Automobile Industry*. New York: Harcourt, Brace & World, 1964.

———, ed. *Managerial Innovation at General Motors*. New York: Arno Press, 1979.

Chandler, Alfred D., Jr., and Salsbury, Stephen. *Pierre S. du Pont and the Making of the Modern Corporation*. New York: Harper & Row, 1971.

Chaplin, Charles. *My Autobiography*. New York: Simon and Schuster, 1964.

Chapman, Captain Charles. *The Ocean Waves: Travels by Land and Sea*. London: George Berridge & Co., 1875.

Clair, René. *A nous la liberté and entra' acts*. Translated by Richard Jacques and Nicola Hayden. New York: Simon and Schuster, 1978.

Clymer, Floyd. *Henry's Wonderful Model T, 1908–1927*. New York: Bonanza Books, 1955.

Colvin, Fred H. *60 Years with Men and Machines*. New York: McGraw-Hill, 1947.

Cooper, Grace Rogers. *The Sewing Machine: Its Invention and Development*. Washington, D.C.: Smithsonian Institution Press, 1976.

Copley, Frank B. *Frederick W. Taylor*. 2 vols. New York: Taylor Society, 1922.

Davies, Robert Bruce. *Peacefully Working to Conquer the World: Singer Sewing Ma-*

chines in Foreign Markets, 1854–1920. New York: Arno Press, 1976.

Dewing, Arthur S. *Corporate Promotions and Reorganizations.* Cambridge, Mass.: Harvard University Press, 1914.

Deyrup, Felicia Johnson. *Arms Makers of the Connecticut Valley: A Regional Study of the Economic Development of the Small Arms Industry, 1798–1870.* Smith College Studies in History, vol. 33. Northampton, Mass.: Smith College, 1948.

Dos Passos, John. *The Big Money.* New York: Washington Square, 1936.

Drucker, Peter. *Concept of the Corporation.* Rev. ed. New York: John Day, 1972.

80 Years Ferracute. Bridgeton, N.J.: Ferracute Machine Company, 1943.

Erskine, Albert Russel. *History of the Studebaker Corporation.* South Bend, Ind.: Studebaker Corporation, 1924.

Ewen, Stuart. *Captains of Consciousness.* New York: McGraw-Hill, 1976.

Ferguson, Eugene S. *Bibliography of the History of Technology.* Cambridge, Mass.: MIT Press, 1968.

————. *Oliver Evans.* Greenville, Del.: Hagley Museum, 1980.

Filene, Edward A. *The Way Out: A Forecast of Coming Changes in American Business and Industry.* Garden City, N.Y.: Doubleday, Page, 1925.

Filene, Edward A., with Wood, Charles W. *Successful Living in This Machine Age.* New York: Simon and Schuster, 1932.

Flink, James J. *America Adopts the Automobile, 1895–1910.* Cambridge, Mass.: MIT Press, 1970.

Ford, Henry, with Crowther, Samuel. *Moving Forward.* Garden City, N.Y.: Doubleday, 1930.

————. *My Life and Work.* Garden City, N.Y.: Garden City Publishing Co., 1922.

Fuller, Claud E. *The Breech-Loader in the Service.* Topeka, Kan.: Arms Reference Club of America, 1933.

Gamel, Joseph. *Description of the Tula Weapon Factory in Regard to Historical and Technical Aspects.* Moscow, 1826.

General Motors Corporation. *Annual Report of General Motors Corporation for 1925.* N.p.: N.p., 1926.

Giedion, Siegfried. *Mechanization Takes Command: A Contribution to Anonymous History.* Oxford: Oxford University Press, 1948.

Gilbert, K. R. *The Portsmouth Block-making Machinery: A Pioneering Enterprise in Mass Production.* London: HMSO, 1965.

Goodrich, Carter. *The Miner's Freedom: A Study of Working Life in a Changing Industry.* Boston: Marshall Jones, 1925.

Green, Constance, M. *Eli Whitney and the Birth of American Technology.* Boston: Little, Brown, 1956.

Gustin, Lawrence R. *Billy Durant: Creator of General Motors.* Grand Rapids, Mich.: Eerdmans, 1973.

Habakkuk, H. J. *American and British Technology in the Nineteenth Century.* Cambridge: Cambirdge University Press, 1962.

Haber, Samuel. *Efficiency and Uplift: Scientific Management in the Progressive Era, 1890–1920.* Chicago: University of Chicago Press, 1964.

Haven, Charles T., and Belden, Frank A. *A History of the Colt Revolver.* New York: Morrow, 1940.

Herbert, Gilbert. *Pioneers of Prefabrication: The British Contribution in the Nineteenth Century.* Baltimore: Johns Hopkins University Press, 1978.

Hutchinson, William T. *Cyrus Hall McCormick: Harvest, 1856–1884.* New York: D. Appleton-Century, 1935.

————. *Cyrus Hall McCormick: Seedtime, 1809–1856.* New York: Century, 1930.

Huxley, Aldous. *Brave New World.* 1932. Perennial Classic. New York: Harper & Row, 1969.

Jefferson, Thomas. *Papers of Thomas Jefferson.* Edited by Julian P. Boyd. 60 vols. Princeton: Princeton University Press, 1950–.

Jerome, Chauncey. *History of the American Clock Business for the Past Sixty Years.* New York: F. C. Dayton, Jr., 1860.

Kettering, Charles F., and Orth, Allen. *The New Necessity.* Baltimore: Williams & Wilkins, 1932.

Knight, E. Vernon, and Wulpi, Meinrad, eds. *Veneers and Plywood: Their Craftsmanship and Artistry, Modern Production Methods and Present-Day Utility.* New York: Ronald, 1927.

Knight, Edward H. *Knight's American Mechan-*

ical Dictionary. 3 vols. Boston: Houghton, Mifflin, 1876.

Leland, Ottilie M. *Master of Precision: Henry M. Leland*. Detroit: Wayne State University Press, 1966.

Lewis, David L. *The Public Image of Henry Ford*. Detroit: Wayne State University Press, 1976.

London. Great Exhibition of the Works of Industry of All Nations, 1851. *Official Descriptive and Illustrated Catalogue*. 3 vols. London: Spicer Brothers, 1851.

Lyons, Timothy James. *Charles Chaplin: A Guide to References and Resources*. Boston: G. K. Hall, 1979.

McCabe, John, *Charlie Chaplin*. Garden City, N.Y.: Doubleday, 1978.

McCormick Harvesting Machine Co. *La Compagnie MacCormick de Chicago a l'Exposition de 1900*. N.p., n.d.

———. *Triumph Throughout all Nations: The Standard of the World*. Chicago: McCormick Harvesting Machine Co., 1896.

McKenzie, Frederick A. *The American Invaders: Their Plans, Tactics and Progress*. New York: Street & Smith, 1901.

May, Earl Chapin. *The Canning Clan*. New York: Macmillan, 1937.

Mayr, Otto, and Post, Robert C., eds. *Yankee Enterprise: The Rise of the American System of Manufactures*. Washington, D.C.: Smithsonian Institution Press, 1982.

Maxim, Hiram Percy. *Horseless Carriage Days*. New York: Harper & Brothers, 1937.

Mazur, Paul M. *American Prosperity: Its Cuases and Consequences*. New York: Viking, 1928.

Meikle, Jeffrey L. *Twentieth Century Limited: Industrial Design in America, 1925–1939*. Philadelphia: Temple University Press, 1979.

Meyer, Stephen, III. *The Five Dollar Day: Labor Management and Social Control in the Ford Motor Company, 1908–1921*. Albany: State University of New York Press, 1981.

Moffett, Cleveland. *Marvels of Bicycle Making*. New York: McClure's Magazine, 1897.

Montgomery, Davis. *Workers' Control in America*. Cambridge: Cambridge University Press, 1979.

Müller-Freienfels, Richard. *The Mysteries of the Soul*. Translated by Bernard Miall. New York: Alfred A. Knopf, 1929.

Mumford, Lewis. *City Development*. New York: Harcourt, Brace & World, 1945.

———. *Technics and Civilization*. New York: Harcourt, Brace & World, 1934.

Nelson, Daniel. *Managers and Workers: Origins of the New Factory System in the United States, 1880–1920*. Madison: University of Wisconsin Press, 1975.

Nevins, Allan, and Hill, Frank Ernest. *Ford: Decline and Rebirth, 1933–1962*. New York: Charles Scribner's Sons, 1962.

———. *Ford: Expansion and Challenge, 1915–1933*. New York: Charles Scribner's Sons 1957.

———. *Ford: The Times, the Man, the Company*. New York: Charles Scribner's Sons, 1954.

Niebuhr, Reinhold. *Leaves from the Notebook of a Tamed Cynic*. 1929. Reprint. Hamden, Conn.: Shoe String Press, 1956.

North, S. N. D., and North, Ralph H. *Simeon North: First Official Pistol Maker of the United States*. Concord, N.H.: Rumford Press, 1913.

Noyes, William. *World and Forest*. Peoria: Manual Arts Press, 1912.

Nutter, Waldo. E. *Manhattan Firearms*. Harrisburg, Pa: Stackpole Press, 1958.

Official Catalogue of the New York Exhibition of the Industry of all Nations. Rev. ed. New York: Association for the Exhibition, 1853.

Ozanne, Robert. *A Century of Labor-Management Relations at McCormick and International Harvester*. Madison: University of Wisconsin Press, 1967.

Perry, Thomas D. *Modern Plywood*. New York: Pitman Publishing Corporation, 1942.

Pope Manufacturing Company. *Columbia Bicycle, Catalog for January, 1881*. Boston: Pope Manufacturing Co., 1881.

———. *Columbia Bicycles, 1895*. Hartford, Conn.: Pope Manufacturing Co., 1895.

———. *Columbia Bicycles, 1896*. Hartford, Conn., Pope Manufacturing Co., 1896.

———. *An Industrial Achievement*. Hartford, Conn.: Pope Manufacturing Co., 1907.

Potter, David M. *People of Plenty: Economic Abundance and the American Character*. Chicago: University of Chicago Press, 1954.

Pye, David. *The Nature and Art of Workmanship.* Cambridge: Cambridge University Press, 1968.

Rae, John B. *American Automobile Manufacturers.* Philadelphia: Chilton, 1959.

———, ed. *Henry Ford.* Englewood Cliffs, N.J.: Prentice-Hall, 1969.

Ransom, Frank Edward. *The City Built on Wood: A History of the Furniture Industry in Grand Rapids, Michigan, 1850–1950.* Ann Arbor: Frank E. Ransom, 1955.

Rauschenbush, Carl. *Fordism.* New York: League for Industrial Democracy, 1937.

Richards, John. *Treatise on the Construction and Operation of Wood-Working Machines.* London: E. & F. N. Spon, 1872.

Roberts, Kenneth D. *The Contributions of Joseph Ives to Connecticut Clock Technology, 1810–1862.* Bristol, Conn.: American Clock & Watch Museum, 1970.

———. *Eli Terry and the Connecticut Shelf Clock.* Bristol, Conn.: Ken Roberts Publishing Co., 1973.

———. *Some Observations Concerning Connecticut Clockmaking, 1790–1850. Bulletin of the National Association of Watch and Clock Collectors,* Suppl. 6 (1970).

Roe, Joseph W. *English and American Tool Builders.* New Haven: Yale University Press, 1916.

Roe, Joseph W., and Lytle, Charles W. *Factory Equipment.* Scranton, Pa.: International Textbook Co., 1935.

Rohan, Jack. *Yankee Arms Maker.* New York: Harper & Bros., 1935.

Rolt, L. T. C. *A Short History of Machine Tools.* Cambridge, Mass.: MIT Press, 1965.

Rosenberg, Nathan, ed. *The American System of Manufactures: The Report of the Committee on the Machinery of the United States 1855 and the Special Reports of George Wallis and Joseph Whitworth 1854.* Edinburgh: Edinburgh University Press, 1969.

Rothschild, Emma. *Paradise Lost: The Decline of the Auto Industrial Age.* New York: Vintage Books, 1973.

Sargent, C. S. *Woods of the United States.* New York: D. Appleton, 1885.

Scott, John. *Genius Rewarded; or the Story of the Sewing Machine.* New York: John J. Caulon, 1880.

Sellers, George Escol. *Early Engineering Reminiscences (1815–40) of George Escol Sellers.* Edited by Eugene S. Ferguson. Washington, D.C.: Smithsonian Institution Press, 1965.

Sheldon, Roy, and Arens, Egmont. *Consumer Engineering: A New Technique for Prosperity.* New York: Harper & Brothers, 1932.

Shlakman, Vera. *Economic History of a Factory Town: A Study of Chicopee, Massachusetts.* Smith College Studies in History, vol. 20. Northampton, Mass.: Smith College, 1934–35.

Sinclair, Upton. *The Flivver King.* Pasadena, Calif.: Upton Sinclair, 1937.

Singer Manufacturing Co. *Catalogue of Singer Sewing Machines, Illustrating Their Construction, Their Variety, and Their Special Uses by Manufactures.* New York: Singer Manufacturing Co., 1896.

———. *Report of the Proceedings on the Occasion of Breaking Ground for the Singer Manufacturing Company's New Factory.* Glasgow: Singer Manufacturing Co., 1882.

Singer Sewing Machine Co. *Mechanics of the Sewing Machine.* New York: Singer Sewing Machine Co., 1914.

Sloan, Alfred P., Jr. *My Years with General Motors.* New York: MacFadden-Bartell Books, 1965.

Smallzried, Kathleen A., and Roberts, Dorothy J. *More Than You Promise.* New York: Harper & Brothers, 1942.

Smith, Merritt Roe. *Harpers Ferry Armory and the New Technology.* Ithaca, N.Y.: Cornell University Press, 1977.

Smith, Oberlin. *Press-Working of Metals.* New York: John Wiley & Sons, 1896.

Smith, Robert A. *A Social History of the Bicycle.* New York: American Heritage Press, 1972.

Sorensen, Charles F. *My Forty Years with Ford.* New York: Norton, 1956.

Stern, Philip Van Doren. *Tin Lizzie: The Story of the Fabulous Model T Ford.* New York: Simon and Schuster, 1955.

Storck, John, and Teague, Walter Dorwin. *Flour for Man's Bread: A History of Milling.* Minneapolis: University of Minnesota Press, 1952.

Studebaker Brothers Manufacturing Company.

Illustrated Catalogue of Studebaker Brothers Manufacturing Company Wagons, Buggies, and Carriages. South Bend, Ind.: Studebaker Brothers Manufacturing Company, 1892–93.

————. *Illustrated Souvenir of the Studebaker Brothers Manufacturing Company, South Bend, Ind., U.S.A.* South Bend, Ind.: Studebaker Brothers Manufacturing Company, 1893.

Studebaker Corporation. *100 Years on the Road.* South Bend, Ind.: Studebaker Corporation, 1952.

Sward, Keith. *The Legend of Henry Ford.* New York: Antheneum, 1972.

Taylor, Frederick W. *The Principles of Scientific Management.* New York: Harper & Brothers, 1911.

Terry, Henry. *American Clock Making: Its Early History and Present Extent.* Waterbury, Conn.: J. Giles & Son, 1872.

Times (London). *American Engineering Competition.* New York: Harper & Brothers, 1901.

Tomlinson, Charles, ed. *Rudimentary Treatise on the Construction of Locks.* London: John Weale, 1853.

Troitsch, Ulrich, and Weber, Wolfhard, eds. *Die Technik.* Braunschweig: Westermann, 1982.

Van Slyck, J. D. *New England Manufacturers and Manufactories.* 2 vols. Boston: Van Slyck, 1879.

Western Toy Company. *Price List for Spring 1887, The Western Toy Company.* Chicago: Western Toy Company, 1887.

Western Wheel Company. *Crescent Bicycles.* Chicago: Western Wheel Co., 1898.

[Wheeler and Wilson Manufacturing Company]. *The Sewing Machine: Its Origin, Introduction in General Use, Progress and Extent of its Manufacture [and] A Great Machine-Shop Described.* New York: Wm. W. Rose, 1863.

Williams, Raymond. *Keywords: A Vocabulary of Culture and Society.* New York: Oxford University Press, 1976.

Woodbury, Robert S. *History of the Grinding Machine.* Cambridge, Mass.: MIT Press, 1959.

————. *History of the Milling Machine.* Cambridge, Mass.: MIT Press, 1960.

Woodworth, Joseph V. *American Tool Making and Interchangeable Manufacturing.* New York: N. W. Henley, 1905.

Articles

Abell, Oliver J. "Making the Ford Motor Car." *Iron Age* 89 (1912): 1383–90; 1454–60; 92 (1913): 1–7; 93 (1914): 902–4.

Alford, L. P. "Duplicate and Interchangeable Manufacture." In *Organizing for Production and Other Papers on Management 1912–1924,* edited by Israel Mayer, pp. 331–44. Easton, Pa.: Hive Publishing Co., 1981.

"American Industries—No. 10: Sewing Machines." *Scientific American* 40 (1879): 271, 274–75.

"American Industries—No. 35: The Manufacture of Sewing Machines." *Scientific American* 42 (1880): 179, 181.

American Society of Mechanical Engineers. Sub-Committee on Machine Shop Practice. "Developments in Machine Shop Practice during the Last Decade." *Transactions of the American Society of Mechanical Engineers* 34 (1912): 847–65.

Anderson, John. "On the Application of Machinery in the War Department." *Journal of the Society of Arts* 5 (1857): 155–66.

————. "On the Application of the Copying or Transfer Principle in the Production of Wooden Articles." *Proceedings of the Institution of Mechanical Engineers* (1858): 237–48.

————. "On the Application of the Copying Principle in the Manufacture and Rifling of Guns." *Proceedings of the Institution of Mechanical Engineers* (1862): 125–45.

Ardrey, R. L. "The Harvest Machine Industry." *Scientific American Supplement* 54 (1902): 22544–47.

Arnold, Horace L. [Hugh Dolnar]. "Bicycle Brazing." *American Machinist* 19 (1896): 1077–81.

————. "Bicycle Tools." 31-part series. *American Machinist* 18 (1895): 781–82; 801–2; 821–22; 842–44; 863–64; 909–11; 924–25; 941–42, 963–64; 1001–2; 1021–22; 19 (1896): 1–2; 50–52; 79–81; 104–5; 152–54; 182–84; 205–7; 231–33; 252–55; 275–76; 325–26; 348–50; 474–76; 495–97;

517–19; 617–19; 657–60; 677–79; 736–39; 871–73; 894–97.

———. "Cycle Stampings." *American Machinist* 19 (1896): 1163–67.

Arnold, Horace L. [Henry Roland]. "The Revolution in Machine-Shop Practice." 6-part series. *Engineering Magazine* 18 (October 1899–March 1900); 41–58; 177–200; 369–88; 530–49; 729–46; 903–6.

———. "Six Examples of Successful Shop Management—V." *Engineering Magazine* 12 (October 1896–March 1897): 994–1000.

"Assembling Motor Cars in the Packard Plant." *Iron Age* 96 (1915): 873–76.

[Avery, Clarence]. "How Mass Production Came into Being." *Iron Age* 123 (1929): 1638.

Bacon, Elizabeth M. "Marketing Sewing Machines in the Post–Civil War Years." *Bulletin of the Business History Society* 20, no. 6 (1946): 90–94.

Basalla, George. "Keaton and Chaplin: The Silent Film's Response to Technology." In *Technology in America: A History of Individuals and Ideas,* edited by Carroll W. Pursell, Jr., pp. 192–201. Cambridge, Mass.: MIT Press, 1981.

Battison, Edwin A. "Eli Whitney and the Milling Machine." *Smithsonian Journal of History* 1 (1966): 9–34.

———. "Searches for Better Manufacturing Methods, Section Two." *Tools and Technology* 3 (1979): 13–14, 16–18.

"Bennett, E. H." *Transactions of the American Society of Mechanical Engineers* 19 (1897–98): 979– 80.

"Bicycle Engineering." *American Machinist* 9 (September 4, 1886): 8.

"Bicycle Manufacturing." *Bicycling World* 1 (1880): 204, 206.

Beckett, Sir Edmund. "Lock." *Encyclopaedia Britannica,* 14: 750–58. Philadelphia: J. M. Stoddart & Co., 1882.

Blackmore, Howard L. "Colt's London Armoury." In *Technological Change: The United States and Britain in the Nineteenth Century,* edited by S. B. Saul, pp. 171–95. London: Methuen, 1970.

Bornholdt, Oscar C. "Continuous Manufacturing by Placing Machines in Accordance with Sequence of Operations." *Journal of the American Society of Mechanical Engineers*

35 (1913): 1671–78; reprinted in *Iron Age* 92 (1913): 1267–77.

Bourne, Frederick G. "American Sewing-Machines." In *One Hundred Years of American Commerce,* edited by Chauncey M. Depew, 2: 524–39. New York: D. O. Haines, 1895.

Brophy, J. P. "Interchangeability." *Machinery* 21 (1915): 967.

Burn, D. L. "The Genesis of American Engineering Competition, 1850–1870." *Economic History* 2 (1931): 292–311.

Buttrick, John. "The Inside Contract System." *Journal of Economic History* 12 (1952): 205–21.

Calkins, Earnest Elmo. "The New Consumption Engineer and the Artist." In *A Philosophy of Production,* edited by Justis George Frederick, pp. 107–29. New York: Business Bourse, 1930.

Callender, Harold. "Gandhi Dissects the Ford Idea." *New York Times Magazine,* November 8, 1931, pp. 2–3, 19.

"Carriages and Coaches." *Eighty Years' Progress of the United States . . . , by Eminent Literary Men.* Hartford, Conn.: L. Stebbins, 1866.

Chase, Stuart. "Henry Ford's Utopia." *Nation,* July 21, 1926, pp. 53–55.

———. "Play." In *Whither Mankind: A Panorama of Modern Civilization,* edited by Charles A. Beard, pp. 332–53. London: Longmans, Green, 1930.

"Chevrolet Begins Production of New Six." *Automotive Industries* 59 (1928): 734–35.

"Chevrolet Production Change Over to 6-Cylinder Car Held Industrial Feat." *Automotive Daily News,* February 8, 1929, p. 12.

"The Chicago Trade." *Wheel* 6 (1890): 472–74.

Cleveland, Reginald McIntosh. "How Many Automobiles Can America Buy?" *World's Work* 27 (1914): 679–89.

Clewett, Richard C. "Mass Marketing of Consumers' Goods." In *The Growth of the American Economy,* edited by Harold F. Williamson, pp. 766–84. 2d ed. Englewood Cliffs, N.J.: Prentice-Hall, 1951.

Cole, Arthur C. "American System." In *Dictionary of American History,* 1: 113–14. Rev. ed. New York: Scribner's, 1976.

"Colonel Albert A. Pope." *Bicycling World* 3 (1881): 129–30.

Colt, Samuel. "On the Application of Machinery to the Manufacture of Rotating Chambered-Breech Fire-Arms, and the Peculiarities of Those Arms." *Proceedings of the Institution of Civil Engineers* 11 (1851–52): 30–68.

Colvin, Fred H. "Assembling Magnetos, Motors and Transmissions." *American Machinist* 42 (1915): 557–58.

———. "Building an Automobile Every 40 Seconds." *American Machinist* 38 (1913): 757–62.

———. "Continuous Assembling in Modern Automobile Shops." *American Machinist* 43 (1915): 365–70.

———. "Continuous Pouring in the Ford Foundry." *American Machinist* 39 (1913): 910–12.

———. "Ford Crank Cases and Transmission Covers." *American Machinist* 39 (1913): 49–53.

———. "Ford Radiators and Gasoline Tanks." *American Machinist* 39 (1913): 393–96.

———. "Machining the Ford Cylinders—I." *American Machinist* 38 (1913): 841–46.

———. "Machining the Ford Cylinders—II." *American Machinist* 38 (1913): 971–76.

———. "Making Rear Axles for the Ford Auto." *American Machinist* 39 (1913): 143–49.

———. "Methods Employed in Making the Ford Magneto." *American Machinist* 39 (1913): 311–16.

———. "Methods Used in Assembling Sweepers." *American Machinist* 58 (1923): 281–83.

———. "Special Machines for Auto Small Parts." *American Machinist* 39 (1913): 439–43.

———. "Special Machines for Making Pistons." *American Machinist* 39 (1913): 349–53.

"Continuous Assembling Frame." *American Machinist* 46 (1917): 92.

Cooper, Carolyn C. "The Production Line at Portsmouth Block Mill." *Industrial Archaeology Review* 6 (Winter 1981–82): 28–44.

"Country Learns of Ford Plans for 1932." *Ford News* 12 (March 1932): 3, 17.

"Cutlery at Sheffield." *Journal of Domestic Appliances and Sewing Machine Gazette* 11 (March 1, 1883): 19–20.

Davis, Harvey N. "Spirit and Culture under the Machine." In *Toward Civilization,* edited by Charles Beard, pp. 282–96. London: Longmans, Green, 1930.

"Dealer-Opinion on Filene's Mass-Production Forecast." *Furniture Record and Journal,* June 1931, pp. 44–47, 55–56.

"Decidedly the Greatest." *Wheel* 18 (January 29, 1897): 94, 96.

Dibble, L. C. "Slow Progress in Plant Changes Delays New Ford Model." *Automotive Industries* 57 (1927): 404–5.

Douglas, Diane, "The Machine in the Parlor: A Dialectical Analysis of the Sewing Machine." *Journal of American Culture* 5 (1982): 20–29.

Dowd, Albert A., and Curtis, Frank W. "Saving Time in Assembling." *Machinery* 28 (1921): 296–300.

Downs, Linda. "The Rouge in 1932: The *Detroit Industry* Frescoes by Diego Rivera." In *The Rouge: The Image of Industry in the Art of Charles Sheeler and Diego Rivera,* pp. 47–52. Detroit: Detroit Institute of Arts, 1978.

Duggan, Edward. "Machines, Markets, and Labor: The Carriage and Wagon Industry in Late-Nineteenth Century Cincinnati." *Business History Review* 51 (1977): 308–25.

Durfee, William F. "The First Systematic Attempt at Interchangeability in Firearms." *Cassier's Magazine* 5 (1893–94): 469–77.

———. "The History and Modern Development of the Art of Interchangeable Construction in Mechanism." *Transactions of the American Society of Mechanical Engineers* 14 (1893): 1225–57.

Earl, Polly Anne. "Craftsmen and Machines: The Nineteenth Century Furniture Industry." In *Technological Innovation and the Decorative Arts,* edited by Ian M. G. Quimby and Polly Anne Earl, pp. 307–29. Charlottesville: University Press of Virginia, 1974.

"Edward Clark." *Obituary Record of Donors and Alumni of Williams College, 1882–3* 18 (1883): 304–6.

"Electric Welding." *Iron Age* 43 (1889): 589.

Ettema, Michael J. "Technological Innovation and Design Economics in American Furniture Manufacture of the Nineteenth Century." *Winterthur Portfolio* 16 (1981): 197–223.

"Extracts from Chordal's Letters." *American*

Machinist 6 (May 5, 1883): 3–4; (June 16, 1883): 2–3; (June 23, 1883): 1–3.

Faurote, Fay Leone. "Equipment Makes Possible the Ford Model A." 11-part series. *American Machinist* 68 (1928): 679–84; 721–24; 761–62; 805–7; 874–79; 917–21; 1034–39; 69 (1928): 15–19; 61–66; 153–60; 269–75.

———. "Final Assembly." *American Machinist* 69 (1928): 269–75.

———. "Ford Shop Changes Estimated at $25,000,000." *Iron Age* 121 (1928): 1080–83.

———. "Henry Ford Still on the Job with Renewed Vigor." *Industrial Management* 74 (1927): 193–202.

———. "How Ford Plans His Layout of Grounds, Buildings, and Plant." *Factory and Industrial Management* 75 (1928): 1196–99.

———. "Machining and Welding Operations on the Rear Axle Housings. . . ." *American Machinist* 69 (1928): 15–19.

———. "Make Time and Space Earn Their Keep." *Factory and Industrial Management* 75 (1928): 541–45.

———. "Packing and Shipping Ford Models and Parts." *Factory and Industrial Management* 76 (1928): 269–72.

———. "Planning and Mass Production Coordinated." *Factory and Industrial Management* 75 (1928): 984–87.

———. "Planning Production through Obstacles, Not around Them." *Factory and Industrial Management* 75 (1928): 302–6.

———. "Preparing for Ford Production." *American Machinist* 68 (1928): 635–39.

———. "Producing the New Ford Model." *Factory and Industrial Management* 75 (1928): 62–65.

———. "Research is Back of all Ford Manufacturing." *Factory and Industrial Management* 75 (1928): 74–77.

———. "Single-Purpose Manufacturing." *Factory and Industrial Management* 75 (1928): 769–73.

———. "Splitting an Inch a Million Ways." *Factory and Industrial Management* 76 (1928): 510–13.

———. "What Is Going on Behind the Scenes at the Ford Plants." *Industrial Management* 74 (1927): 257–62.

Ferguson, Eugene S. "The American-ness of American Technology." *Technology and Culture* 20 (1979): 3–24.

———. "Elegant Inventions: The Artistic Component of Technology." *Technology and Culture* 19 (1978): 450–60.

Filene, Edward A. "Foresees Ford Prosperity for Furniture Industry." *Furniture Record and Journal*, April 1931, pp. 14–17.

———. "Mass Production Makes a Better World." *Atlantic* 143 (1929): 625–31.

"Final Assembly Line Ready. . . ." *Ford News* 7 (October 1, 1927): 1, 8.

Fitch, Charles H. "The Rise of a Mechanical Ideal." *Magazine of American History* 11 (1884): 516–27.

"Flexibility in Making Automobile Frames." *Iron Age* 122 (1928): 19–23.

Ford, Henry, "Mass Production." In *Encyclopaedia Britannica*, 13th ed., Suppl. Vol. 2 (1926); pp. 821–23.

"Francis's Corrugated Metallic Boats." *Artizan* 14 (1856): 253–55.

Fries, Russell I. "British Response to the American System: The Case of the Small-Arms Industry after 1850." *Technology and Culture* 16 (1975): 377–403.

"Furniture Men Flay Filene's Proposal of Mass Production." *Furniture Record and Journal*, May 1931, pp. 47–49, 59.

Gamble, William B. "List of Works in the New York Public Library Relating to Electric Welding." *Bulletin of the New York Public Library* 17 (1913): 375–95.

Gartman, Davis. "Origins of the Assembly Line and Capitalist Control of Work at Ford." In *Case Studies on the Labor Process*, edited by Andrew Zimbalist, pp. 193–205. New York: Monthly Review Press, 1979.

"General Motors II: Chevrolet." *Fortune*, January 1939, pp. 39–40, 46, 103–4.

Goodyear, S. W. "Personal Recollections." *American Machinist* 6 (October 13, 1883): 4–5; (December 1, 1883): 4–5; 7 (January 12, 1884): 7.

———. "Working to Standards in Large and Small Shops—Early Sewing-Machine Economies." *American Machinist* 6 (August 4, 1883): 1–2.

"A Great American Manufacture." *Bicycling World* 2 (1880–81): 326–32.

Green, Constance M. "Light Manufactures and the Beginnings of Precision Manufacture."

In *The Growth of the American Economy*, edited by Harold F. Williamson, pp. 190–210. 2d ed. Englewood Cliffs, N.J.: Prentice-Hall, 1951.

Greenwood, Thomas. "On Machinery for the Manufacture of Gunstocks." *Proceedings of the Institution of Mechanical Engineers* (1862): 328–40.

Gregory, Charles Hutton. "Address of the President." *Proceedings of the Institution of Civil Engineers* 27 (1867–68): 180–203.

Harmond, Richard. "Progress and Flight: An Interpretation of the American Cycle Craze of the 1890s." *Journal of Social History* 5 (1971): 235–57.

Harrison, A. E. "The Competitiveness of the British Cycle Industry, 1890–1914." *Economic History Review*, 2d ser., 22 (1969): 287–303.

Hayward, Charles Harold. "Furniture Industry." In *Encyclopaedia Britannica*, 15th ed., 7: 807–11.

Heitkamp, Frederick B. "Using More Semi-Special Tools." *Iron Age* 119 (1927): 1072–74.

Henry, Leslie R. "The Ubiquitous Model T." In *Henry's Wonderful Model T, 1908–1927*, edited by Floyd Clymer, pp. 102–37. New York: Bonanza Books, 1955. First published in *Antique Automobile* 16 (1952).

Higgins, Carter C. "The Pressed Steel Industry." *Mechanical Engineering* 63 (1941): 534–35.

Higham, John. "The Reorientation of American Culture in the 1890s." In *The Origins of Modern Consciousness*, edited by John Weiss, pp. 25–48. Detroit: Wayne State University Press, 1965.

"Hobbs' Lock Manufactory." *Engineer* 7 (1859): 188–90.

Hounshell, David A. "The Bicycle and Technology in Late Nineteenth Century America." In *Transport Technology and Social Change*, edited by Per Sörbom, pp. 173–85. Stockholm: Tekniska Museet, 1980.

———. "Gunnison, Foster." In *Dictionary of American Biography*. 7th suppl. New York: Scribner's, 1981.

———. "Public Relations or Public Understanding? The American Industries Series in *Scientific American*." *Technology and Culture* 21 (1980): 589–93.

"How Precision Is Being Built into the Model A." Dealers Supplement, *Ford News* 11 (March 15, 1929): 69.

Howard, Robert A. "Interchangeable Parts Reexamined: The Private Sector on the Eve of the Civil War." *Technology and Culture* 19 (1978): 633–49.

Hubbard, Guy. "Development of Machine Tools in New England." 23-part series. *American Machinist* 59 (1923): 1–4; 139–42; 241–44; 311–15; 389–92; 463–67; 541–44; 579–81; 919–22; 60 (1924): 129–32; 171–73; 205–9; 255–58; 271–74; 437–41; 617–20; 875–78; 951–54; 61 (1924): 65–69; 195–98; 269–72; 313–16; 453–55.

"'I Will': Chicago True to Her Motto in the Carrying out of Her Cycle Show. . . ." *Wheel* 16 (January 10, 1896): 64.

"Influence of the Bicycle upon Machine Tools." *Iron Age* 55 (1895): 496–97.

"Installment Selling to the Front." *Review of Reviews* 73 (1926): 469–70.

"Institution of Civil Engineers." *Civil Engineer and Architect's Journal* 14 (1851): 611.

Jack, Andrew B. "The Channels of Distribution for an Innovation: The Sewing-Machine Industry in America, 1860–1865." *Explorations in Entrepreneurial History* 9 (1951): 113–41.

Kettering, Charles F. "Keep the Consumer Dissatisfied." *Nation's Business* 17 (January 1929): 30–31.

Klein, Julius. "Business." In *Whither Mankind: A Panorama of Modern Civilization*, edited by Charles A. Beard, pp. 83–109. London: Longmans, Green, 1930.

Knudsen, William S. "'For Economical Transportation': How the Chevrolet Motor Company Applies Its Own Slogan to Production." *Industrial Management* 76 (August 1927): 65–68.

Lathrop, W. G. "The Development of the Brass Industry in Connecticut." In *Tercentenary Commission of the State of Connecticut*, vol. 49. New Haven: Yale University Press, 1934.

Leslie, Stuart W. "Charles F. Kettering and the Copper-cooled Engine." *Technology and Culture* 20 (1979): 752–76.

"Life and Work of Philip Diehl." *Sewing Machine Times*, n.s., 23 (April 25, 1913): 1.

"Looking Back: A Bird's Eye View of Our

Trade." *Sewing Machine Times*, n.s. 18 (October 25, 1908): 8.

Lyon, Peter. "Isaac Singer and His Wonderful Sewing Machine." *American Heritage* 9 (October 1958): 34–39, 103–9.

McCormick, Anne O'Hare. "The Future of the Ford Idea." *New York Times Magazine*, May 22, 1932, pp. 1–2.

[McCormick, Cyrus H., Jr.]. "American Industries—No. 73: Harvesting Machines." *Scientific American* 44 (1881): 303, 307–8.

McCurdy, Charles W. "American Law and the Marketing Structure of the Large Corporation, 1875–1890." *Journal of Economic History* 38 (1978): 631–49.

McGee, WJ. "Fifty Years of American Science." *Atlantic Monthly* 87 (1898): 307–20.

Madigan, Mary Jean Smith. "Influence of Charles Lock Eastlake on American Furniture Manufacture, 1870–1890." In *Winterthur Portfolio 10*, edited by Ian M. G. Quimby, pp. 1–22. Charlottesville, Va.: Henry Francis du Pont Winterthur Museum, 1975.

Maier, Charles S. "Between Taylorism and Technocracy: European Ideologies and the Vision of Industrial Production in the 1920s." *Journal of Contemporary History* 5 (1970): 27–51.

"Making a Bicycle." *Iron Age* 48 (1891): 1070–72.

"Making Balls for Bearings." *American Machinist* 19 (1896): 917–24.

"Making Shop Layout Efficient." Dealers Supplement, *Ford News* 11 (March 15, 1929): 70.

"The Manufacture of Bicycle Tubing." *Iron Age* 54 (January 7, 1897): 1–2; (January 14, 1897): 1–5; (March 4, 1897): 1.

"The Manufacture of Sewing Machines." *Scientific American* 42 (1880): 175, 181.

"Many Highly Specialized Jobs Now Being Performed with Standard Machine Tools." *Automotive Industries* 59 (1928): 694–705.

Mazur, Paul M. "Mass Production: Has It Committed Suicide?" *Review of Reviews* 77 (1928): 476–79.

Meckier, Jerome. "Debunking Our Ford: *My Life and Work* and *Brave New World*." *South Atlantic Quarterly* 78 (1979): 448–59.

"Miller, Lebbeus B." *Transactions of the American Society of Mechanical Engineers* 44 (1922): 927–28.

"Mr. Ford Doesn't Care." *Fortune*, December 1933, pp. 62–69, 121–34.

Murphy, John Joseph. "Entrepreneurship in the Establishment of the American Clock Industry." *Journal of Economic History* 24 (1966): 169–86.

"Norton's Automatic Can Making Machinery." *American Machinist* 6 (July 14, 1883): 1–2.

"Obituary of John Anderson." *Proceedings of the Institution of Civil Engineers* 86 (1885–86): 346–53.

"Walter E. Flanders Dies in Virginia." *Automobile Topics* 70 (1923): 517–25.

Ordway, Warren. "Assembling by Conveyor." *Machinery* 29 (1922): 103–6.

Parks, B. A. "Engineering in Furniture Factories." *Mechanical Engineering* 43 (1921): 85–90.

Perrine, Frederick A. C. "Practical Aspects of Electric Welding." *Iron Age* 47 (1891): 1114; also in *Transactions of the American Institute of Electrical Engineers* 8 (1891): 246–65.

Perry, Thomas D. "The Engineer and the Woodworking Industry: A Great Industry in Which There Exists an Urgent Need of Engineering Skill." *Mechanical Engineering* 42 (1920): 448–50.

———. "The Wood Industries." *Mechanical Engineering* 52 (1930): 429–34.

Peterson, Charles E. "Early American Prefabrication." *Gazette des Beaux-Arts*, January 1948, pp. 37–46.

Pope, Albert A. "The Bicycle Industry." In *One Hundred Years of American Commerce*, edited by Chauncey Depew, 1: 549–53. New York: D. O. Haines, 1895.

Porter, Harry Franklin. "Four Big Lessons from Ford's Factory." *System: The Magazine of Business* 31 (1917): 639–46.

"Preparations for Building New Ford Car to Cost Many Millions." *Ford News* 7 (September 1, 1927): 1.

"Press and Die Work on Bicycles." *American Machinist* 19 (1896): 1097–1100.

"Pressed Steel Automobile Parts." *Horseless Age* 24 (1909): 263–64.

Rezneck, Samuel. "Mass Production and the Use of Energy." In *Growth of the American Economy*, edited by Harold F. Williamson, pp. 718–38. 2d ed. Englewood Cliffs, N.J.: Prentice-Hall, 1951.

Richards, Archer W. "Mass Production of Radio Cabinets." *Transactions of the American Society of Mechanical Engineers* 52 (1930): 65–66.

Rigby, John. "The Manufacture of Small Arms." *Engineering* 55 (1893): 757–58.

Roe, Joseph W. "Development of Interchangeable Manufacture." *American Machinist* 40 (1914): 1079–84.

———. "Interchangeable Manufacture." *Mechanical Engineering* 59 (1937): 755–58.

Rosenberg, Nathan. "America's Rise to Woodworking Leadership." In *America's Wooden Age,* edited by Brooke Hindle, pp. 37–55. Tarrytown, N.Y.: Sleep Hollow Restorations, 1975.

———. "Technological Change in the Machine Tool Industry, 1840–1910." *Journal of Economic History* 23 (1963): 414–43.

"The Royal Small-Arm Manufactory, Enfield." *Engineer* 7 (1859): 204–5.

Rumely, Edward A. "The Manufacturer of Tomorrow." *World's Work* 28 (1914): 106–12.

Russell, Jack. "The Coming of the Line." *Radical America* 12 (May–June 1978): 28–45.

Saul, S. B. "The Market and the Development of the Mechanical Engineering Industries in Britain, 1860–1914." In *Technological Change: The United States and Britain in the Nineteenth Century,* edited by S. B. Saul, pp. 141–170. London: Metheun, 1970

Sawyer, John E. "The Social Basis of the American System of Manufacturing." *Journal of Economic History* 14 (1954): 361–79.

"Scientific Methods Applied to Bicycle-making." *Leslie's Illustrated Weekly* 80 (1895): 364.

"A Scrap of History." *Sewing Machine Advance* 4 (January 1882): 3.

"Sewing Machines." In *Eighty Years' Progress of the United States . . . by Eminent Literary Men,* pp. 421–23. Hartford, Conn.: L. Stebbins, 1866.

Sharpe, Henry Dexter. "Joseph R. Brown, Mechanic, and the Beginnings of Brown & Sharpe." New York: Newcomen Society of England, American Branch, 1949.

"Short Cuts on Standard Machines for Common Automotive Jobs Are Being Developed to Reduce Use of Special Tools." *Automotive Industries* 61 (1929): 508–13.

Slauson, Harold Whiting. "Efficient System for the Rapid Assembly of Motor Cars." *Machinery* 16 (1909): 114–16.

———. "A Ten-Million Dollar Efficiency Plan." *Machinery* 21 (October 1914): 83–87.

Smith, Merritt Roe. "Eli Whitney and the American System of Manufacturing." In *Technology in America: A History of Individuals and Ideas,* edited by Carroll W. Pursell, Jr., pp. 49–65. Washington, D.C.: Voice of America, 1979.

———. "John H. Hall, Simeon North, and the Milling Machine: The Nature of Innovation among Antebellum Arms Makers." *Technology and Culture* 14 (1973): 573–91.

Spalding, B. F. "The 'American System' of Manufacture." *American Machinist* 13 (November 6, 1890): 2–4.

———. "The 'American' System of Manufacture." *American Machinist* 13 (November 13, 1890): 2–3; (November 20, 1890): 11–12.

Stevens, Joshua. "Sixty Years a Mechanic." *Machinery* 1 (October 1894): 3–4.

"The Story of a Success." *Wheel* 18 (August 18, 1897): 48.

"Testing the Parts of a Modern Bicycle." *Scientific American* 75 (1896): 18, 23.

Thomas, Robert Paul. "Style Change and the Automobile Industry during the Roaring Twenties." In *Business Enterprise and Economic Change,* edited by Louis P. Cain and Paul J. Uselding, pp. 118–38. Kent, Ohio: Kent State University Press, 1973.

Thomson, Elihu. "Electric Welding." *Iron Age* 46 (1890): 988.

Tobin, Gary Allan. "The Bicycle Boom of the 1890s: The Development of Private Transportation and the Birth of the Modern Tourist." *Journal of Popular Culture* 7 (1974): 836–49.

Trescott, Martha Moore. "The Bicycle: A Technical Precursor of the Automobile." *Business and Economic History,* 2d ser., 5 (1976): 51–75.

Uselding, Paul. "Elisha K. Root, Forging, and the 'American System.'" *Technology and Culture* 15 (1974): 543–68.

———. "Studies of Technology in Economic History." In *Recent Developments in the Study of Economic and Business History: Es-*

says in Memory of Herman E. Krooss*, edited by Robert E. Gallman, pp. 159–219. Greenwich, Conn.: JAI Press, 1977.

Van Deventer, John H. " 'Dat ole Debil,' Machinery." *Iron Age* 128 (1931): 923–25.

Waldo, Leonard. "The American Bicycle: Its Theory and Practice of Construction." *Journal of the Society of Arts* 46 (December 3, 1897): 46–56.

Westbrook, Robert B. "Tribune of the Technostructure: The Popular Economics of Stuart Chase." *American Quarterly* 32 (1980): 387–408.

"The Westinghouse Foundry, Near Pittsburg, Pa.," *Scientific American* 62 (1890): 369–70.

Whorton, James C. "The Hygiene of the Wheel: An Episode in Victorian Sanitary Science." *Bulletin of the History of Medicine* 52 (1978): 61–88.

Wilkinson, Norman B. "The Forgotten 'Founder' of West Point." *Military Affairs* 24 (1960–61): 177–88.

Williamson, Harold F. "Mass Production, Mass Consumption, and American Industrial Development." In *First International Conference of Economic History, Contributions*, pp. 137–47. The Hague: Mouton, 1960.

"Wilson Improved Patent Sewing Machine." *Scientific American* 8 (1853): 298.

Woodbury, Robert S. "The Legend of Eli Whitney and Interchangeable Parts." *Technology and Culture* 1 (1960): 235–53.

Woodward, Henry P. "Manufactures in Hartford." In *Hartford in History*. Hartford, Conn.: Board of Trade, 1899.

Newspapers and Directories

Boston Almanac
Boston Directory
Detroit Free Press
Detroit News
Elizabeth Daily Journal
Los Angeles Times
Louisville Courier-Journal
New York Times
Providence Daily Journal
South Bend News Times
South Bend Sun News

Unpublished Material

Basalla, George. "Science, Technology, and Popular Culture." Manuscript, 1982.

Brown, Howard Francis. "The Saga of Brown and Sharpe (1833–1968)." Master's thesis. University of Rhode Island, 1971.

Cesari, Gene S. "American Arms-Making Machine Tool Development, 1789–1855." Ph.D. dissertation, University of Pennsylvania, 1970.

Davies, Robert B. "International Operations of the Singer Manufacturing Co., 1854–1895." Ph.D. dissertation, University of Wisconsin, 1966.

Dunham, Norman Leslie. "The Bicycle Era in America." Ph.D. dissertation, Harvard University, 1957.

Fries, Russell I. "A Comparative Study of the British and American Arms Industries, 1790–1890." Ph.D. dissertation, Johns Hopkins University, 1972.

Hallock, James Lindsey. "Woodworking Machinery in Nineteenth Century America." Master's thesis, University of Delaware, 1978.

Hounshell, David A. "From the American System to Mass Production: The Development of Manufacturing Technology in the United States, 1850–1920." Ph.D. dissertation, University of Delaware, 1978.

Mason, Philip Parker. "The League of American Wheelmen and the Good-Roads Movement, 1880–1905." Ph.D. dissertation, University of Michigan, 1957.

Meyer, Stephen, III. "Mass Production and Human Efficiency: The Ford Motor Company, 1908–1921." Ph.D. dissertation, Rutgers University, 1977.

Molloy, Peter M. "Technical Education and the Young Republic: West Point as America's Ecole Polytechnique, 1802–1833." Ph.D. dissertation, Brown University, 1975.

Murphy, John Joseph. "The Establishment of the American Clock Industry: A Study in Entrepreneurial History." Ph.D. dissertation, Yale University, 1961.

Schmitt, John B. "The Invention of Electric Resistance Welding: Elihu Thomson as Pioneer Inventor-Entrepreneur." Research paper. Division of Electricity, Smithsonian Institution, Washington, D.C., 1974.

Smith, Peter Haddon. "The Industrial Archeology of the Wood Wheel Industry in America." Ph.D. dissertation, George Washington University, 1971.

Thomas, Selma. "The Greatest Economy and the Most Exact Precision: The Work of Honoré Blanc." Paper presented at Twentieth Annual Meeting of the Society for the History of Technology, October 21, 1977, Washington, D.C.

White, Edward Hartwell, Jr. "The Development of Interchangeable Mass Manufacturing in Selected American Industries from 1795 to 1825." Ph.D. dissertation, University of Maryland, 1973.

Index

—as art, 304
—changeover to, 12; announcement of, 279; chaos of, 289–92; contrasted with Chevrolet, 266–67; cost of, 379 n. 85; lessons of, 293–94; problems in, 281–92; and recordkeeping at Ford, 271–72
—changes in, 295–96
—cost of, 281
—design of, 279–84
—drop in sales of, 296
—first public showing of, 282, **283**
—manufacture of: end of, 296; final assembly in, 289–91, **290–91**; first engine, **282**; Henry Ford's plans for, 283; machine tools acquired for, 288; new parts production departments for, 286–87; precision in, 286, 379 n. 80; problems in, 281, 283–86, 292; record output achieved in, 295; tooling up for, 281–82
—public response to, 292, 295–96, 304
Ford Model B, 297
Ford Model N, 218, 220–21, **222**
Ford Model T
—as a "car for the masses," 9, 218–19
—in *Brave New World*, 316
—changes in, 273–75
—decline of, 12, 264, 276–77, 279
—design of, 218–19
—engine of, compared to Model N, 224
—Henry Ford's decision to manufacture only, 227
—Henry Ford's views on, 277
—history of, 1915–1927, 267–80
—idea of, 218–19, 275
—influence of 1926 changes in, on Ford machine tools, 294
—manufacture of: accommodation of change in, 273; "all-metal" body and, 274–75, 378 n. 40; before the assembly line, 228–38; cost accounting methods for, 272; cost and price trends in, 11, 224, 377 n. 29; crankshaft grinding machines, **232**; drilling and reaming engine block, **233**; end of, 279; experimentation with techniques of, 234; fifteen millionth, 279, **280**; likened to tin can factory, 229, 241; machine tool acquisition for, 231–33; machine tool placement for, 229; machining engine blocks, **231**; principles of economy in, 229; profit margin on, 378 n. 49; quick-change fixtures for, 230–31, **230–31**; special-purpose machinery for, 233; standards of accuracy in, 229; system to increase output in, 268, 270–71; testing methods in, 229;

tooling up for, 223; work-scheduling for, 229
—1913 touring car, **219**
—output of, 11, 223, 224, 268
—placement of gasoline tank in, 284
—reintroduction of colors for, 275
Ford Motor Company
—ability of, to accommodate changes in Model T, 274–75
—adoption of armory practice, 10
—adoption of sheet steel stamping, 10, 224–25
—analyses of, 276–77
—and annual model change, 186–87, 263, 277, 300–301
—assembly techniques of: and body drop station, **254**; and delay of Model A production, 292; development of moving line, 237–56; influence of breweries and flour mills on, 241, **243**; influence of canning on, 241, **243**; influence of foundry conveyors on, 239–40; influence of slaughterhouses on, 241, **242**; for Model A chassis, 289, **290, 291**; for Model N (static), **222**; for Model T (static), 234–38; for Model T chassis (line), 249, **252, 253**, 254, **255, 256, 257**; for Model T chassis (static), **237**; for Model T contrasted with Model A, 289; for Model T dashboards (line), **259**; for Model T dashboards (static), **236**; for Model T engines (line), 248–49, **250, 251**; for Model T engines (static), 234, **235**; for Model T magneto coils (static), **234**; for Model T magneto coils (line), 244–48, **246, 247**; for Model T rear axles (line), **258**; for Model T rear axles (static), 234, **235**, 236; for Model T transmissions (line), **247**, 248; for Model T upholstery (line), **260**; origin of moving line, 212, 237; at Piquette Avenue factory, 220; productivity gains of moving line, 248, 249, 254–56; role of Taylorism in, 249–53; for V-8 engine, **299**
—backward integration of, 268
—changeover from Model A to V-8, 296–301
—changeover from Model T to Model A, 12–13, 279–95
—contrasted with Pope Manufacturing Company, 206
—cost accounting at, 272, 377 n. 32
—design of machine tools at, 227, 230–33, 372 n. 43
—early history of, 218–20
—as educator on mass production, 11, 260–61

—engineering department of, 268, 270–73, 380 n. 124
—factory letters of, 273
—gaugemaking operations of, 286
—Highland Park factory of: aerial photograph of, **227**; day's output of, **2**; decline of, 268; description of, 227; employees at, **3**; foundry at, **238–39**, 239–40; initial production at, 228; labor turnover at, 11, 257–58; planning and erection of, 225–27; power plant at, 228; punch press operations at, **225–26**; role of Taylorism at, 249–53
—institutes credit financing, 293
—introduces Victoria, 296
—managerial structure of, 293
—marketing strategy of, 9, 275, 278, 292–93, 295–96
—market share of, 11, 12, 264, 276, 295, 296
—mechanical drawing system of, 224, 272–73, 286, 377 n. 34
—operations sheets at, 224, 270–71, 273, 376–77 n. 24
—performance of, in 1929, 295
—performance of, in 1931, 296
—Piquette Avenue factory of, 220, **222**
—reasons for success of, 220
—recordkeeping at, 271–72
—and rise of mass production, 9, 215
—River Rouge factory of: assembly operations at, 289–91; conveyor belt at, **298**; development of, 267–68; Diego Rivera's frescoes of, 323, **324–26**, 327; furlough of workers from, 296; high overhead costs of, 274; Model A frame production at, 292; Model T body production at, 174; in 1930, **269**; soybean operations at, 309, **310**; visitors at, 383 n. 61
—sociological department of, 259
—Sorensen's purge of Model T men from, 289–92
—strategy of purchasing from outside suppliers, 272, 300
—and support for soybean development, 309, **310**
Ford News, 282, 289
Fordson tractor, 268
Ford V-8: changeover to, from Model A, 296–301; cost of changeover to, 298; delays in introduction of, 296–97; design and production of, as an engineering feat, 297; production problems with, 297–98; serial number one, **300**
Ford X-car, 278
Forging. *See* Drop forging
Fortune, 300, 314
Foundry: at Ford's Highland Park

—and sequential operation of special-purpose machinery, 15, 35, 38, 64, 350 n. 77, 352 n. 144
—in sewing machine industry, 71, 80–81, 91, 92, 93, 96, 109
—in small-arms industry: Blanchard's contribution to, 35; British study of, 19–21, 23–24, 61–62, 64; at Colt's armories, 47–50, 350 n. 77; Hall's work, 41–43; North's work, 28–29; at Springfield Armory, 35, 38, 44; War Department's pursuit of, 27–35, 38–45; and Whitney's work, 30–32
—in wagonmaking, 147, **148**, 149
—in woodworking, 127–29
Meida, Charles, 279
Meikle, Jeffrey L., 304, 308–9
Metal spinning machines, hot, 285–86
Meyer, Stephen, III, 249–50
Military, 25, 27, 330
Miller, Lebbeus B.: career of, at Singer, 92–93, 96, 120; and Chapman tour, 104–6; and inside contracting, 93, 109, 114; and Montreal factory, 117–18; reports of, to McKenzie, 110, 112–20; role of, in expansion of Singer output, 95, 104, 109, 120, 122, 358 n. 135
Miller, Logan, 374 n. 87, 376 n. 20
Miller, William H., 47
Milling machine: dependence on, at Colt's armory, 50; first owned by Singer, 92; at Ford, **231**, 288; Hall's development of, 41; importance of, in American system, 29; improvements to tools at Brown & Sharpe, 80; North's, 29; universal, 80
Mitchell, George, 54
Model, 6, 34, 42, 164–65
Model changes
—American fetish for, 261
—annual: at Chevrolet, 264–66; discussed in *Fortune*, 300–301; at Ford, 264, 295; and General Motors, 13, 263–64, 376 n. 6; impact of, on machine tool acquisitions, 380 n. 106; at McCormick, 8, 185–86; and mass production, 264
—Ford's views on, 277
—in furniture industry, 145, 150
—impact of, on market share, 278
—at McCormick, 159, 166–67, 169, 170, 178
"Model T dilemma," 315
Modern Times (Chaplin), 318–20, **318, 319**, 323
Molding machinery, 105, **239, 240**
Monroe, James, 33, 39

Morgan, Henry, 32
Morgana, Charles, 225, 231, 270–71
Motion study, 236, 249–53
Müller-Freienfels, Richard, 320–21
Mumford, Lewis, 25, 314–15, 330, 382 n. 30
Murphy, John J., 54, 60–61
My Life and Work (Ford), 305, 316, 321, 373 n. 79
Mysteries of the Soul, The (Müller-Freienfels), 321

Nasmyth, James: argues for superiority of Colt arm, 21; on British origin of the American system, 24, 348 n. 32; as expert witness before Select Committee on Small Arms, 19; maintains interchangeability of Colt revolvers, 23; and origin of term "American system of manufactures," 17, 334–35; testimony of, about machinery production, 19; testimony of, on labor inputs, 20; on ultimate result of mechanization, 23; views of, on Colt's gauging methods, 351 n. 113
National armories, 17–18, 25, 34. *See also* Hall's Rifle Works; Harpers Ferry Armory; Springfield Armory
Needle manufacture, **74**, 75, 92
Nevins, Allan, 10; corroborates Sorensen's account of conveyors, 244; on Ford organization in 1930s, 267; on Ford production technology, 220, 223, 233; on Ford's work with soybeans, 309; on Kanzler memorandum, 277–78; on Model T announcement, 219; on origin of Ford assembly line, 237–38, 244–46, **246**; on River Rouge factory, 267–68; on role of Taylorism at Ford factory, 249–51
New York Times: article of, on mass production, 1, 303, 306; on decline of Ford Motor Company, 276; editorial on Ford's article, "Mass Production," 306–7; editorial on Ford's proposal on mass production in agriculture, 310; editorial on lack of planning at Ford, 301; Ford's article, "Mass Production," in, 1, 303; report of, on Ford changeover, 279, 281, 282, 288; report of, on Model A changes, 295
New York Times Magazine, 322
Niebuhr, Reinhold, 381 n. 133
North, Simeon: contracts to make Hall breechloader, 42; contracts to make pistols with interchangeable parts, 28; decline of, 45; Hall's influence on, 43; and mechanization of arms production, 29, 41, 349 n.

51; Model 1813 pistol made by, **30**; and uniformity system, 29; use of division of labor by, 28; and use of receiver gauge, 29, 34, 41; work of, compared to Whitney's, 32
Norton, Edwin, **243**, 373 n. 70
Nyssa, 138–41

Oak, 143
O'Connor, James, **254**, 290–91
Official Descriptive and Illustrated Catalogue (London Crystal Palace Exhibition), 331
Operations sheets: Henry Leland's development of, at Brown & Sharpe, 82; Singer's blue book of, 120; use of, at Ford, 224, 270–71, 273, 376–77 n. 24
Ordnance Department. *See* United States Ordnance Department
Orr, Robert, 32
Overman, A. H., 200–201, 369 n. 64

Parks, B. A., 126, 145
Parliamentary committee on small arms. *See* Select Committee on Small Arms
Parrish, Maxfield, 203, 368 n. 44
Patent armsmakers, 45–46
Patent Arms Manufacturing Company (Colt's), 47. *See also* Colt, Samuel
Patent pool, sewing machine industry, 68
Peabody Rifle Company, 333
Peddler system, 54
Pelletier, LeRoy, 223
Pentz, Albert D., 110, 112–17, 119, 120, 358 n. 134
Perkins, Jacob, 51
Perry, Thomas, 126
Perry, William H., 69–73
Phelps, Orson C., 82–83
Piece rate system, 32, 81, 118, 206, 248
Pierce Foundation, 311
Pine, Leighton: advocates use of bentwood covers, 142; advocates use of gumwood, 138–39; as cabinet maker for Singer, 132; explores production costs, 142; leaves and returns to Singer, 134; as manager of Singer cabinetmaking operations, 133; on the meaning of quantity production, 140; work of, on all-veneer cabinets, 136–39, 142, 361 n. 54
Pinmaking, 51, **52**
Pioch, William: on changeover to Ford Model A, 284, 287, 293–94; on changeover to V-8, 297; on Ford's abandonment of stampings, 280; as head of Ford tool depart-

THE JOHNS HOPKINS UNIVERSITY PRESS

From the American System to Mass Production

*This book was set in Goudy Old Style display and Times
Roman text type by The Composing Room of Michigan, Inc.,
from a design by Susan P. Fillion. It was printed on S. D.
Warren's 50-lb. Olde Style Wove paper and bound in Holliston
Roxite A by The Maple Press Company.*